Advanced Digital Signal Processing: From Concepts to Applications

Advanced Digital Signal Processing: From Concepts to Applications

Edited by **Edmond Thor**

C WILLFORD PRESS

New York

Published by Willford Press,
118-35 Queens Blvd., Suite 400,
Forest Hills, NY 11375, USA
www.willfordpress.com

Advanced Digital Signal Processing: From Concepts to Applications
Edited by Edmond Thor

International Standard Book Number: 978-1-68285-088-6 (Hardback)

Printed in the United States of America.

Contents

Preface

In my initial years as a student, I used to run to the library at every possible instance to grab a book and learn something new. Books were my primary source of knowledge and I would not have come such a long way without all that I learnt from them. Thus, when I was approached to edit this book; I became understandably nostalgic. It was an absolute honor to be considered worthy of guiding the current generation as well as those to come. I put all my knowledge and hard work into making this book most beneficial for its readers.

The topics covered in this extensive book deal with the core areas of digital signal processing. It is compiled in such a manner, that it will provide in-depth knowledge about the theory and practices of signal processing through detailed discussions of concepts such as time and space domains, wavelet, discrete signals, etc. There has been rapid progress in this field and its applications are finding their way across multiple industries. This book compiles significant researches contributed by scientists and engineers. It will prove beneficial for students of engineering. Academicians and research scholars will also find this book useful.

I wish to thank my publisher for supporting me at every step. I would also like to thank all the authors who have contributed their researches in this book. I hope this book will be a valuable contribution to the progress of the field.

Editor

Accurate modeling of high frequency microelectromechanical systems (MEMS) switches in time- and frequency-domain

F. Coccetti[1], **W. Dressel**[1], **P. Russer**[1], **L. Pierantoni**[2], **M. Farina**[2], and **T. Rozzi**[2]

[1]Technische Universität München, Lehrstuhl für Hochfrequenztechnik, Munich, Germany
[2]Dipartimento di Elettronica ed Automatica, University of Ancona, Ancona, Italy

Abstract. In this contribution we present an accurate investigation of three different techniques for the modeling of complex planar circuits. The em analysis is performed by means of different electromagnetic full-wave solvers in the time-domain and in the frequency-domain. The first one is the Transmission Line Matrix (TLM) method. In the second one the TLM method is combined with the Integral Equation (IE) method. The latter is based on the Generalized Transverse Resonance Diffraction (GTRD). In order to test the methods we model different structures and compare the calculated S-parameters to measured results, with good agreement.

1 Introduction

The goal of our outgoing joint effort is the development and the application of efficient numerical tools for the analysis and modeling of complex open planar circuits, such as antennas, filters and more complex structure as Micro-Electro-Mechanical-Systems (MEMS).These latter structures exhibit usually several geometrical details, finite dielectric layers, losses and thick metals and also strongly critical "aspect-ratios". Typically, it is very useful to deal with all these structures using different methods or solvers. The use of semi-analytical methods like the integral equation method (IE) in connection with the method of moments (MoM) is usually restricted to strictly planar structures, Harrington (1982). However GTRD allows setting up an integral equation for truly 3-D structures, complementing the known advantages of MoM techniques (such as speed and reliability) with the flexibility of 3-D full-wave approach in the frequency domain. Its disadvantage lies in the need for some hypothesis on the structure, as it relies on knowledge of the Green Function describing the structure under test.

In (Farina and Rozzi, 2001) a 3-D GTRD formulation for boxed multilayer structures was presented that exploited the Green's function of a loaded box and was shown to be especially suited for MMIC and MEMS analysis. Space discretizing methods like the TLM method allow the numerical field modeling of structures with nearly arbitrary geometry (Johns, 1987; Russer, 2000). Their disadvantages appear when dealing with free space regions which increases considerably the 3-D-spatial domain of computation, thus increasing the number and the size of the elementary cells for the field modeling. The hybrid TLM-IE method combines the advantages of the TLM method in modeling nearly arbitrary complex structures and the advantages of the IE method in dealing with wide homogeneous regions, (Pierantoni et al., 1999). A minor drawback is the need of storing the time-evolution of the tangential field where TLM is coupled to the Green's function-based Integral Equation.

Three full-wave numerical tools were developed based to the aforementioned techniques:

(i) A solver based on the TLM method, which involves computer visualization (Mangold and Russer, 1999).

(ii) A solver based on the TLM-IE method (Pierantoni et al., 1999).

(iii) A general-purpose commercial program, including tools for pre and post processing, EM3-DS, distributed by MEM Research, based on GTRD method (Farina and Rozzi, 2001).

In order to compare accuracy and efficiency of the above three methods we have modeled a MEMS structure. In this contribution we discuss a MEMS capacitive switch, known to be challenging structure because of its severe aspect-ratio constrains (Sheen et al., 1990; Coccetti et al., 2000). Theoretical S-parameters are compared to experimental ones with very good agreement.

Correspondence to: F. Coccetti

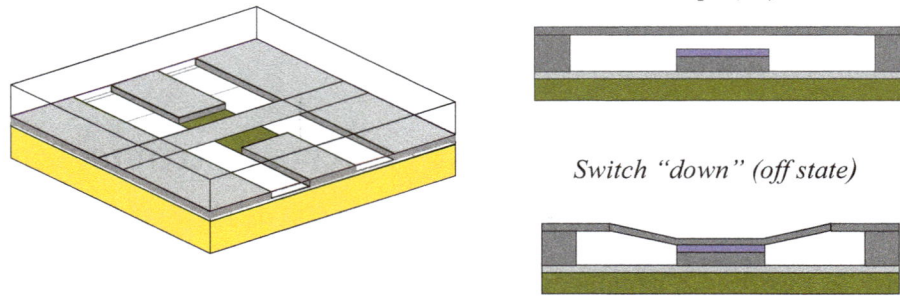

Fig. 1. MEMS switch and the cross section in the two working states.

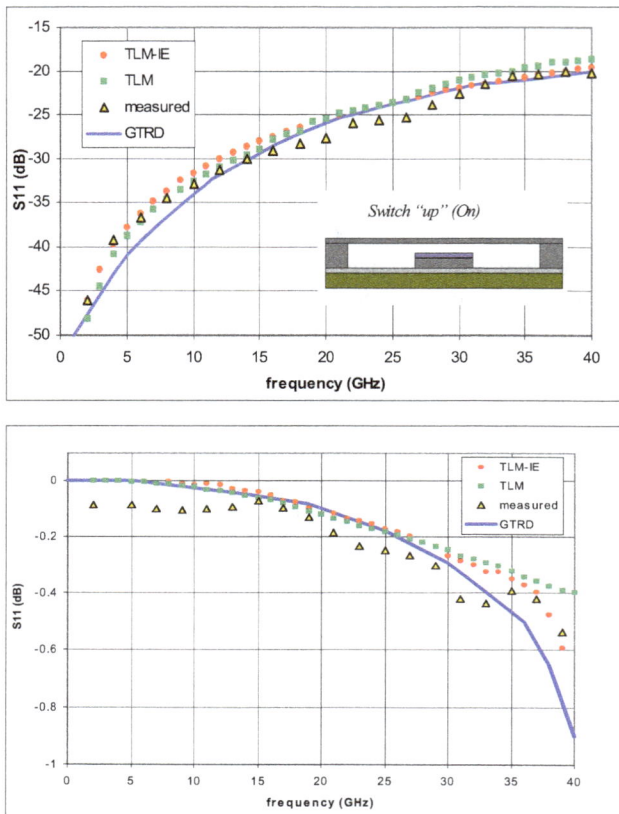

Fig. 2. On State: comparison of magnitude in dB for S_{11} (left) and S_{21} (right) vs. freq. (GHz).

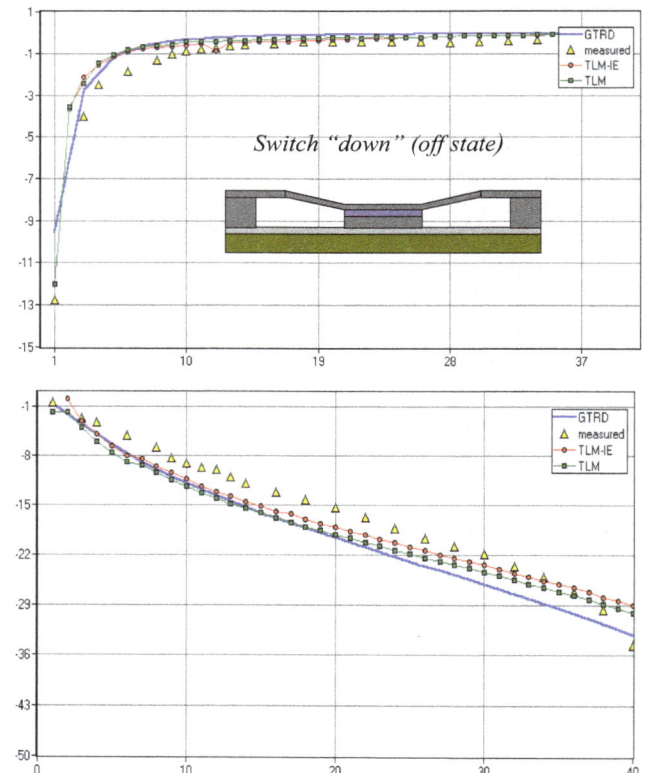

Fig. 3. Off State: comparison of magnitude in dB for S_{11} (left) and S_{21} (right) vs. freq. (GHz).

2 Theory

In the TLM method the evolution of the discretized electromagnetic field is modeled by wave pulses propagating on a mesh of transmission lines and scattered at the mesh nodes (Johns, 1987; Russer, 2000). In the TLM-IE method the 3-D space is segmented into different sub-regions, where the best suited method, be it TLM or IE is applied. Inside the TLM-regions, the e.m. field is modeled by the TLM method. In IE-regions the e.m. field is analytically by means of the appropriate Green's function. The continuity of the field is

applied at the interfaces between regions, providing appropriate integral equations for the tangential field.

The tangential field solution represents the exact boundary condition for the TLM algorithm (Pierantoni et al., 1999). In the GTRD approach the Green's function of a multilayer dielectric stack is calculated (Farina and Rozzi, 2001): the Green's function links fields within the stack to arbitrary current source distributions. Currents are defined in volumes describing lossy conductor regions, and by imposing Ohm's law to hold, an eigenvalue equation is obtained. The final step is to select appropriate excitation so as to transform the

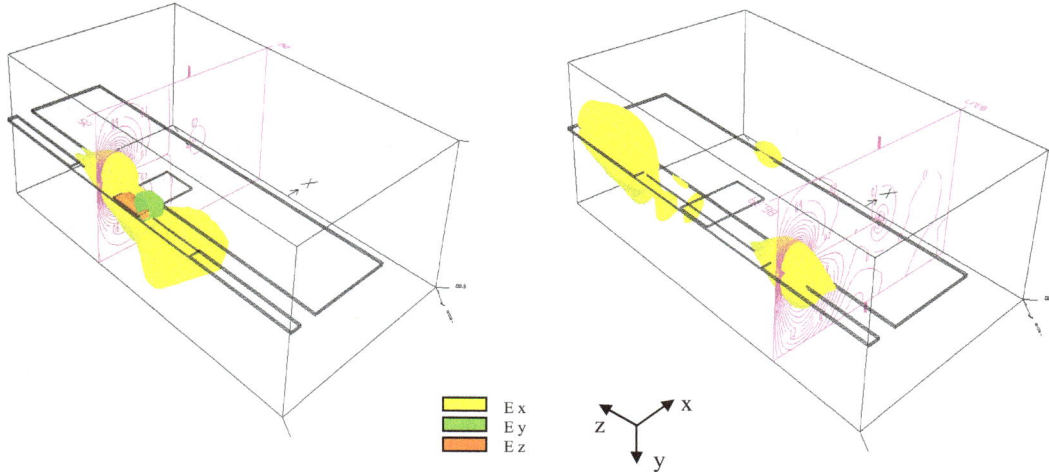

Fig. 4. On State 3-D field representation at two time instants: $t = t_0$ (left) and $t = t_0 + 2.5\,\text{ps}$ (right).

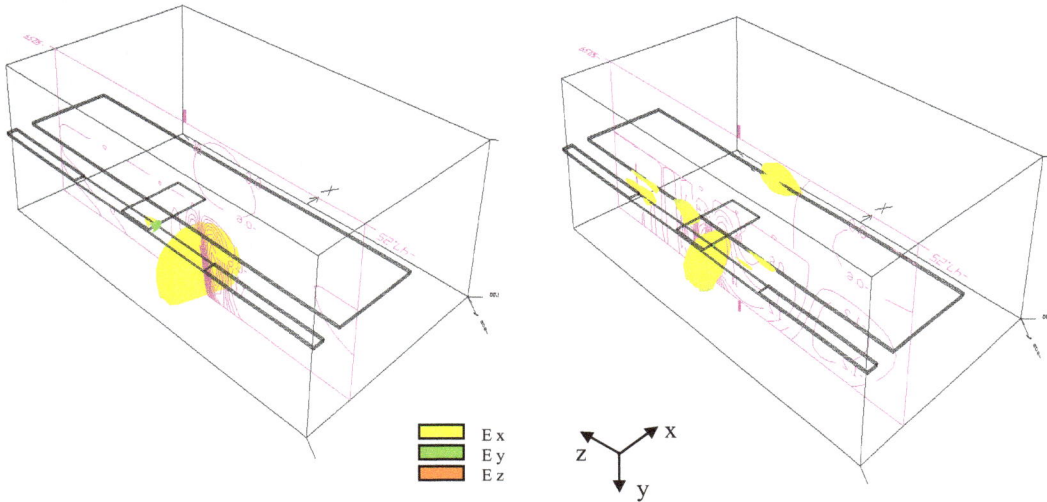

Fig. 5. Off State 3-D field representation at two time instants: $t = t_0$ (left) and $t = t_0 + 2.5\,\text{ps}$ (right).

eigenvalue equation into a deterministic one. In our case, excitations were selected to be standard delta-gap field sources, while the source discontinuity was removed by appropriate de-embedding.

3 Results

The three methods show high accuracy and agreement with experimental data. The TLM-IE and GTRD simulations have been performed by a 512 Mb-RAM 300 MHz PC, while TLM over a HP-9000 C360.

It has to be mentioned that TLM and TLM-IE have their strongest point in the ability to model very complex structures, with nearly arbitrary shaped object in space. It should be remarked, however, that GTRD is a frequency-domain approach, so that the simulation time is dependent of the number of frequency points required, while TLM and TLM-IE obtain the frequency-domain response as FFT of a time-domain evolution, implying advantages for broad band simulations.

Figure 1 shows the MEMS switch reported in Strohm et al. (1999). Figure 2 shows a comparison between TLM, TLM-IE, GTRD and experimental data for the "on" state and the same comparison for the "off" state. In both cases GTRD computation required roughly 3 h CPU-time; the TLM-IE simulation requires about 2 h CPU-time. It is remarkable to observe that by using the TLM-IE method both the bulk Si-region and the free-space regions are modeled by means of the appropriate Green's function, thus drastically reducing the 3-D spatial domain of computation for the TLM algorithm.

A comparison of TLM method, TLM-IE method and GTRD method with experimental data shows a very good agreement in any analyzed structure. Slight differences on the accuracy are mostly due to the selection running parameters (mesh size, number of time steps for TLM/TLM-IE, number of expansion functions for GTRD).

Any method has its own advantages and drawbacks. TLM-IE, due to its hybrid nature, seems to offer a good trade-off between flexibility, accuracy and computation time.

Figures 4 and 5 show the time domain simulation of the electric field in the MEMS switch according to Fig. 4. The structure is excited with a Gaussian pulse 1.6 ps of width. Due to the symmetry of the problem only half of the structure is depicted. The field distribution of the traveling pulse in the "on" state is depicted in Fig. 4. The shaded surfaces are the isopheres of E_x, E_y and E_z. In the xz-plane and in the yz-plane respectively the isoclinal lines of E_x are depicted.

The pulse is a snapshot of the time propagation in two time instant; t_0 just before the MEMS bridge, and at $t = t_0 + 2.5$ ps just after it. In a similar fashion the field distribution of the traveling pulse in the "off" state is depicted in Fig. 5.

Is pretty evident from these time domain plots, how the bridge in the two different state acts differently, producing a completely dual behavior, highlighted already in the given scattering parameters, with only 0.1 dB insertion loss and almost 20 dB isolation at 20 GHz.

4   Conclusions

The performances of the three full-wave approaches, TLM, TLM-IE and GTRD, have been compared for the case of planar and quasi-planar structures. Comparison with experimental results shows very good agreement. Besides the high accuracy a further advantage of TLM is its high flexibility with respect to general structures. A reduction of computation time by up to one order of magnitude with pure TLM can be achieved using system identification methods (Chtchekatourov et al., 2001).

References

Harrington, R. F.: Field Computation by Moment Methods, Krieger, 1982.

Farina, M. and Rozzi, T.: A 3-D integral equation-based approach to the analysis of real-life MMICs: application to microelectromechanical systems, IEEE Trans. Microwave Theory and Tech., December 2001.

Johns, P. B.: A symmetrical condensed node for the TLM method, IEEE Trans. Microwave Theory Tech., 35, 370–377, 1987.

Russer, P.: The transmission line matrix method, in: Applied Computational Electromagnetics, NATO ASI Series, pp. 243–269, Springer, Berlin, New York, 2000.

Mangold, T. and Russer, P.: Full-wave modeling and automatic equivalent-circuit generation of millimeter-wave planar and multilayer structures, IEEE Trans. Microwave Theory Tech., 47, 851–858, 1999.

Pierantoni, L., Lindenmeier, S., and Russer, P.: Efficient analysis and modeling of the radiation of microstrip lines and patch antennas by the TLM-integral equation (TLM-IE) method, Int. J. Numerical Modelling: Electronic Networks, Devices and Fields, 12, 4, 329–340, 1999.

Sheen, D. M., Ali, S. M., Abouzahra, M. D., and Kong, J. A.: Application of the three-dimensional finite-difference time-domain method to the analysis of planar microstrip circuits, IEEE Trans. Microwave Theory Tech., 38, 7, 849–857, 1990.

Coccetti, F., Vietzorreck, L., Chtchekatourov, V., and Russer, P.: A numerical study of MEMS capacitive switches using TLM, Proc. 16th Annual Review of Progress in Applied Computational Electromagnetics, Monterey 2000, 580–586, 2000.

Strohm, K., Rheinfelder, C., Schurr, A., et al.: SIMMWIC capacitive RF switches, 29th European Microwave Conference, Munich 1999, 2, 411–414, 1999.

Chtchekatourov, V., Coccetti, F., and Russer, P.: Direct Y-parameters estimation of microwave structures using TLM simulation and Prony's method, Proc. 17th Annual Review of Progress in Applied Computational Electromagnetics, Monterey 2001, 461–467, 2001.

Study of heterogeneous and reconfigurable architectures in the communication domain

H. T. Feldkaemper, H. Blume, and T. G. Noll

Chair of Electrical Engineering and Computer Systems, RWTH Aachen University, Schinkelstr. 2, 52062 Aachen, Germany

Abstract. One of the most challenging design issues for next generations of (mobile) communication systems is fulfilling the computational demands while finding an appropriate trade-off between flexibility and implementation aspects, especially power consumption. Flexibility of modern architectures is desirable, e.g. concerning adaptation to new standards and reduction of time-to-market of a new product. Typical target architectures for future communication systems include embedded FPGAs, dedicated macros as well as programmable digital signal and control oriented processor cores as each of these has its specific advantages. These will be integrated as a System-on-Chip (SoC). For such a heterogeneous architecture a design space exploration and an appropriate partitioning plays a crucial role.

On the exemplary vehicle of a Viterbi decoder as frequently used in communication systems we show which costs in terms of *ATE* complexity arise implementing typical components on different types of architecture blocks. A factor of about seven orders of magnitude spans between a physically optimised implementation and an implementation on a programmable DSP kernel. An implementation on an embedded FPGA kernel is in between these two representing an attractive compromise with high flexibility and low power consumption. Extending this comparison to further components, it is shown quantitatively that the cost ratio between different implementation alternatives is closely related to the operation to be performed. This information is essential for the appropriate partitioning of heterogeneous systems.

1 Introduction

Today's mobile communication standards like GPRS, EDGE, UMTS or CDMA2000 enable high-performance wireless applications as they offer mobile high-speed data rates. For those systems it is required to provide a high degree of flexibility and highest computational capabilities. But the computational demands are beyond the capacities of today's programmable platforms. For example in (Hausner, 2001) the increase in computational demands evolving from one standard to the next one has been compared to the increase in performance of digital signal processor kernels (DSPs). At the time of the standard release the computational requirements are constantly beyond the available performance of on-chip DSP kernels (Fig. 1). Even more severe, future generations of communication standards tend to strengthen this computational gap.

In addition to these computational demands a high degree of flexibility is required because of:

- Avoidance of the design of a new platform for future products within the design cycles as sufficient flexibility is provided for an adaptation to changing demands e.g. for the integration of new features from one product generation to the next one, or for required adaptations to standard updates (e.g. variation of chiprate with evolution of a communication standard like UMTS). By this, short innovation cycles and longer product lifetimes can be achieved.

- Runtime adaptivity due to switching between several cells/standards (e.g. handover between standards), respectively adaptation to channel quality.

Therefore, the underlying architecture of a communication system has to include architecture blocks that support these aspects of flexibility. Dedicated hardware implementations offer orders of magnitude better performance with respect to throughput and power dissipation. But flexibility of those implementations is restricted to weak programmability (e.g. switching of coefficient sets) considered at design time. Altogether, a well-balanced architecture of a communication system has to include different types of architecture blocks in order to provide the required performance at reasonable costs (e.g. area and power dissipation) on one hand and ensuring sufficient flexibility on the other. Future communication systems will consist of a variety of system blocks. These

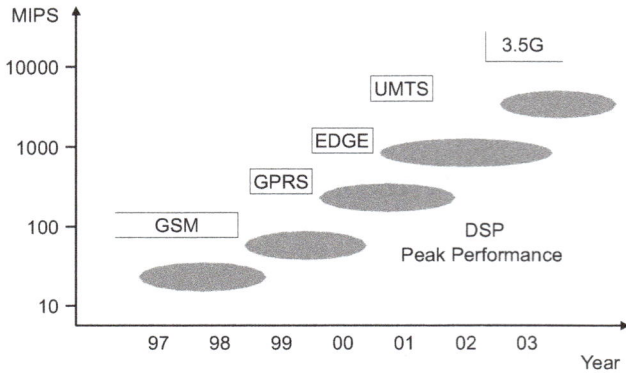

Fig. 1. Yearly increase in computational complexity and DSP-performance (Hausner, 2001).

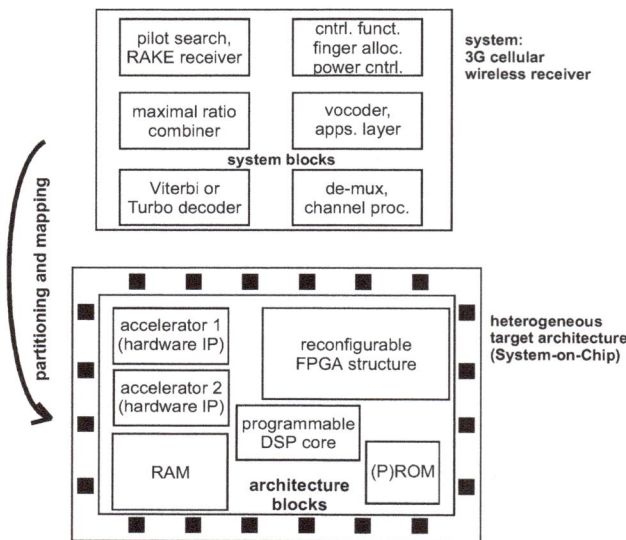

Fig. 2. Partitioning and mapping from system to architecture.

systems have to be partitioned and mapped to the architecture blocks of such heterogeneous target architectures (Fig. 2). In order to meet the challenging demands sketched above it is important to elaborate methodologies which assist designers with metrics and with an early assessment of the capabilities of a given platform.

The next section lists classical quantitative metrics. An exemplary implementation of a key component in digital communication systems – the Viterbi decoder – is shown in the following section considering principle architecture blocks. Finally, the results for the different implementation alternatives are discussed.

2 Possible evaluation metrics for quantitative optimisation

For the evaluation of basic operations in the field of digital signal processing several aspects have to be considered. Various metrics considering silicon area A and symbol rate

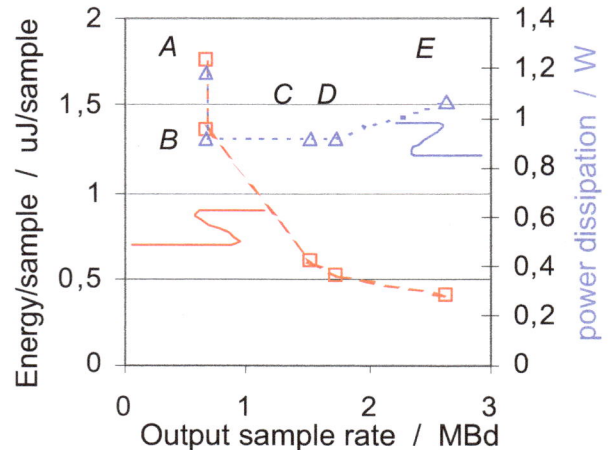

Fig. 3. Software optimisations for the TM 1300.

$1/T$ have been proposed in the past e.g. (De Hon, 2000). Due to the importance of low-power operation especially in the communication domain the energy per output sample $E_{per sample}$ has to be taken into account. Therefore, a combined cost function is taken exemplarily in the following as evaluation metric

$$cost = A \cdot T \cdot E_{per\,sample}.$$

A common primary design challenge is to implement a system achieving a specified throughput rate at minimised energy per sample and silicon area. According to scaling theory (general scaling for short channel devices) all parameters are normalised with respect to the minimal feature size L_{min}

$$T_{norm} = T \cdot \left(\frac{1 \mu m}{L_{min}} \right), \quad A_{norm} = A \cdot \left(\frac{1 \mu m}{L_{min}} \right)^2,$$

$$E_{per\,sample,norm} = E_{per\,sample} \cdot \left(\frac{1V}{V_{DD}} \right)^2 \cdot \left(\frac{1 \mu m}{L_{min}} \right)^{0.75}.$$

Overhead for time-sharing e.g. of free computational resources of a DSP or parallelisation is neglected here.

3 Exemplary vehicle: Viterbi decoder

In order to perform a fair comparison between the costs of an algorithm mapped on different architecture blocks, the implementations need to be optimised individually to the specific architecture block, e.g. by including algorithmic transformations. This will be shown in the following applying the exemplary vehicle of a Viterbi decoder.

The architecture of the Viterbi decoder can be divided into three basic units: the branch metric unit (BMU), the path metric unit (PMU) and the survivor memory unit (SMU). The following subsections review optimisation techniques and present the costs of a rate-1/2 64-state Viterbi decoder with a survivor memory length of 128 and a path-metric word length of eight bits mapped to a DSP, an FPGA and a physically optimised macro.

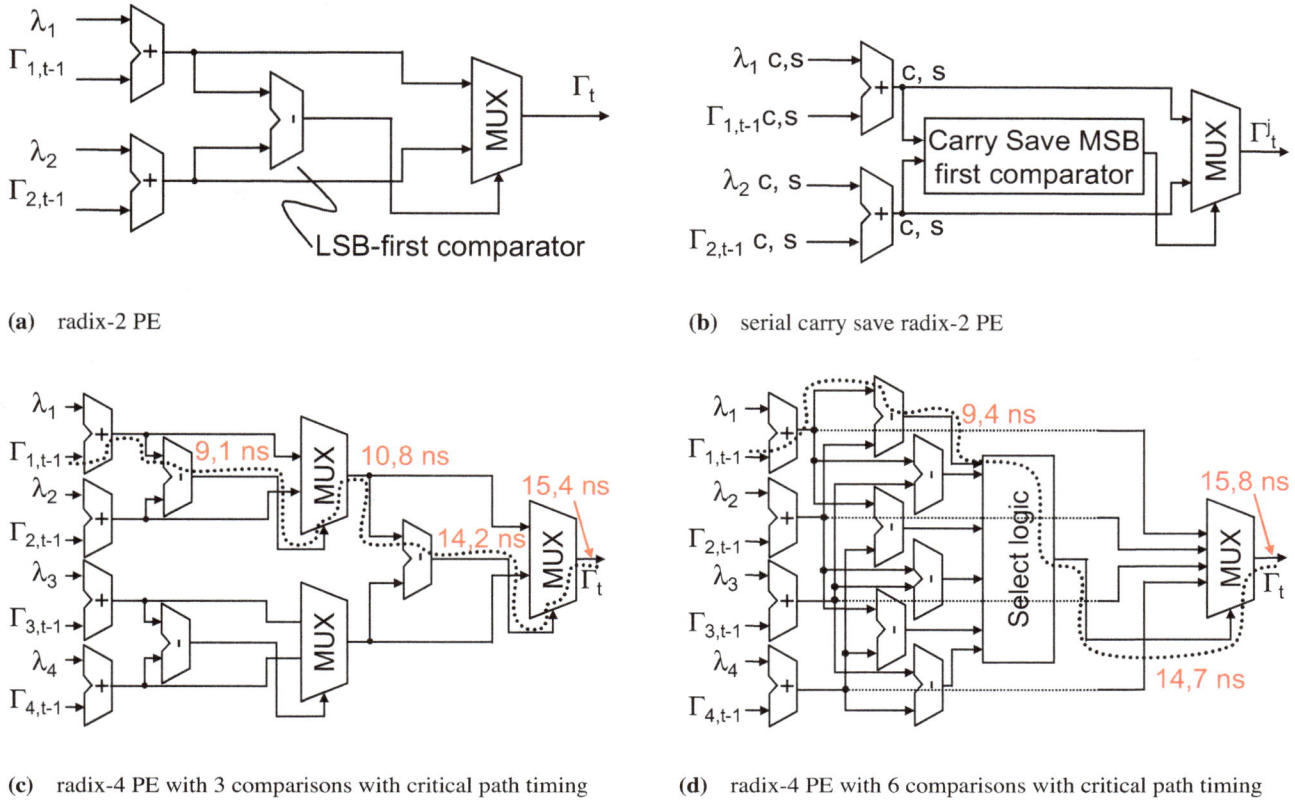

(a) radix-2 PE

(b) serial carry save radix-2 PE

(c) radix-4 PE with 3 comparisons with critical path timing

(d) radix-4 PE with 6 comparisons with critical path timing

Fig. 4. Simplified Signal flow graphs (SFG) of different PEs for the PMU.

3.1 Optimisation for a DSP implementation

Due to high data rates in the communication domain, very long instruction word (VLIW) DSPs like the TM 1300 processor are well suited for Viterbi decoding (Ahmad Khan et al., 2000). This processor features five issue slots (i.e. at max. five instructions/cycle) and 27 functional units dedicated to frequently used basic operations ranging from simple ALU operations to square root or division. It has been applied here in order to implement the Viterbi decoder. Special attention has been directed to the power dissipation as this is one of the most decisive factors for wireless applications and therefore is examined here for different kinds of software optimisation. For several optimisation steps power dissipation has been measured within continuous operation of the program code and is sketched in Fig. 3 with the achieved symbol rate. Furthermore, the resulting energy per output sample is depicted over the symbol rate.

The reference software taken from (Karn, 1996) is denoted by the symbol A. In the first step, unused blocks in the DSP for example video I/O units are powered down, resulting in B. Compiler options were adjusted appropriately within the next optimisation (C). Custom operations were applied in order to get linear program code (D). All of these optimisation steps were provided for the PMU, since it requires about 79% of the total execution time. The variable type of the

path metrics were changed from an array type to a scalar one resulting in the most energy efficient implementation (E). Though the power dissipation is increasing, from B to E, the energy per sample is decreased as the symbol rate features at the same time a more significant increase.

3.2 Optimisation for FPGA implementations

As an exemplary SRAM-based reconfigurable FPGA an Altera Apex 20KE device (EP20K200EQC240-2) was applied (Altera 2002). The available logic resources allow an implementation on a single device. This FPGA offers arrays of logic cells and additional embedded system blocks as memories. The logic cells include a 4-bit look-up table, a subsequent register and additional logic e.g. for the implementation of fast carry ripple adders.

The PMU and the SMU were examined and optimised for the FPGA implementation because of their significance for the overall area and speed of the design (Black, 1993). For the SMU a trace back method was chosen, because the register exchange method requires too many registers and routing resources for this specification. In order to use a simple clocking scheme, the two pointer trace back method was chosen, since hybrid architectures with a trace forward method again result in inefficient usage of FPGA resources.

In order to find an optimal solution different implementa-

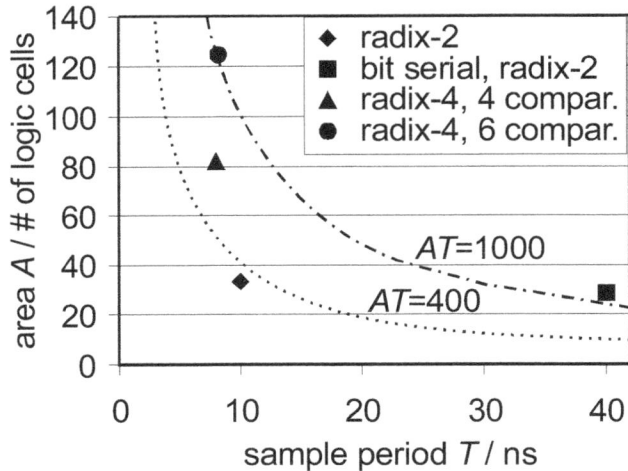

Fig. 5. Comparison of different PEs.

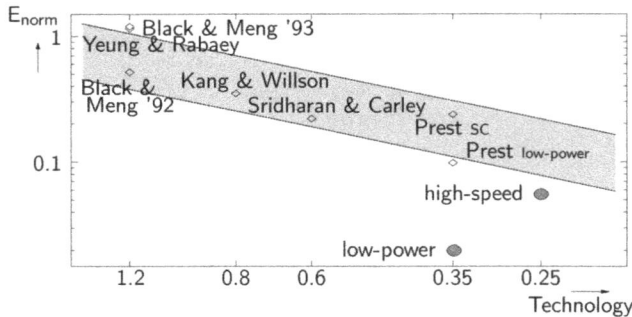

Fig. 6. Normalised energy conversion (E_{norm}) per symbol of Viterbi decoder implementations (Gemmeke et al., 2001); (E_{norm} is defined as power dissipation normalised to the Viterbi decoder specifications of DVB-S and throughput rate).

tions of the add-compare-select processing element (PE) of the PMU were examined regarding the area (assumed to be proportional to the number of required logic cells) and the maximal symbol rate $1/T$ (Fig. 4). The radix-2 butterfly PE has been presented several times in literature (Kivioja et al., 1999; Pandita et al., 1999). With an LSB first comparator, it utilises the mentioned fast carry chain efficiently (Fig. 4a). It requires 33 logic cells with a symbol rate of 99 Megabit/s. A bit serial carry save MSB first implementation was further examined providing the advantage, that the path metric does not need to be stored. For this implementation a carry save MSB first comparator is required which is rather complex.

In a radix-4 butterfly PE two succeeding stages of a radix-2 add-compare-select operation are performed in a single PE. For a hierarchical comparison three comparators are required (Fig. 4c). One would expect that a doubled throughput rate should be achievable with a radix-4 approach, but due to the underlying hardware structure, the resulting sample rate is only almost as fast as in the case of a radix-2, but much more area is required. For a parallel, pairwise comparison six comparators are necessary (Fig. 4d). The critical path is shown

in both PEs with the corresponding timing values. However, due to the inefficient placement of the logic cells in the select logic, the critical path is even worse although the number of logic cells was increased for the intention of speedup. For a dedicated macro, this implementation is often chosen, because of better implementation results.

The AT complexities of different PEs are shown in the diagram in Fig. 5. The radix-2 element is by a factor of 2.5 more efficient in terms of AT complexity than a bit serial radix-2 PE. Overall, here a radix-2 PE is the most cost efficient implementation. This is due to the use of fast carry chains in the logic cells, which can be applied to efficiently implement the add-compare operation (implemented as a subtraction).

3.3 Physically optimised implementation

A dedicated hardware implementation offers higher throughput rates and less power dissipation compared to any programmable/reconfigurable solution. A physically oriented design style allows to fully exploit the throughput potential of a technology, and to reach the lowest possible energy conversion per operation.

To reduce the design effort of a hardware implementation the regularity of common digital signal processing datapaths can be exploited. Common arithmetic algorithms inherently contain a high degree of locality. Preserving this down to the layout ensures high throughput at low silicon area and even more important low power dissipation. The use of a datapath generator (Weiss et al., 2001) automates the macro generation.

Applying quantitative optimisation on all levels of design hierarchy two Viterbi decoders were designed (Gemmeke et al., 2002): one optimised for high speed operation, the other for low power dissipation.

A comparison of these two designs to the trend of other published leading edge implementations is shown in Fig. 6. Apparently, the designs fall into a band of decreasing power dissipation. Its slope indicates an exponential trend according to the minimum feature size in VLSI technology. The physically optimised low-power implementation disrupts the common trend by approximately one decade. Whereas, the high performance design traded some power dissipation for high-speed operation.

4 Final comparison

The preceding implementation results are compared by means of the ATE cost function. As a consequence Fig. 7a can be derived. Besides the three implementation alternatives discussed before also the cost value for a dedicated standard cell implementation is depicted. The cost values are normalised to that of the physically optimised implementation. It is shown that the normalised cost of a DSP and a physically optimised implementation differ by about seven orders of magnitude. This is compared to the costs of further basic operations. In Fig. 7b a comparison for a variety of digital signal processing operations is depicted (Blume et al., 2002).

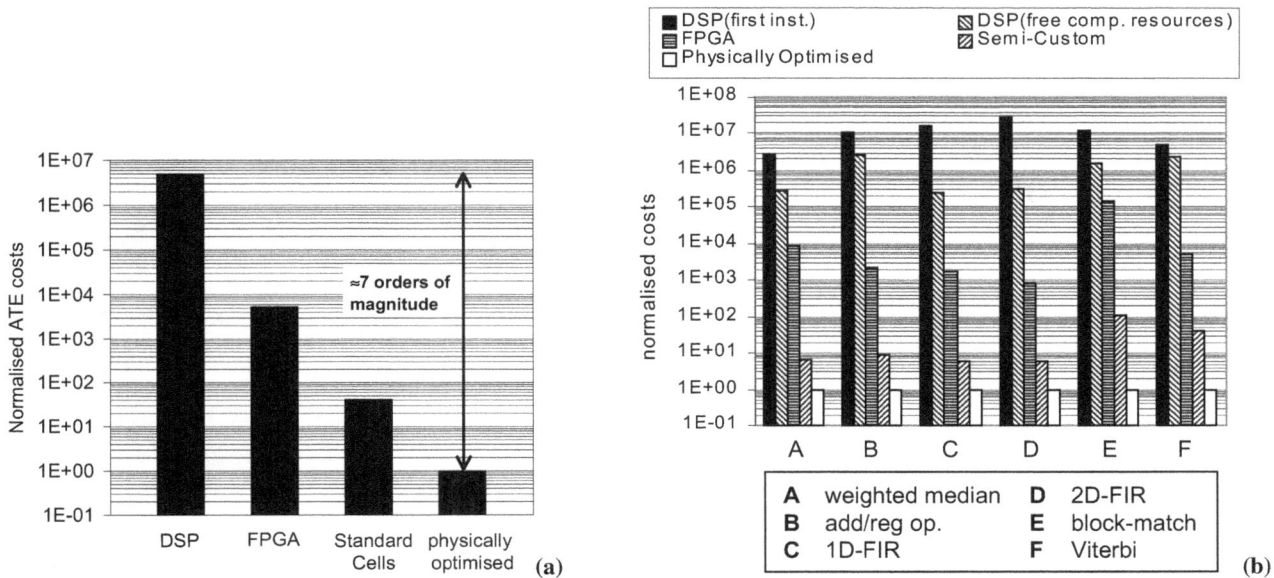

Fig. 7. Comparison by means of normalised costs, (a) Comparison of a Viterbi decoder, (b) Comparison for different operations.

It is distinguished between differential and absolute power consumption of an operation on a given architecture block. Differential power consumption applies to the availability of free computational resources. Therefore, only the operation-dependent power consumption is considered. If the computational resources are exhausted, a new device needs to be instantiated. This is referred to as absolute power consumption, where additionally operation-independent power consumption (overhead e.g. for the clock system) has to be considered.

For the case of a DSP with free computational resources the cost ratio between a physically optimised and a DSP based implementation spans from at least four to six orders of magnitude. FPGA based implementation costs mostly lie between the physically optimised and the DSP implementations. Considering the absolute power consumption, the cost ratio between DSP and physically optimised implementation increases up to two additional orders of magnitude. All the investigations discussed here were based on discrete devices. Future SoCs will embed several architecture blocks of the types described before on one chip.

5 Conclusion

As the cost differences between dedicated and programmable implementations of system blocks are rather huge and flexibility is required for future SoCs, partitioning of a system to a target architecture is most important. It demands for an analysis of implementation specific parameters for basic operations. This is required in order to choose the optimum implementation for basic operations with demanding specifications. Cost modelling of basic operations on different architecture blocks like DSPs, FPGA like structures, semi-

custom and physically optimised macros is required for this partitioning.

Using the Viterbi decoder as an exemplary component a quantitative cost function based analysis for heterogeneous SoCs has been shown. Future work has to be directed to refined partitioning strategies utilising approved models for key elements allowing an early assessment of possible implementation alternatives.

References

Ahmad Khan, S., Saqib, M., and Ahmed, S.: Parallel Viterbi algorithm for a VLIW DSP, Proc. ASSP, pp. 3390–3393, 2000.

Altera: APEX 20K Programmable Logic Device Family Data Sheet, Version 4.3, Feb. 2002.

Black, P.: Algorithms and Architectures for High Speed Viterbi Decoding, PhD dissert., Stanford, March 1993.

Blume, H., Huebert, H., Feldkaemper, H., and Noll, T. G.: Model based exploration of the design space for heterogeneous Systems on Chip, ASAP, pp. 29–40, 2002.

De Hon, A.: The Density Advantage of Configurable Computing, IEEE Computer, pp. 41–49, 4/2000.

Gemmeke, T., Gansen, M., and Noll, T. G.: Scalable, Power and Area Efficient High Throughput Viterbi Decoder Implementations, IEEE, Journal of Solid-State Circuits, pp. 941–948, July 2002.

Hausner, J.: Integrated Circuits for Next Generation Wireless Systems, Proc. ESSCIRC, pp. 26–29, 2001.

Karn, P.: http://people.qualcomm.com/karn/code/index.html, 1996.

Kivioja, M., Isoaho, J., and Vanska, L.: Design and Implementation of a Viterbi Decoder with FPGAs, Journal of VLSI Signal Processing, pp. 5–14, Kluwer, Netherlands, May 1999.

Pandita, B. and Roy, S.: Design and Implementation of a Viterbi Decoder using FPGAs, Proc. VLSI Design, pp.611–614, 1999.

Weiss, O., Gansen, M., and Noll, T. G.: A flexible Datapath Generator for Physical Oriented Design, ESSCIRC, pp. 408–411, 2001.

3

Resonance circuits for adiabatic circuits

C. Schlachta and M. Glesner

Darmstadt University of Technology, Institute of Microelectronic Systems, Germany

Abstract. One of the possible techniques to reduces the power consumption in digital CMOS circuits is to slow down the charge transport. This slowdown can be achieved by introducing an inductor in the charging path. Additionally, the inductor can act as an energy storage element, conserving the energy that is normally dissipated during discharging. Together with the parasitic capacitances from the circuit a LC-resonant circuit is formed.

1 Introduction

In conventional CMOS logic the load capacitor is charged to V_{DD} via the PMOS block (pull up network) and discharged to GND via the NMOS block (pull down network). During the charging of the output, the power supply has to deliver a charge $Q = C_L V_{DD}$ at the voltage V_{DD} resulting in a supplied energy of $Q V_{DD} = C_L V_{DD}^2$. that dissipats as heat.

To decrease the dissipated energy, the charge transport can be slowed down. Using a ramp-like charging voltage, the output capacitance is charged at a constant current. This current source delivers the charge $C_L V_{DD}$ over a period of time T. The dissipation through the resistance of the charging path R is then (Chandrakasan and Brodersen, 1995):

$$E_{diss} = P \cdot T = I^2 RT = \left(\frac{C_L V_{DD}}{T}\right)^2 RT$$
$$= \left(\frac{RC_L}{T}\right) C_L V_{DD}^2 \ . \tag{1}$$

This shows that it is possible to charge and discharge a capacitance through a resistance while dissipating less energy than $C_L V_{DD}^2$. By introducing an energy storage element, the energy that is normally dissipated as capacitances are discharged can be conserved for later use. This can be observed in LC resonant circuits, where the energy oscillates between the capacitance and the inductance.

To determine the efficiency of a resonant circuit, the amount charge per charging/discharging cycle of an ideal abrupt charge/discharge circuit can be compared to the average charge required by the oscillator.

$$E_{oscillator, percycle} = \frac{i_{avg,osc}}{f_{osc}} \tag{2}$$

$$E_{abrupt, percycle} = C_{load} \cdot V_{dd} \tag{3}$$

In the equations C_{load} is the sum of all capacitances in the resonance circuit and f_{osc} the frequency of the oscillator. From this equations we can determine the efficiency of an oscillator

$$\eta = \frac{E_{oscillator, percycle}}{E_{abrupt, percycle}} = \frac{\frac{i_{avg,osc}}{f_{osc}}}{C_{Load} \cdot V_{dd}} \tag{4}$$

This means, that at a resonant circuit with an efficiency of 70% dissipates only 30% of the power of an abrupt charging circuit.

2 Multi-stage charging

A very simple approach to implement a charging/decharging circuit is to use multi-stage charging (Saas et al., 2001). Multi-stage charging is not a LC oscillator due to its missing inductor however it is presented here as comparision to the other resonant oscillators. The basic concept is to charge/discharge the capacitance stepwise in n steps each with the voltage $V_{step} = \frac{V_{supply}}{n}$.

Multi-stage charging also reuses the charge stored in the load capacity. Under the assumption that the pull-up transistors charge the load to the full supply voltage, the power dissipation for discharging is the same as for charging. It is not necessary to use power supplies for the intermediate voltages because the average current is zero in the ideal case.

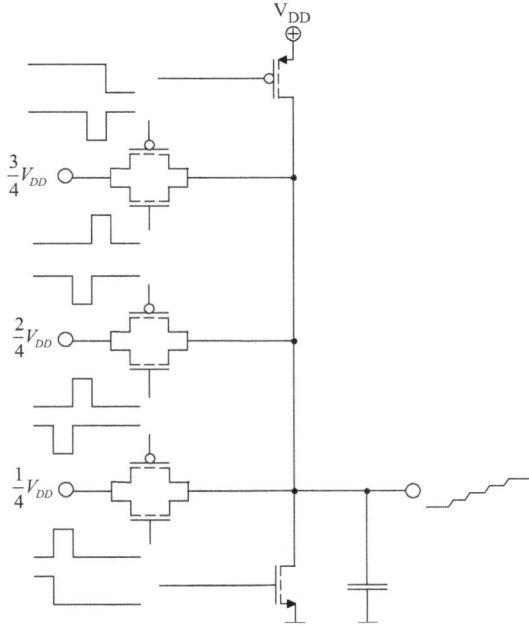

Fig. 1. Principle of multi-stage charging.

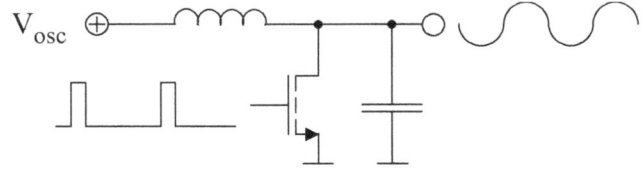

Fig. 3. Schematic of the one transistor oscillator.

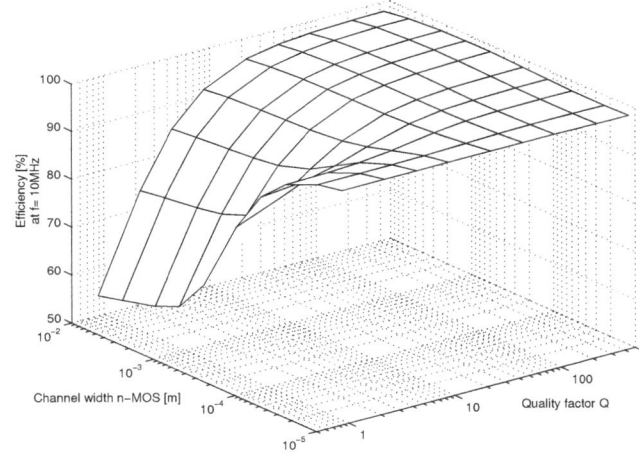

Fig. 4. Efficiency for one transistor oscillator.

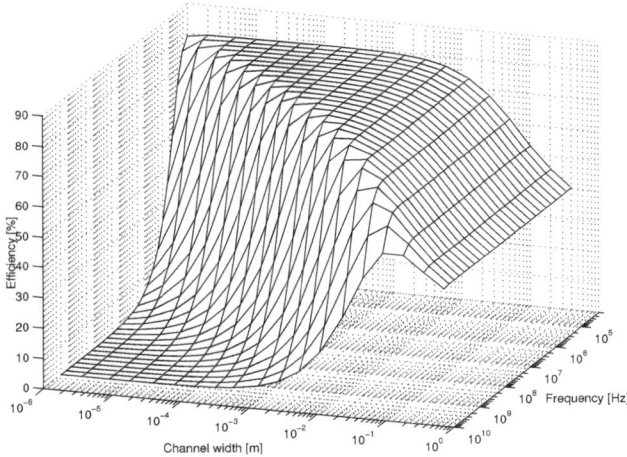

Fig. 2. Efficiency for 5-stage (4 intermediate voltages) charging/discharging.

Instead, large buffering capacitors are sufficient. These capacitors must be charged to the desired voltage just at startup and afterwards it is only necessary to compensate losses.

The efficiency of the multi-stage charging is assessed through a simulation that is depicted in Fig. 2. This simulation was done for a five step circuit, this means four intermediate voltages are used. The current consumption is calculated for a load capacity of 1500 pF, the current consumed by the control logic is estimated to be equal to the current needed for driving the pass transistor gates. This means, that the current is twice the current consumed by driving the pass transistor gates. The channel length of the transistors is 0.35 μm and the circuit is supplied with 3.3 V. The top area

of the graph shows the range for usefull operation. Increasing the channel width leads to bad efficieny because of the rising current consumption for the control logic. The other limitation is the charging time, increasing the operating frequency reduces the time period for charging/discharging the load capacity. If the time interval is getting to small the load capacity can not be completely charged or discharged. Consequently case multi-stage charging is not applicable in this case.

3 Continuously oscillating circuits

3.1 One transitor resonant circuit

The simplest attempt to design a resonant circuit is the 1-transistor circuit depicted in Fig. 3 (Voss, 2001). The gate of the transistor is driven by a control signal that must be derived from an other oscillator, i.e. a crystal oscillator. Besides the control of the frequency the gate driver logic has also to control the duty cycle of the transistor. The efficiency diagram of this oscillator is depicted in Fig. 4. This diagram results from a simulation with a transistor channel length of $w_n = 0.35\,\mu$m, a resonant capacity of $C_{res} = 1500$ pF, an inductance with $L_{res} = 170$ nH and a supply voltage of $V_{osc} = 1.65$ V. The oscillation frequency is about 10 MHz. The gate of the MOSFET is driven with a pulse width $pw = 10$ ns (equiv. to a duty cycle of $\approx 10\%$) and a periode of $per = 100$ ns (resonance frequency). The power consumption of the driver logic for the transistor gate was also taken

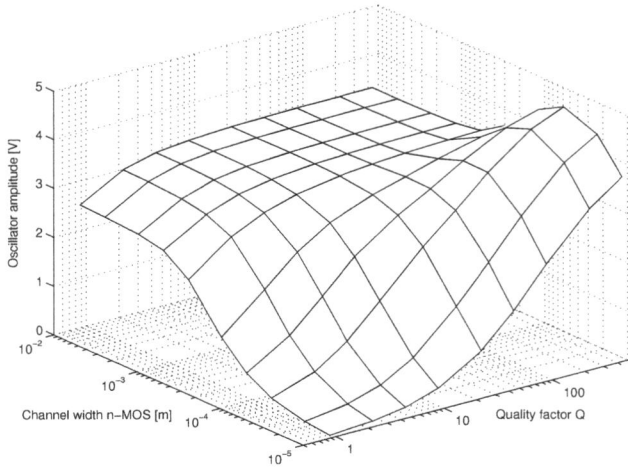

Fig. 5. One transistor oscillator: Variation of the quality factor of the inductor.

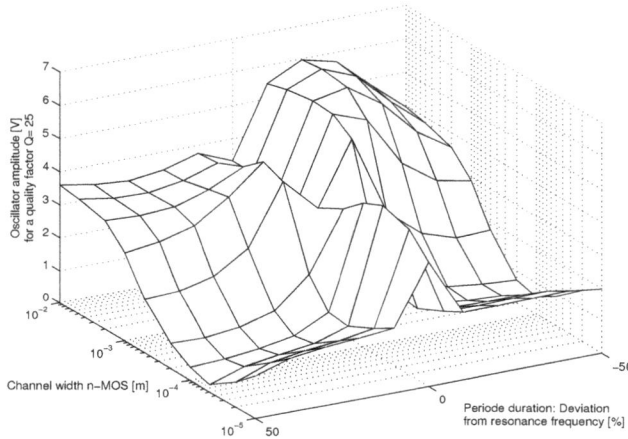

Fig. 6. One transistor oscillator: Amplitude as function of variation of the control frequency.

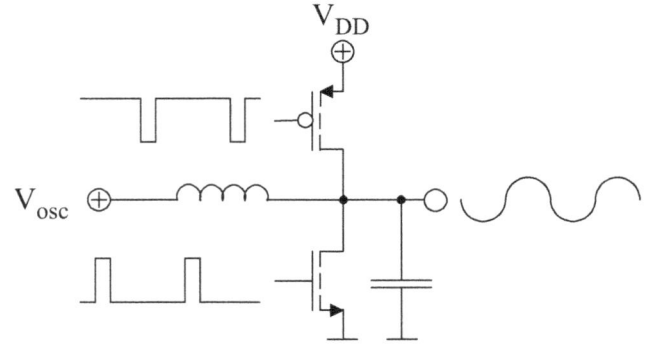

Fig. 7. Two transistor oscillator.

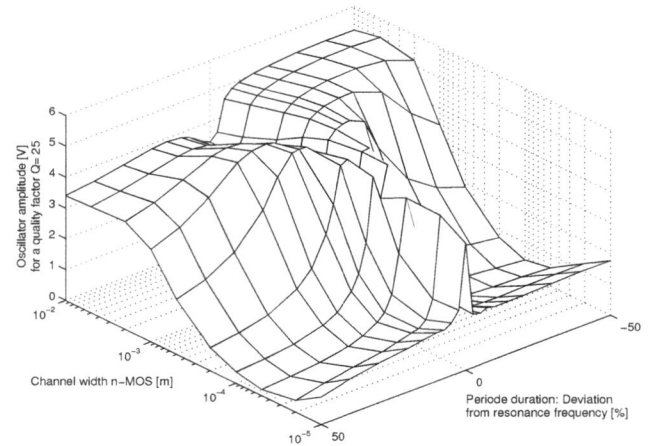

Fig. 8. Two transistor oscillator: Amplitude as function of variation of the control frequency.

the resonance frequency:

$$\frac{f_{resonance,actual}}{f_{resonance,nominal}} = \sqrt{\frac{C_{actual}}{C_{nominal}}} \tag{5}$$

This equation shows, that the deviation in the load capacitance must also be in the range of some percent around the nominal value for a stable operation.

3.2 Two-transistor oscillator

The one transistor oscillator can be expanded to a push-push stage by adding a p-MOSFET and the necessary control logic. This stage behaves mostly like the single transistor oscillator previously presented. It can be clearly seen in Fig. 8 that this circuit is also not stable against mismatch between control and resonance frequency or varying the load capacity.

3.3 H-bridge oscillator

If the circuitry needs a sine signal with as well 0° as 180° phase shift, the two-transistor oscillator can be extended to the H-bridge shown in Fig. 9 (Voss, 2001). The main advantages of the H-bridge are the stable phase shift between

into account. It's value was estimated as twice the power needed for driving only the gate of the MOSFET.

In Fig. 3.1 the amplitude of the oscillator as a function of the channel width of the MOSFET and the quality factor of the inductor is shown. It can be seen that only for very high quality factors of the inductor and small channel width of the MOSFET the amplitude of the oscillation differs from the nominal value (3.3 V, twice the oscillator voltage).

In Fig. 3.1 the periode duration of the gate drive signal is varied from the period duration at resonance frequency (frequency mismatch) and the resulting oscillator amplitude is drawn. It can be seen very clearly that the oscillator is only working well within in the range of some percent around the resonance frequency. This limits also changes in the load capacity because these changes are equal to changes in the resonance frequency as it can be derived from the calculation of

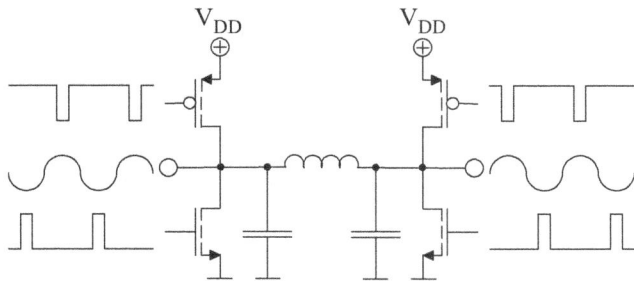

Fig. 9. H-bridge with four transistors.

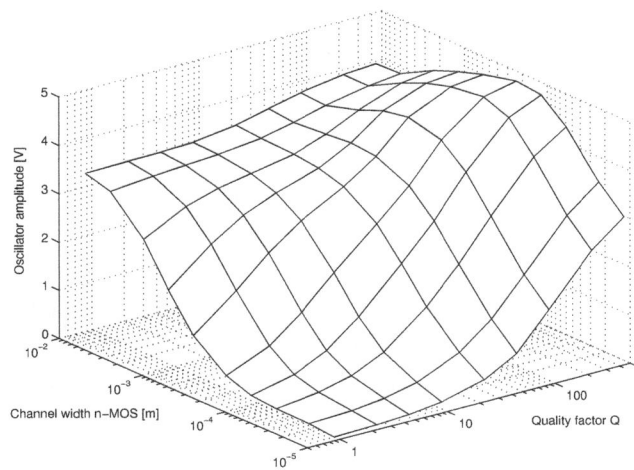

Fig. 10. H-bridge: Amplitude for varying quality factor of the inductor.

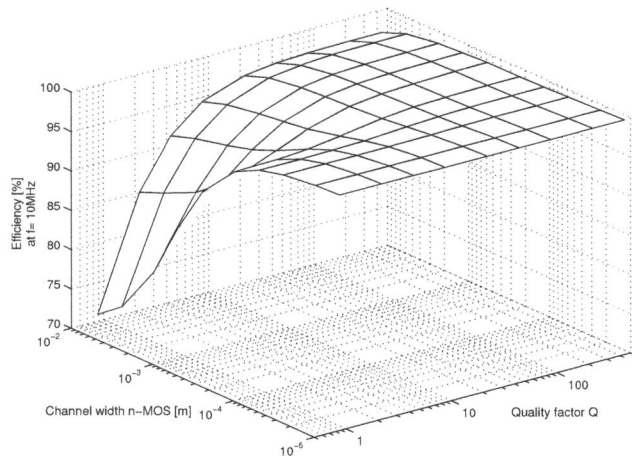

Fig. 11. H-bridge: Efficiency for varying quality factor of the inductor.

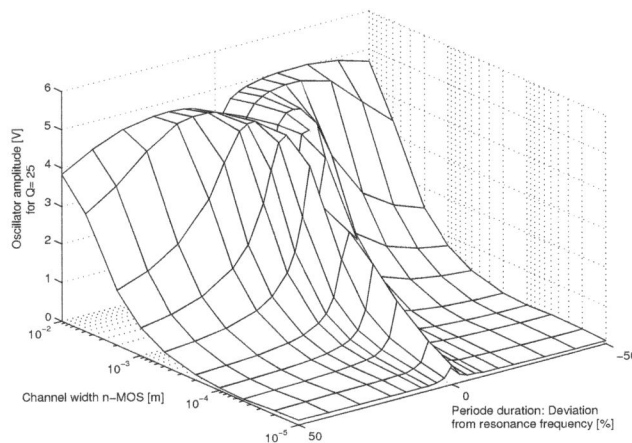

Fig. 12. H-bridge:

the two outputs and the use of only one inductor. The stability against mismatch between control and resonance frequency or variation of the capacitances is even worse compared to both the one-transitor and the two-transistor oscillator (Fig. 8).

4 Controlled resonance circuit

Another approch is the oscillator circuit shown in Fig. 13 (Schlachta et al., 2002; Athas et al., 1996). Contrary to the previous discussed oscillators this resonane cell doesn't oscillate continuously.

We start the discussion of the circuit by assuming that the connected load has to change its state. That triggers the circuit, which behaves like a clocked latch. The actual voltage over the resonance capacitor determines the needed transition and is saved in an RS flip flop in the control logic during the whole transition. The power transistor (depicted as ideal switch) is switched on and starts the resonant process by connecting the inductor to $\frac{V_{supply}}{2}$ This starts a resonant oscillation with an amplitude of also $\frac{V_{supply}}{2}$, achieving a voltage swing between $\frac{V_{supply}}{2} - \frac{V_{supply}}{2} = 0$ and $\frac{V_{supply}}{2} + \frac{V_{supply}}{2} = Vdd$. Therefore it is only necessary to stop the oscillation at the desired voltage level to get the demanded change in the

voltage of the load. If these levels are the minimum or maximum voltages of the resonant circuit, the inductor contains no energy so that the free wheeling diodes are not needed. A comparator is used to determine the threshold at which the power transistor is switched off. Since there are two thresholds, one for switching from LOW to $HIGH$ and one for switching from $HIGH$ to LOW, two comparators adjusting the switching thresholds are used where the active one is selected depending on the initial value of the resonance capacitor. To ensure constant levels after switching, a tristate buffer is used. This buffer is enabled when the switching cycle is finished and pulls up or down the resonant capacitor, thus compensating the energy and the consequent voltage losses in the resonant ciruit.

Because the resistance of the charging/discharging path influences the energy dissipation in the design (see Eq. 1), the power transistors had to be made very wide. On the other hand, the width of the switching transistors is responsible for the gate capacitance that has to be charged and discharged conventionally each switching cycle, thence dissipating en-

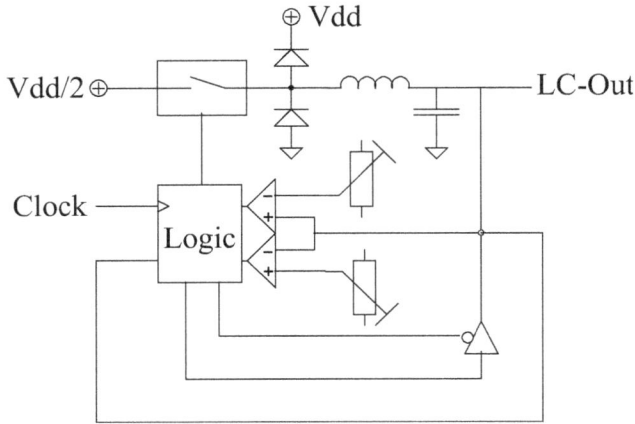

Fig. 13. Controlled oscillator cell.

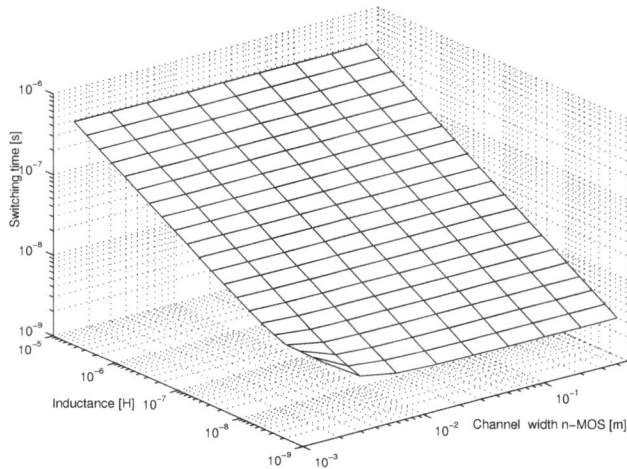

Fig. 14. Controlled oscillator: Switching as function of inductance.

ergy. Thereby, the channel widths of the transistors are a tradeoff between the resistive loss in the resonant circuit especially at full load as well as the switching losses due to the gate capacitance at small load.

Another subject to tradeoff is the value of the resonant capacitor. The resonant capacitor together with the switchable and thereby variable load capacitances forms the resonant capacitance and determines the switching time and thus the slew rate. To achieve a small range of switching time between no load and full load it is necessary and advantageous to use a large additional capacitor (in the range of the maximum load capacitance). The drawback of a large resonant capacitor is that it significantly increases switching losses especially with less load capacitance. As a compromise, the resonant capacitor is sized equal to one third of the maximum load capacitance, achieving a change of $1 : \sqrt{3+1} \approx 1 : 2$ in the switching time between minimum (no load) and full load.

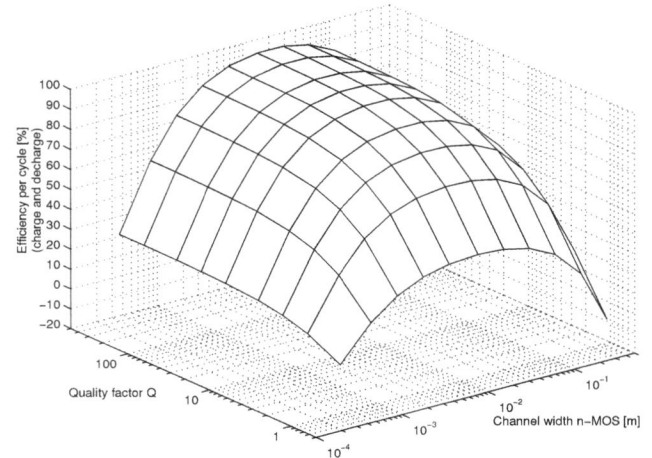

Fig. 15. Controlled osciolltor: efficiency as function of quality factor.

5   Conclusions

Several oscillator designs have been presented. All of these designs have their advantages and drawbacks:

- Multi stage charging is only applicable for limited frequencies; the design without an inductor and the high efficiency makes it a good choice for circuits with lower clock frequencies.

- The one- and two-transistor oscillator and H-bridge have a good efficiency but they all require a nearly constant load capacitance. This limits the application to nets with a constant capacitvie load, e.g. the clock net on a chip.

- The controlled oscillator contains some difficult to implement components like the analog comparators, and the operating speed is limited because of the difficult turn-off detection. On the other hand the ability to drive loads with varying capacity makes it very flexible. Consequently it is a good choice for I/O-buffers and similiar applications.

As a result can be said, that none of the proposed concepts is an optimal choice for all application. Nevertheless, through a careful selection of the oscillator type according to the demands of the load it is possible to find very good solutions for a broad range of applications.

Acknowledgements. This work is granted by the Deutsche Forschungsgemeinschaft as part of the VIVA project under the label GL 144/18-2.

References

Athas, W., Svenson, J., and Tzartzanis, N.: A resonant signal driver for two-phase, almost-non-overlapping clocks, in Proceedings of the ISCAS 1996, pp. 129–132, 1996.

Chandrakasan, A. P. and Brodersen, R. W.: Low Power Digital CMOS Design, chap. 7: Minimizing Switched Capacitance, Kluwer Academic Publishers, 1995.

Saas, C., Schlaffer, A., and Nossek, J. A.: An adiabatic multi stage driver, in Proc. of ECCTD'01 European Conference on Circuit Theory and Design, 2001.

Schlachta, C., Voss, B., and Glesner, M.: A low-power line driver using resonant charging, in Design Automation and Test Europe (DATE), 2002.

Voss, B.: Resonantes Umladen als Schaltungstechnik zur Verlustleistungsreduktion in digitalen CMOS-Schaltungen, Ph.D. thesis, TU Darmstadt, Shaker Verlag, ISBN: 3-8265-9800-8, 2001.

4

Active miniature radio frequency field probe

A. Glasmachers

Fachbereich Elektrotechnik und Informationstechnik, Universität Wuppertal, Fuhlrottstrasse 10, 42097 Wuppertal, Germany

Abstract. For the measuring of the electromagnetic interference (e.g. on men) of RF fields produced by mobile communication equipment field probes are required with high spatial resolution and high sensitivity. Available passive probes show good results with respect to bandwidth and low field distortion, but do not provide the required sensitivity and dynamic range. A significant limitation for active miniature probes is the power supply problem, because batteries cannot be used. Therefore the effect of high impedance connection lines is examined by a numerical field simulation. Different approaches for the design of an active probe are discussed, a favourable solution with a logarithmic demodulator is implemented and measuring results are presented.

1 Introduction

Radio frequency field probes were originally designed for measurements of antenna systems, since some decades additional applications arise in the field of electromagnetic compatibility. New requirements were caused by the introduction of mobile communication. Radio frequency field probes described in this paper are used to examine health hazards among other measurements. These measurements require following points:

- measuring of the electrical field

- frequency range 800 MHz–2,4 GHz

- spatial resolution 10 mm

- high sensitivity

- high dynamic range

- compatible with new modulation techniques (e.g., CDMA)

Because of the required spatial resolution the admissible size of the dipole antenna is limited to $2\,h<10\,mm$. Therefore for the power supply of the probe a battery which is much larger then the antenna cannot be used because it will produce inadmissible field distortions. For an external power supply and for the measuring signal transmission two solutions are applicable:

a) An optical method (which is very difficult to implement for miniature probes) or

b) a connection via high impedance lines which will not significantly influence the RF field.

Most of the passive probes and the active probe presented in this paper use the connection via high impedance lines.

2 Principles and limits of passive field probes

Passive RF filed probes, which fulfill the frequency range and the spatial resolution requirements, are commercially available (Fig. 1).

The design is based on a long insulating carrier material with the antenna and a detection unit at the measuring end and a box with battery and post processing electronics at the other end. For the detection of the RF signal either a rectification by a Schottky diode or the conversion into heat by a resistor are used (Fig. 2). With respect to sensitivity the rectification gives the better result.

For a short dipole antenna with the length $2\,h$ placed in parallel to the electrical field E the open circuit voltage v_0 is given by

$$v_0 \approx 2\mathrm{h}E \tag{1}$$

and the source impedance by

$$Z_0 \approx -j37\Omega\frac{\lambda}{2\mathrm{h}} \tag{2}$$

For a frequency of $f = 1\,GHz$ ($\lambda = 300\,mm$) and an antenna length of $2\,h = 10\,mm$ this impedance of $Z \approx -j1.1k\Omega$ represents a capacitance of $C_0 \approx 0.14pF$.

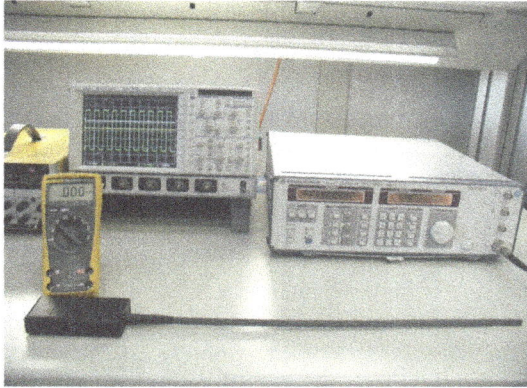

Fig. 1. Commercially available passive field probe.

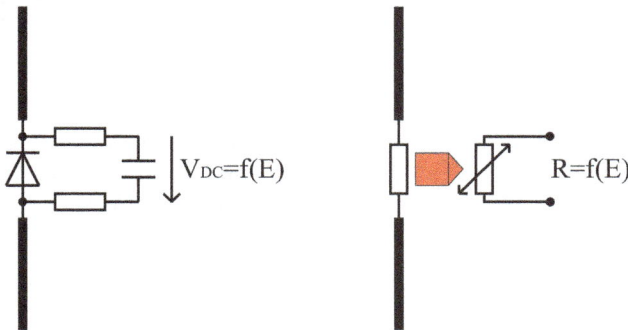

Fig. 2. Principles of passive field probes.

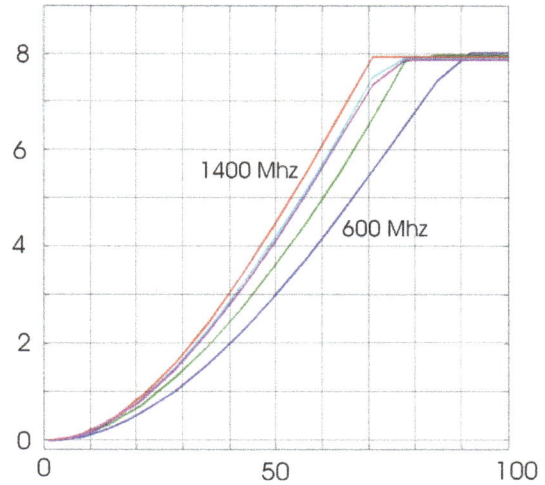

Fig. 3. Properties of a passive field probes with a Schottky diode, x-axis: field strength E [V/m], y-axis: dc output voltage [V].

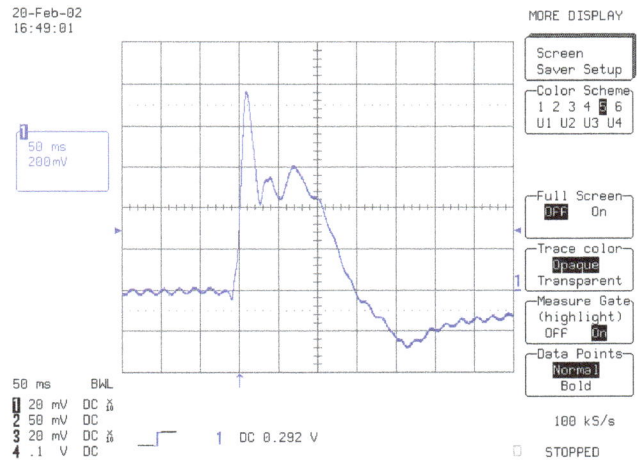

Fig. 4. Effect of low frequency electrostatic noise signals.

For the Schottky diode the relation of diode current I_D und diode voltage V_D is given by

$$I_D = I_S\{\exp(V_D/V_T) - 1\}$$

with I_S : saturation current,

    V_T : temperature voltage   (3)

which for small voltages V_D can be approximated by

$$I_D \approx I_S\{V_D/V_T + 1/2(V_D/V_T)^2\} \tag{4}$$

The detection effect at low levels is given by the quadratic term in Eq. (4), therefore an approximately quadratic interrelationship of field strength and rectifier voltage with low sensitivity (useful only for field strength $E > 10$ V/m) and low dynamic range (less than one decade) is achieved.

Another problem of a passive field probe is the high cross sensitivity against low frequency electrostatic noise signals. The large noise signal given in Fig. 4 was produced when the shoe of the operator slipped over the floor covering.

3 Power supply of an active electric field probe

The power supply of an active electric field probe by optical transmission is a significant problem for miniature probes. Batteries cannot be used because of their large dimensions which will disturb the rf field. A better solution is to supply power over high resistance lead wires. A MAFIA simulation (Fig. 5) shows that a positive result is achieved with a lead wire resistance of $RL' > 1$ kΩ/cm. A lead wire length of approximately 50 cm is shown in Fig. 6 with a resistance of 50 kΩ of the supply and signal lead. The electrical supply current Isupply must not exceed a few mA to make sure that the supply voltage stays in a range with no need of special safety measures (e.g. <500 V).

4 Active circuits for electric field probes

The market volume of electric field probes is very low. For this financial reason it is not possible to design a special integrated circuit. Integrated circuits for measuring the transmit signal in mobile radio devices can be used for electric field probes. There are 4 classes of circuits available:

   a) RF preamplifier in front of a passive rectifier

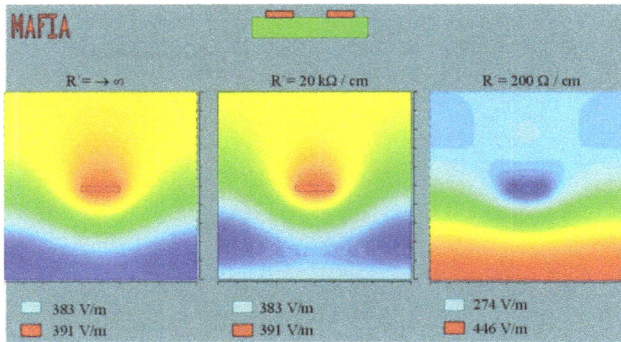

Fig. 5. Field simulation for different lead wire resistances.

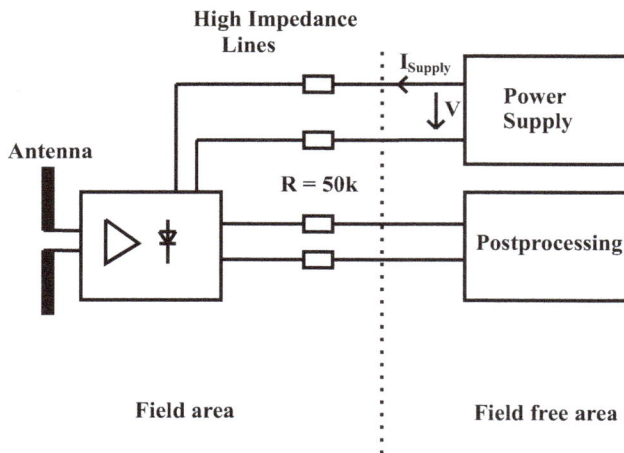

Fig. 6. Power supply of the electric field probe.

b) rectifier with Schottky diodes with bias current and DC buffer amplifier (e.g. LTC5505, Linear Technology)

c) RMS rectifier (e.g. AD8361, Analog Devices)

d) logarithmic demodulator (e.g. AD8314/MAX4000, Analog Devices / Maxim)

a) RF preamplifier

An increase of sensitivity can be achieved e.g. by use of an RF amplifier in front of the Schottky rectifier. RF amplifiers are available in small packages, but there is a need for extra circuitry and for a very high supply current of $I_{supply} > 15$ mA.

b) Rectifier with Schottky diodes with bias current and DC buffer amplifier

At the input of these integrated circuits a Schottky diode is used as rectifier. The efficiency is improved by a DC no-load current (bias current). The additional circuitry produces a higher input capacitance which reduces the signal level (capacitive voltage divider). Therefore for this application there is no advantage

in sensitivity compared to the passive circuitry. The internal dynamic compression improves the usable dynamic range significantly.

c) RMS rectifier

RMS rectifiers (e.g. AD8361) do not measure the peak of the HF field like the other methods but the root mean square value. RMS rectifiers qualify to measure complex signals (e.g. Code Division Multiple Access used with UMTS). Currently those components have a low power consumption ($I_{supply} < 2$ mA) but a small dynamic range of 25 dB and a low sensitivity. A newer version (AD8362) has a much higher dynamic range of 60 dB, but a very high supply current of $I_{supply} > 20$ mA which is not acceptable for electric field probes.

d) Logarithmic demodulator

A favorable alternative are logarithmic demodulators on a chip. The block diagram of the chip AD8314 is shown in Fig. 7. The supply current is approximately 4 mA. This circuit provides high sensitivity and high dynamic range.

5   Example for an active electric field probe

The circuit of a electric field probe with a dipole of $2h = 10$ mm length with a logarithmic demodulator AD8314 is shown in Fig. 8 on a printed circuit board made out of FR4 (still without high resistant lead wires). The high resistant lead wires for power supply are placed on one side of the PCB, the high resistant wires for the signals on the opposite side. Graphite spray is used for the lead wires.

The measurement of the active field probe was done in a TEM cell. The measurement results compared to a commercial passive probe with the same antenna length are shown in Fig. 9. The advantages are the logarithmic characteristic, higher sensitivity and the enhanced dynamic range.

6   Summary and outlook

Integrated circuits for measuring transmit power in mobile radio devices are particularly suitable for use in active electric field probes. Both applications have several identical requirements as frequency range, small package and low power consumption.

Currently available integrated circuits make it possible to build electric field probes which provide remarkably better sensitivity and dynamic range compared to passive probes. In the near future newer components are expected, respectively announced, to have more improvements in supply current and the ability to measure RMS values of complex signals.

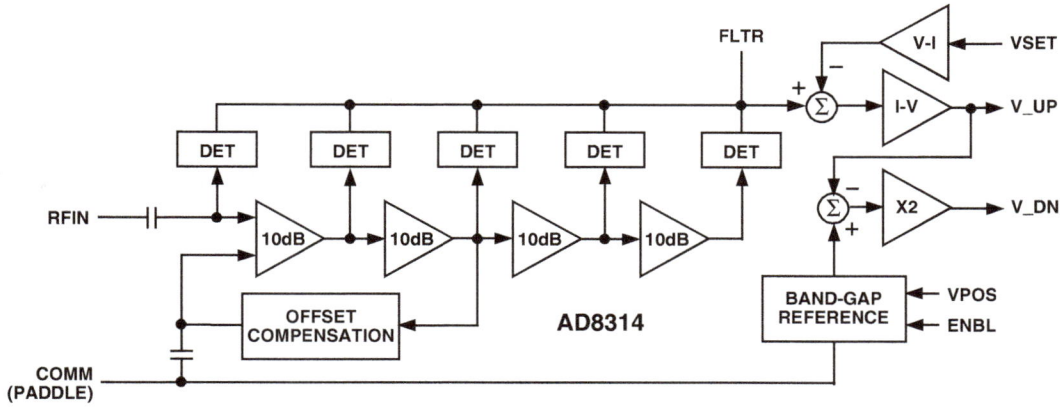

Fig. 7. Block diagram of the logarithmic demodulator AD8314.

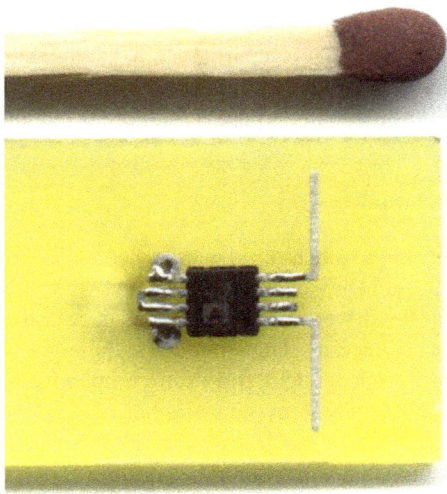

Fig. 8. Layout of an active electric field probe.

The technique of using high resistance lead wires for power supply has a very small effect on the field to be measured. The technological realization of these wires is still insufficient; a better replacement for the graphite spray to form the high resistant lead wires is needed.

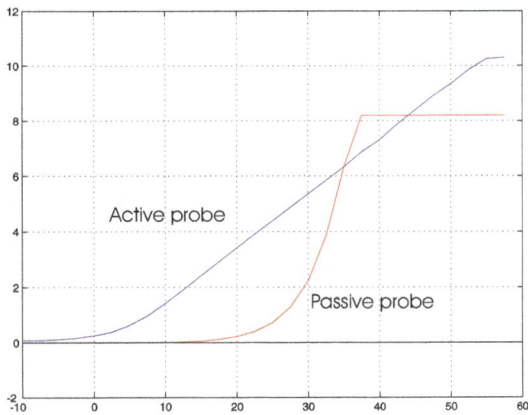

Fig. 9. Characteristics of field probes with x-axis: field strength E [dBV/m], y-axis: dc output voltage [V].

5

Consideration of parasitic effects on buses during early IC design stages

J. Rauscher, M. Tahedl, and H.-J. Pfleiderer

Department of Microelectronics, University of Ulm, Germany

Abstract. Complex integrated systems contain more and more on-chip components which exchange data and access memories via buses. To consider parasitic effects on bus structures during the early design phase appropriate models and wiring methods must be available. A RC-Pi model is proposed to model the load of a bus driver cell that also considers capacitive coupling between the signal lines. On the basis of simple estimations it is shown under what conditions the influence of inductance on the bus is negligible. Furthermore a method to consider inductive effects is shown. An early consideration of parasitic effects on global interconnect structures and an assertion which effects have to be considered is mandatory to effectively estimate the behavior of wires. Hence it also helps to avoid wrong assumptions.

1 Introduction

In novel technologies the performance and speed of integrated systems is increasingly dominated by on-chip interconnects. Hierarchical design approaches are used to deal with the complexity of whole Systems-on-Chip. Several system parts are developed independent. This is also necessary for enabling the possibility to reuse system parts or modules from previous developed systems. Bus systems are used for inter block communication. Physically a bus is a bundle of parallel routed wires. The bus structure, i.e. the two dimensional cross section of a bus should be defined while the logical top level design is done. The topology of a bus can be extracted from a floorplanning step. Hence, an early performance estimation of a bus structure is possible. Such an approach requires accurate bus models which consider also capacitive coupling effects. This results from the fact that the capacitance of minimal dimensioned buses is dominated by the coupling capacitances. The validity and reliability of the model has been investigated. An easy method to decide

if inductance has to be considered is shown. To accurately simulate inductive effects a partial element equivalent circuit (PEEC) model (Ruehli, 1972) is used. But unlike in Gala et al. (2000) our proposal is to use model order reduction.

2 RC-load model

A modeling approach is used which decouples the typical nonlinear driver behavior and the linear network of bus wires. The load for the driver caused by the bus is described in a π-model for this purpose. The structure and complexity of the load model is independent from the complexity of the bus structure, i.e. the topology and number of branches. Thus the model permits an integrative algorithm to determine the drivers' output signals. It should be possible to determine the output signals of all branches of the linear network after the drivers' output signals, i.e. the bus input signals are known.

A π-model which describes the load admittance of the driver was presented in Tahedl and Pfleiderer (2003). The model is based on the first three moments of the bus admittance where an estimation of the bus length for all branches and the positions of nodal points is required. The model is an extension to the model presented in O'Brien and Savarino (1989) from single wires to capacitive coupled interconnect buses. Compared to interconnect models where a Thevenin model is used for driver modeling, every model for the driver's nonlinear behavior is possible. Because the Thevenin resistance is often unknown, it is preferred to use timing libraries in early stages of design (Sheehan, 2002), especially when analog simulations with Spice or Spectre are not applicable.

2.1 Model computation

The derivation of the model was presented in Tahedl and Pfleiderer (2003). The used π-model is depicted in Fig. 1. The model is fully described with three parameter matrices C_n, C_f and R_π. To compute these matrices the first three admittance moments of the bus structure are matched to the

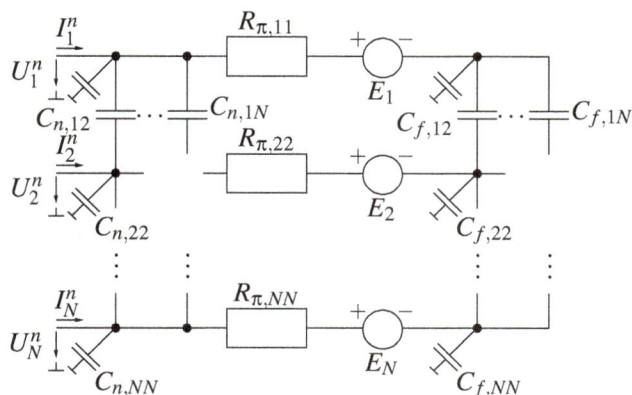

Fig. 1. Capacitive coupled RC-π-model.

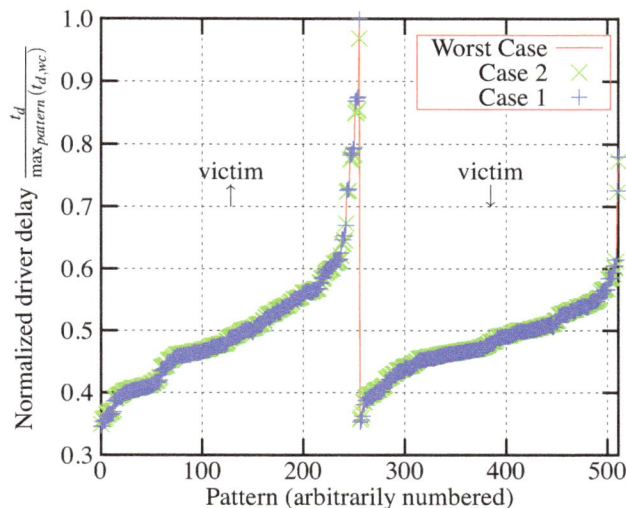

Fig. 2. Dependency of the driver delay on the switching condition in adjacent wires of a bus. The delay is plotted for two cases of wire densities in adjacent metal layers where the technology specific limits for the cases of metal density were chosen. The patterns are sorted by the worst case driver delay for each switching direction on the victim wire.

first three admittance moments of the model. Consequently C_n, C_f and R_π depend on the bus structures first three admittance moments. A bus is modeled by piecewise homogenous segments and lumped elements, e.g. vias between metal layers. The first three admittance moments of a lumped element or a distributed segment depends on the respective capacitance and resistance matrices as well as the first three admittance moments of the load to the element under consideration. Additionally, of course, the moments of a segment depends on the segment length. This dependencies cause a recursive computation of the first three admittance moments of a bus structure. The computation is started at every far-end of the structure. At a nodal point the respective admittance moments of the branches are summed up.

A mathematical simulation approach is suggested instead of the more physical representation depicted in Fig. 1. The physical representation may result in negative capacitances. Additionally some current driven voltage sources are necessary. However, the passivity of the model is proven mathematically.

2.2 Model validity and accuracy

The π-model is valid as long as the bus is sufficiently short. It is assumed that this is true due to the application of repeater insertion algorithms for performance optimization. Anyhow, if reversely scaled buses are used where inductive effects are still negligible or repeater insertion is not applied, an approach for computing a feasible model is presented in Tahedl and Pfleiderer (2003). Therefore a specific saturation length is required where the driver delay saturates. This length strongly depends on the driver's output waveform and can differ for each wire in the bus. The bus is modeled as a bus with open ended wires with saturation length. No closed form expression for the saturation length was derived by now.

Model validity depends on the knowledge of the saturation length and on a statement that inductive effects are negligible. If the model validity can be proven, the accuracy of the load estimation is defined by the accuracy of the structure

prediction and by the order of modeling accuracy. For example in a first step the same layer could be assumed for the whole tree without any vias. In some further steps informations on different layers as well as lumped models for vias or bends may be available. The model accuracy depends on the knowledge of the three-dimensional interconnect structure and the model is suitable through all stages of a circuit design process from first performance analysis steps to analysis of final extracted structures and for performance checks before tape-out. As long as the model is valid no differences compared to simulations with long RC-chains are recognizable.

2.3 Bus example

It is possible to estimate the driver delay in early design stages because the influence of wires in the adjacent layers of a bus structure on the driver delay is almost negligible ($< 4\%$). This was shown on the example of a bus in a complex environment. Therefore two cases of wire densities in adjacent layers were assumed. For the two cases the technology specific limits for the metal density were chosen. The dependency of the driver delay from the switching condition in adjacent wires of the bus is depicted in Fig. 2.

3 Inductive effects

Inductive effects have to be considered as a general rule only for long and wide wires which typically can be found in the upper metal layers. Essentially this results from the low resistance and hence the low damping of the wires. A couple of

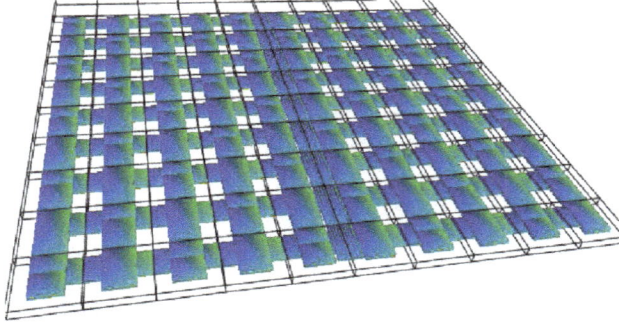

Fig. 3. Three layer power/ground grid with 8 signal lines on the uppermost layer.

formulas to decide when on-chip inductance has to be considered, especially for transmission lines, are well known. In the following we modified a formula (Ismail and Friedman, 2001) to use values of lumped elements.

$$\frac{t_r}{2\sqrt{LC}} < 1 < \frac{2}{R}\sqrt{\frac{L}{C}} \qquad (1)$$

The right side of the inequality considers the damping. The left side in principle the relation between the transition time at the driver input and two times the time of flight. The left side of the condition is only valid if the width of the transistor is chosen to match the impedance of the signal line or to be smaller. This is true in most practical cases, because wider transistors are unwanted and lead to overshoots. Therefore inductance can be neglected as long as this inequality is not fulfilled. To get the parameters of the resistance and inductance ports can be defined at the driver side of the signal line. At the receiver input the signal line can be shorted and the impedance can be obtained. As we will show later this is accurate enough to decide if inductance has to be considered for the early design phase.

For an accurate simulation the actual current loop must be considered. During a switching operation of a driver the coupling capacitors to Vdd or Ground along a signal line will be charged. Respectively the opposite capacitors will be discharged. Hence the current loop will be shorter than assumed before. The effective current loop is frequency dependent. Further it depends on switching events, the resulting voltage fluctuations of nearby gates, the decoupling capacitors and even the package. For this reason we decided to simulate a structure similar to the one depicted in Fig. 3. It's based on the forecasts of a 53 nm copper technology from the ITRS roadmap (ITRS, 2001). We considered a three layer power and ground grid with a bus on the uppermost layer like in Gala et al. (2000). But our approach is to use model order reduction which is considered unsuitable in Gala et al. (2000) for the fully-dense matrix of their model.

3.1 Inductance extraction

In order to get a PEEC model we used a precorrected-fast-Fourier-transform (FFT) approach (Hu et al., 2003) to simulate the on-chip inductance. The merits of this method are that the dense inductance matrix is not calculated explicitly, but a very accurate and fast computation of the product of the inductance matrix with a given vector is provided. Further this method doesn't suffer the problems of other sparsification techniques which are also described in detail in Hu et al. (2003). It considers all mutual partial inductances and is based on accurate partial inductance formulas. For the partial self-inductance we use the formula from Ruehli (1972).

3.2 Model order reduction and simulation

Using a modified nodal analysis (MNA) formulation the linear part of the model is represented as:

$$\mathbf{C}\dot{x}_n = -\mathbf{G}x_n + \mathbf{B}u_N$$
$$i_N = \mathbf{B}^\mathbf{T}x_n.$$

Essentially $\mathbf{C}$ and $\mathbf{G}$ are the susceptance and conductance matrices. The vector x_n consists of the MNA variables (voltages and currents) and the vectors u_n and i_N are the corresponding port voltages and currents. The PRIMA (Odabasioglu et al., 1998) algorithm reduces the MNA matrices to

$$\tilde{\mathbf{C}} = \mathbf{X}^\mathbf{T}\mathbf{C}\mathbf{X} \qquad \tilde{\mathbf{G}} = \mathbf{X}^\mathbf{T}\mathbf{G}\mathbf{X} \qquad (2)$$
$$\tilde{\mathbf{B}} = \mathbf{X}^\mathbf{T}\mathbf{B}.$$

The matrix $\mathbf{X}$ spans a block Krylov subspace and can be calculated for example with the block Arnoldi algorithm as illustrated in Alg. 1. Therefore to obtain a reduced model the explicit inductance matrix which is a submatrix of $\mathbf{C}$ is not necessary. Only the calculation of the product of the matrix $\mathbf{C}$ with some vectors is required in Eq. (2) and Alg. 1. To simulate Eq. (2) together with the nonlinear elements the MNA variables of the reduced system are interpreted as voltages. The symmetric part of the matrices $\tilde{\mathbf{C}}$ and $\tilde{\mathbf{G}}$ will be interpreted as resistors and capacitors in a SPICE netlist. The non-symmetric part will be realized with capacitors and resistors in series with a voltage controlled voltage source to ground. The connection between the ports and the "internal" network will be accomplished by the introduction of voltage controlled current sources at the ports and the internal nodes. There are different realizations imaginable, a similar realization can be found in Heres (2003).

One drawback of the approach is that the expansions point is set to $s_0 = 0$. For a different expansion point s_0 it is possible to replace the $\mathbf{G}^{-1}$ in Alg. 1 by $(\mathbf{G} + s_0\mathbf{C})^{-1}$, but this is numerically expensive. The expansion around zero leads to slightly bigger models than the expansion at higher frequencies to obtain a comparable accuracy.

Algorithm 1 Block Arnoldi with double orthogonalization.

(a) t_r=500ps

(b) t_r=100ps

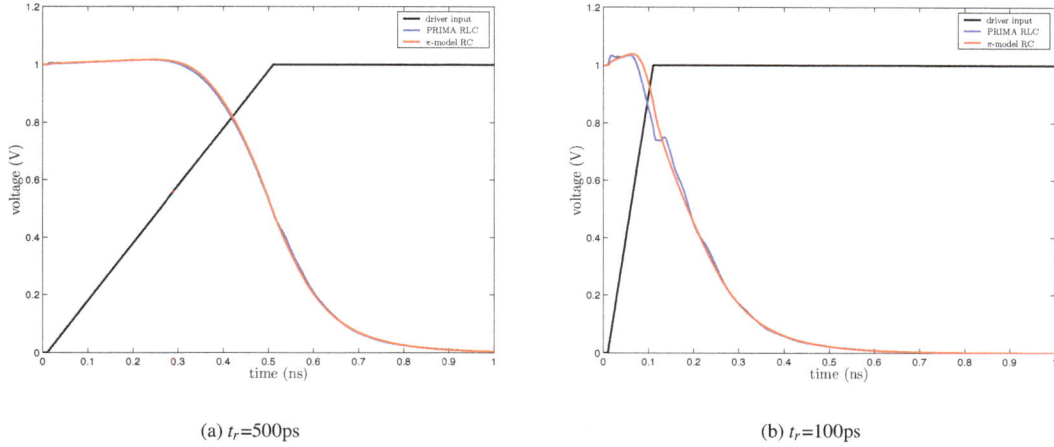

Fig. 4. Simulation results of the driver output for different rise times.

$$qn = ceil\,(q/N)$$
$$[\mathbf{X_0, R}] = qr\,(\mathbf{G^{-1}B})$$
for k=1 to qn
$$\mathbf{X_k = G^{-1}CX_{k-1}}$$
    for np=1 to 2
        for i=1 to k
$$\mathbf{X_k = X_k - X_i\,(X_{i-1}^T X_k)}$$
        end
    end
$$[\mathbf{X_k, R}] = qr\,(\mathbf{X_k})$$
end
$$\widetilde{\mathbf{X}} = \left[\mathbf{X_0, ..., X_{qn-1}}\right]$$
$$\mathbf{X} = \left[\widetilde{X}_0, ..., \widetilde{X}_q\right]$$

3.3 Simulation results

We investigated a bus with a length of 3.5 mm and the surrounding powergrid. For this topology the right side of Eq. (1) is considerably greater than 1. The left side is approximately 1 for a transition time of 100 ps. So for transition times larger than 100ps inductive effects should be negligible. The simulation results in Fig. 4 confirm this. It is also apparent that with a transition time of 100 ps already the first inductive effects arise. Hence our proposed π-load model can be deployed as long as Eq. (1) is not valid. In the other case model order reduction can be used to get a model which considers the inductive effects.

4 Conclusion

To analyze the output of a driver an accurate RC load model that considers capacitive coupling has been used. As long as inductive effects can be neglected and the bus is sufficiently short the model is valid. We have demonstrated that it is possible to use simple conditions to decide whether inductance has to be considered or not. If inductance is not negligible a reduced PEEC model which can be realized as a simple SPICE netlist can be used instead of the RC model.

Acknowledgement. This work was partially supported by Atmel Germany GmbH and the German Bundesministerium für Bildung und Forschung with the indicator 01M3060C. The authors are responsible for the content.

References

Gala, K., Zolotov, V., Panda, R. et al.: On-Chip Inductance Modeling and Analysis, Proc. of the ACM/IEEE DAC, 63–68, 2000.

Heres, P. J., and Schilders, W. A. H.: Reduction and realization techniques in passive interconnect modelling, Proc. IEEE Workshop on Signal Propagation on Interconnects, 157–160, 2003.

Hu, H., Blaunw, D. T., Zolotov, V. et al.: Fast On-Chip Inductance Simulation Using a Precorrected-FFT Method, IEEE Trans. on Computer-Aided Design of Integrated Circuits and Systems, 22, 1, 2003.

International Technology Roadmap for Semiconductors (ITRS): http://public.itrs.net/, 2001.

Ismail, Y. I., Friedman, E. G.: On-Chip Inductance in High Speed Integrated Circuits, Kluwer Academic Publishers, Massachusetts, 2001.

Odabasioglu, A., Celik, M., Pileggi, L. T., and PRIMA: Passive Reduced-Order Interconnect Macromodeling Algorithm, IEEE Trans. on Computer-Aided Design of Integrated Circuits and Systems, 17, 8 August 1998.

O'Brien, P. R. and Savarino, T. L.: Modeling the Driving-Point Characteristic of Resistive Interconnect for Accurate Delay Estimation, IEEE International Conference on Computer-Aided Design, ICCAD, 512–515, 1989.

Ruehli, A. E.: Inductance Calculations in a Complex Integrated Circuit Environment. IBM journal of research and development, 470–481, 1972.

Sheehan, B. N.: Library Compatible Ceff for Gate-Level Timing. Design Automation and Test in Europe Conference, Date Proceedings, 826–830, 2002.

Tahedl, M. and Pfleiderer, H.-J.: A Driver Load Model for Capacitive Coupled On-Chip Interconnect Buses, International Symposium on System-on-Chip, 2003.

6

Application driven evaluation of network on chip architectures for parallel signal processing

C. Neeb, M. J. Thul, and N. Wehn

University of Kaiserslautern, Germany

Abstract. Today's signal processing applications exhibit steadily increasing throughput requirements which can be achieved by parallel architectures. However, efficient communication is mandatory to fully exploit their parallelism. Turbo-Codes as an instance of highly efficient forward-error correction codes are a very good application to demonstrate the communication complexity in parallel architectures. We present a network-on-chip approach to derive an optimal communication architecture for a parallel Turbo-Decoder system. The performance of such a system significantly depends on the efficiency of the underlying interleaver network to distribute data among the parallel units. We focus on the strictly orthogonal n-dimensional mesh, torus and k-ary-n cube networks comparing deterministic dimension-order and partially adaptive negative- first and planar-adaptive routing algorithms. For each network topology and routing algorithm, input- and output-queued packet switching schemes are compared on the architectural level. The evaluation of candidate network architectures is based on performance measures and implementation cost to allow a fair trade-off.

1 Introduction

The network-on-chip approach is a new design paradigm where models and techniques from the computer network community are employed and reevaluated under diversified constraints of the SOC-design approach. Today's technologies allow to integrate vast quantities of functional units on a single chip, where the communication between these components becomes as important as the computations they perform. It is predicted that in the future complex designs with hundreds of functional blocks will be integrated by application-specific networks that offer a high degree of optimization.

In this work we investigate direct network architectures as the most popular interconnection schemes found in many parallel multiprocessor architectures. According to a classification introduced by Duato et al. (2003) we restrict our attention to the subclass of packet switched, strictly orthogonal point-to-point networks. These networks are characterized by their interconnection topology, that determines how nodes are physically interconnected. Each node represents a processing unit realizing the required computation. A communication specific component, the so called router, is attached to it for data transportation. The router implements the routing algorithm which determines the paths on which data is sent through the network.

The efficiency of a communication architecture is strongly application dependent. For a meaningful evaluation, performance measures like network throughput and latency and also area and energy consumption have to be put into the context of the application. A parallel Turbo-Decoder system presented by Thul et al. (2002a) serves as an application example.

Turbo-Codes belong to the iterative channel coding techniques that exhibit an outstanding forward-error correction capability, which made them part of today's communication standards, e.g. 3GPP (Third Generation Partnership Project), and are often found in the outer modem of wireless transmission systems.

For high-throughput, parallel decoder architectures are employed where data distribution due to interleaving makes up the major bottleneck. Therefore our evaluation of the considered networks is based on the specific demands of efficient interleaving architectures in parallel Turbo-Decoders. It should be noted that this topic is not primary specific to Turbo-Codes and thus can be transfered to e.g. LDPC-Decoders where data distribution due to interleaving is even more challenging.

In Sect. 2 we outline the design space for our network-on-chip approach, introducing the network topologies and the applied routing algorithms. Further two packet switched router architectures based on input- and output- queuing

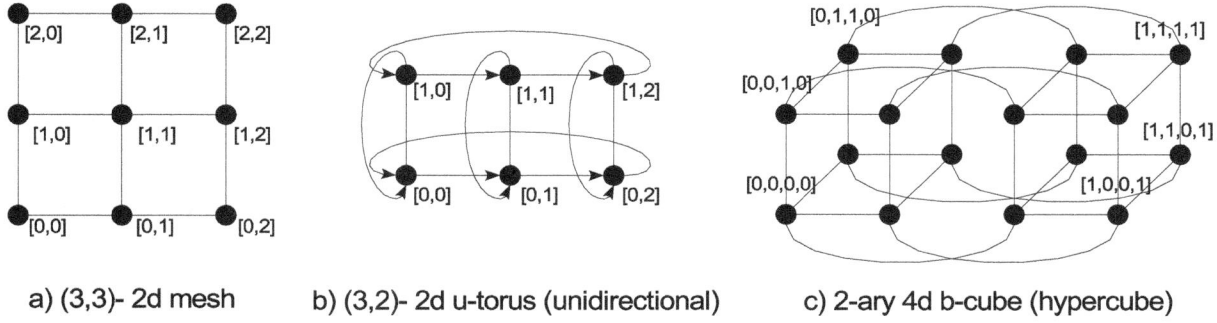

a) (3,3)- 2d mesh   b) (3,2)- 2d u-torus (unidirectional)   c) 2-ary 4d b-cube (hypercube)

Fig. 1. Examples for mesh, torus and k-ary n-cube network topologies.

schemes are discussed in detail. Section 3 presents the special requirements of interleaver networks in parallel Turbo-Decoders, followed by their evaluation in Sect. 4. In Sect. 5 performance and implementation costs are opposed for an exemplary decoder architecture comprised of 16 processing units. Section 6 concludes this paper.

2 Network-on-chip design

2.1 Network topology

Network topologies are traditionally modeled by a strongly connected directed graph $G(N, C)$, where the vertices N represent processing nodes and the edges C the set of connecting physical channels. We focus on n-dimensional meshes, tori and k-ary n-cubes as the most popular representatives of strictly orthogonal networks which are widely employed in commercial multicomputer systems.

They can be defined by a n-dimensional radix vector k, where its components $k_i, 0 \leq i < n$ determine the number of nodes along dimension i. A node x is identified by its coordinates $(x_{n-1}, x_{n-2}, ..., x_1, x_0)$, where $0 \leq x_i < k_i$.

In n-dimensional meshes neighboring nodes are connected by bidirectional channels, where each node has from n to $2n$ neighbors and thus making the mesh an irregular network. An example of a 2d-mesh with $k = (3, 3)$ is given in Fig. 1a. For the sake of simplicity undirected edges are used to symbolize two directed edges leading to opposite directions.

In the torus network wrap-around channels are added to the boundary nodes which gives the torus regularity and symmetry (see Fig. 1b). In contrast to the mesh network adjacent nodes can alternatively be linked by unidirectional channels. We refer to this topology as the u-torus whereas in b-tori nodes are interconnected by bidirectional channels respectively.

The k-ary n-cube represents a special case of the torus, where each dimension accommodates the same number of nodes making up to k^n nodes altogether. For $k = 2$ this topology is also referred to as binary n-cube or hypercube (Fig. 1c).

2.2 Routing algorithm

A predominant advantage of the introduced networks is the regular construction pattern and their high degree of symmetry. This essentially simplifies the employed routing algorithms, which specify the set of allowed paths in the network on which data can travel from an emitting source node to a destined target node. The employed routing algorithm is tightly coupled to the network topology and crucial for efficient communication in parallel architectures.

A routing algorithm is denoted as deterministic if the transmission of a packet, containing target address and data, always follows the same path for a pair of source/target nodes. On the opposite, adaptive algorithms provide alternative paths where local traffic estimates can be exploited to circumvent congested channels and hot-spots in the network. However, to achieve reliable communication, routing freedom is essentially restricted to avoid the case of deadlock (Glass and Ni, 1992). In these situations packets cannot advance towards their destination because requested network resources are already allocated by other packets forming a cyclic dependency. They can systematically be avoided by the introduction of virtual channels as proposed by Dally and Seitz (1987). Virtual channels represent logical packet streams inside a router that share the physical channels for transmission. The impact on router implementation is discussed in Sect. 2.3.

In this paper we restrict our analysis on deadlock-free routing algorithms namely the deterministic dimension-order (Dally and Seitz, 1987), the partially adaptive negative-first (Glass and Ni, 1992) and the planar-adaptive routing proposed by Chien and Kim (1992). In dimension-order routing packets are crossing dimensions in a strictly increasing order. Effectively the address offset of the current node and the target node is computed and incrementally reduced to zero starting with the lowest dimension until the target is reached. This process of clearing dimensions in a fixed order always leads to the selection of the same shortest path between a source and a destination node. The negative-first routing algorithm implies less restrictions on the set of allowed paths and thus provides some adaptivity. At first packets are routed adaptively in negative directions and only when no negative dimension offsets are left then routing proceeds adaptively in positive directions. The planar-adaptive algorithm routes packets in a series of 2d-planes. It provides full adaptivity inside the current plane whereas the transition to the next plane is fixed. Consequently this algorithm can be classified as fully adaptive in 2d-networks whereas only partial adaptivity is provided for higher dimensional networks. To guaranty deadlock-freedom at most three virtual channels for meshes and six virtual channels for tori and cubes are necessary.

2.3 Router architectures

The implementation of the three mentioned routing algorithms is based on two packet switched router architectures. They mainly differ in the way pending packets are internally buffered in the case of contention whereas the appropriate choice does not depend on the algorithm itself.

In the input-queued router (IQ) depicted in Fig. 2a a dedicated input-queue (FIFO buffer) is attached to each inport of the router. Physical channel flow-control (FC) is realized by link-controllers (LC) that stop neighboring nodes sending data packets in case of a full queue. To support virtual channels for deadlock-avoidance extra queues are inserted where each is assigned to a single virtual channel. In this case the link controller must demultiplex incoming packets of the same physical channel on the appropriate virtual channel queue. An additional virtual channel arbiter (VC-Arbiter) selects a packet in one of these queues and forwards it to the address decoder (AD) where the actual routing is done. Dependent on the target address of the packet, one (deterministic routing) or multiple outports (adaptive routing) are requested for further packet delivery. The physical connectivity is realized by a crossbar-switch which is controlled by the scheduler to resolve contending requests for the same outport at the same time. These conflicts are the main reason for the noticeable degradation of the throughput in input-queued routers. We choose the so called SLIP scheduling algorithm (McKeown, 1995) for implementation as it is based on a fair round-robin scheme.

The output-queued router (OQ) circumvents this switching loss by the assignment of a dedicated queue for each pair of in-/out-channels. Hence multiple packets can be forwarded to the same outport at the same time as they go to different out-queues. Again a round-robin scheme is used to multiplex the contents of the queues on the outgoing physical channels. The high throughput and channel utilization of this architecture comes at the expense of the large number of queues needed restricting applicability to low-dimensional routers only. Figure 2b illustrates the output-queued architecture without additional virtual channel queues for the sake of clarity.

3   Interleaving in parallel turbo-decoders

For the evaluation of our network-on-chip approach we refer to concurrent interleaving in a partial parallel Turbo-Decoder

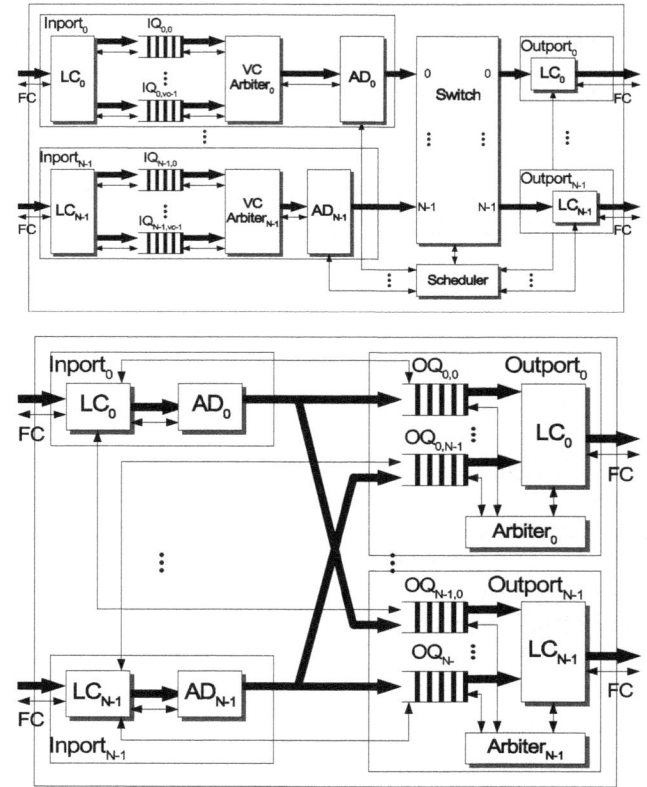

Fig. 2. Router architectures based on (**a**) input-queuing and (**b**) output-queuing.

as the application background. Interleaving is used in many channel coding schemes and essentially impacts the communication performance of Turbo-Codes. It determines a rearrangement of data inside a block to scramble their processing order and thus to break up neighborhood-relations effectively. In parallel Turbo-Decoders each processing unit works on a subblock of data individually where information must be exchanged according to the adopted interleaving scheme.

From a network perspective an all-to-all communication must be established. A fully interconnected architecture where all units are linked by dedicated channels exhibits a high wiring effort which becomes infeasible for increasing degrees of parallelization. Additionally, write conflicts occur when two packets are sent to the same unit at the same time. As only one packet will be accepted by the receiver, the queuing of data becomes inevitable. A dedicated buffer per channel has to be provided which further increases implementation costs drastically.

This issue has already been addressed by Thul et al. (2002c) for the first time where interleaving is identified as the upcoming bottleneck for high-throughput Turbo-Decoding. An optimized architecture based on a ring topology has been presented by Thul et al. (2002b). Both architectures do not employ any flow-control mechanism. They require large queues which are dimensioned by means of

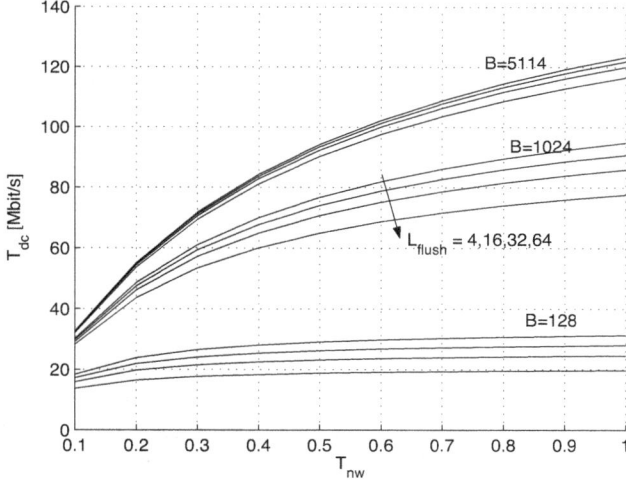

Fig. 3. Impact of the interleaver network throughput and latency on the throughput of a parallel turbo-decoder ($N = 16$, $f = 200\,\text{MHz}$).

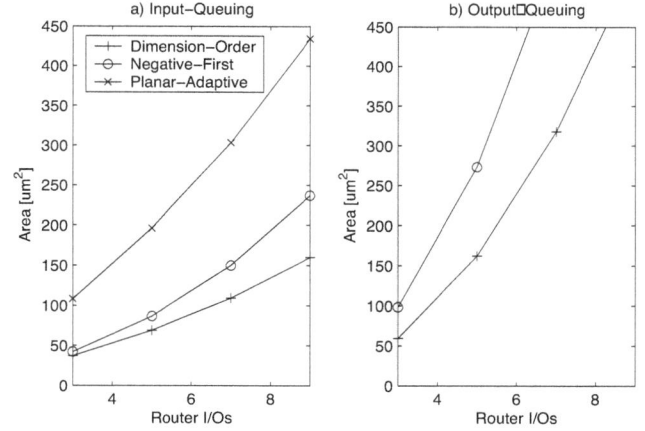

Fig. 4. Quantitative comparison of router complexity for n-dimensional meshes, $Q = 4$.

exhaustive simulations. We will refer to the latter as a benchmark of our approach that requires substantially smaller buffers.

To rate the various network architectures in terms of performance, we derive a model to capture the interleaving specific requirements and to quantify the impact on the overall decoder. The throughput of the considered parallel Turbo-Decoder is given as:

$$T_{dc} = \frac{B \cdot f}{L_{dc}}$$

with B the data block length, f the clock frequency and L_{dc} the decoder latency. Data blocks of length B are partitioned into N subblocks of length B/N where each is decoded separately by one of the N processing units. The decoder latency L_{dc}

$$L_{dc} = B + 2It(\frac{B}{N} + 2W + c + L_{il})$$

amounts to the number of clock cycles needed to decode a complete data block and mainly depends on the interleaver latency L_{il}. According to the given interleaving scheme, N data values have to be communicated per clock cycle. The contribution of the interleaver to the decoder latency can be subdivided into two terms:

$$L_{il} = L_{\text{stall}} + L_{\text{flush}} = \frac{B}{N}(\frac{1}{T_{nw}} - 1) + L_{\text{flush}}(T_{nw}, Q)$$

L_{stall} comprises the fraction of clock cycles a processing unit has to be stalled to avoid overloading the network and hence loosing any packets. The probability of a processing stall particularly depends on the normalized average throughput T_{nw} of the interleaver network. Before the next half-iteration can be started, all data has to be distributed. Thus we define the flush latency L_{flush} as the time span between the arrival of

the last data value at the entrance and the time it leaves the interleaver network. During this time internal data queues are emptied which is mainly influenced by the network throughput and the depth Q of the router queues. The impact on the overall throughput of the Turbo-Decoder is illustrated in Fig. 3 for $N = 16$.

4   Network evaluation

To allow a fair trade-off of the suitability of a network architecture for interleaving, we oppose implementation costs and performance benefits. For further discussions we make the following reasonable assumptions:

1. Data is distributed with equal probability of $1/N$ to any of the target units. This is implied by good interleavers from a communication point of view where data is nearly evenly spread over the address space. We therefore adopt a uniform traffic model where one data value can be delivered by each unit per clock cycle.

2. A physical channel of the network is capable to transfer a single data packet in one clock cycle. We assume a typical channel width of 20 bits for UMTS compliant interleaving.

4.1   Router implementation cost

Figure 4 shows a comparison of the logic area requirements for a single router with respect to the routing algorithm and the number of I/O-ports. For this we assume routing in mesh networks which imply a minimal set of virtual channels and an equal queue-depth to store four packets. The results were obtained through synthesis based on a standard $0.18\,\mu\text{m}$ technology using Synopsis Design Compiler.

All input-queued routers opposed in Fig. 4a are dominated by the size of the queues which make up 50–70% of the overall area. With an increasing degree of adaptivity the routers exhibit a growing complexity of the scheduler (8–22%) and

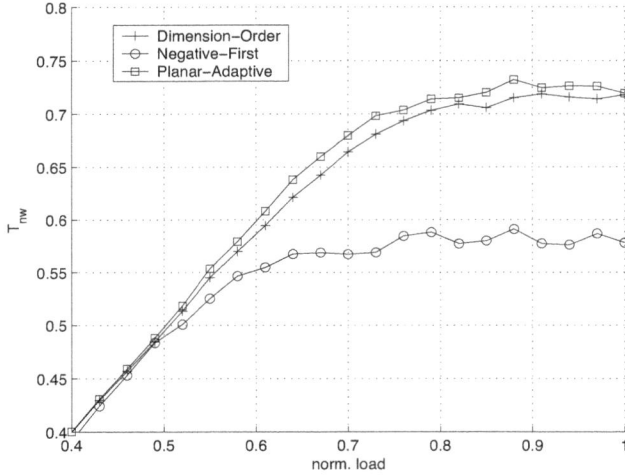

Fig. 5. Normalized throughput of the routing algorithms using input-queuing in a (4,4)-2d-mesh, $Q = 6$.

Table 1. Maximum number of nodes per dimension and in total for 2d- and 3d-networks.

Topology	U_c	BW	k_{max}	N_{max} (2D)	N_{max} (3D)
u-torus	$\frac{N-1}{2 \cdot BW}$	$2k_1 k_2 ... k_{n-1}$	≤ 3	9	27
mesh	$\frac{N}{2 \cdot BW}$	$2k_1 k_2 ... k_{n-1}$	≤ 4	16	64
b-torus	$\frac{N}{2 \cdot BW}$	$4k_1 k_2 ... k_{n-1}$	≤ 8	64	512

the crossbar switch (9–16%). The address decoders which implement the routing algorithm have only a slight influence of about 4–12% on the total area. Particularly for the planar-adaptive router the additional queues and extra arbiters for virtual channel support are expensive to implement.

The output-queued routers opposed in Fig. 4b need substantially more queues compared to the input-queued routers with a contribution of about 85% to the total area. Due to the restrictions of the dimension-order algorithm in contrast to the negative-first, less queues are needed to implement the former. We do not consider the planar-adaptive router for output-queuing as the need for virtual channels requires again a multiple of the already high number of queues.

4.2 Router performance

To quantify the impact of the routing algorithms on network performance for interleaving, simulations are used to measure throughput under varying load. For this, a (4,4)-2d-mesh network using input-queued routers is loaded with uniform traffic. As shown in Fig. 5 negative-first routing achieves the lowest throughput of all for high loads whereas planar-adaptive and dimension-order nearly perform equally. With increasing network dimension, throughput of planar-adaptive routing proved to be even worse compared to deterministic routing. This can be explained by the fact that partially adaptive routing uses local traffic estimates to make a routing decision which distorts the evenness of the global uniform traffic. Opposed to this, deterministic routing maintains this characteristic, spreading traffic more evenly across the network. Similar results are obtained for torus and cube networks as well. Consequently we regard the deterministic dimension-order routing as the best suited algorithm for interleaving. It achieves the highest throughput and comes at lowest implementation cost.

4.3 Topology constraints

Both cost and performance of an interleaver network are heavily affected by the choice of the appropriate topology. In the following we derive a necessary condition for the maximum number of nodes per dimension such that throughput is not degraded by the network topology. For that, we refer to the "bisection" of a topology as the minimum cut to divide the network into two equal sets of nodes. Accordingly the "bisection width" BW comprises the number of channels that have to be cut (see Duato et al., 2003, for further details).

Due to the regular construction and symmetry of the topologies, $(N − 1)/2$ packets in u-torus networks and $N/2$ packets in meshes and b-tori respectively have to cross the bisection during each cycle in average. We can therefore easily derive the normalized channel utilization U_c as the fraction of clock cycles a packet allocates a single channel by the following contemplations: Lets assume that dimension zero contains the largest number $k_0 = k_{max} = \max\{k_i\}$ of nodes in the network with $N = k_0 k_1 ... k_{n-1}$ nodes altogether, then the bisection is orthogonal to this dimension. Since at most one packet can be transferred over a single channel per cycles it follows that $U_c \leq 1$. This condition leads to the maximum number of nodes for any dimension k_{max} depending on the bisection width BW of the topology summarized in Table 1. Consequently the largest number of nodes can be accommodated by cube topologies where the same number of nodes resides in all dimensions. Thus the number of dimensions of the topology sets an upper bound of nodes that can be integrated by the network.

5 Results

The suitability of the presented networks for interleaving is discussed for an exemplary Turbo-Decoder with $N = 16$ parallel processing units. We investigate a (4,4)-2d-mesh and a 4-ary-2d-b-cube using input- and output-queued dimension-order routing. Both of the considered topologies comply to the condition derived in Sect. 4.3 as they do not exceed the maximum number of four nodes per dimension in 2d-meshes and eight nodes for 2d-b-tori (-cubes) respectively. We exclusively employ deterministic dimension-order routing because it achieves the highest throughput (Sect. 4.2) at the lowest cost in all considered topologies (Sect. 4.1).

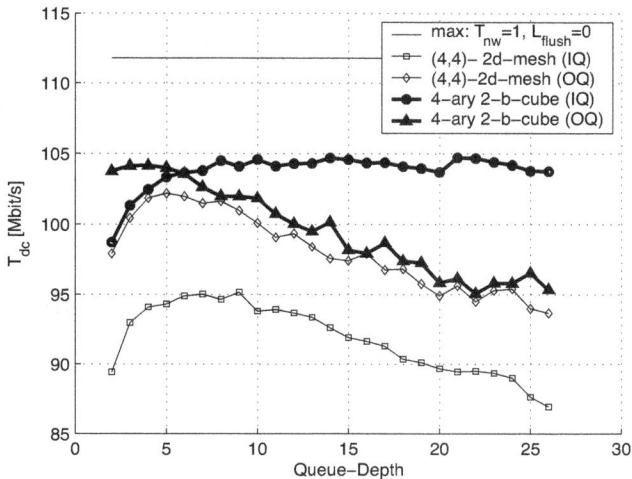

Fig. 6. Impact of router queue-depth on the turbo-decoder throughput using dimension-order routing ($N = 16$, $B = 2048$).

Table 2. Router queue-depths for optimal trade-off of decoder throughput and network area.

Network	Q	T_{dc} [Mbit/s]	Area$_{nw}$ [mm^2]
4-ary-2d-b-cube (IQ)	5	104	1.3
4-ary-2d-b-cube (OQ)	2	104	1.5
(4,4)-2d-mesh (OQ)	4	102	1.7
(4,4)-2d-mesh (IQ)	6	95	1.2

Network throughput and the flush latency finally depend on the depth of the router queues. Thus further performance simulations are carried out with varying queue-depths under full load (norm. load = 1) to quantify their impact on the overall decoder throughput. A fixed number of packets is sent across the network then the source is switched off to measure the flush latency. For small queue sizes, network throughput tends to deteriorate and thus the processing units often have to be stalled (see Fig. 6). With deeper queues throughput of the interleaver network reaches saturation and the flush latency becomes dominant. We choose the minimal queue-depth where at least 90% of the maximum throughput is achieved for comparison. The total network costs are estimated by extrapolation of the single router area with according queue-depths.

Compared to the architecture proposed by Thul et al. (2002b) an area reduction of factor ten for the interleaver network is achieved. Only very small queues are necessary in our approach to implement a high-throughput interleaver network that is capable to handle all possible interleaving schemes.

6 Conclusion and future work

Strictly orthogonal networks facilitate very efficient implementations of interleaver networks. In this context deterministic routing is superior to partial adaptive algorithms in terms of performance and implementation costs. The employment of network flow-control drastically reduces the sizes of the queue buffers and hence logic area of the whole network.

Particularly high-dimensional topologies offer only limited scalability. Future work will have to focus on the exploration of more sophisticated topologies and deadlock-free routing algorithms that can be tailored to an arbitrary number of nodes.

Acknowledgements. This work has been supported by the Deutsche Forschungsgesellschaft (DFG) under grant We 2442/1-3 within the Schwerpunktprogramm "Grundlagen und Verfahren verlustarmer Informationsverarbeitung (VIVA)".

References

Chien, A. and Kim, J.: Planar-adaptive Routing: Low-cost Adaptive Networks for Multiprocessors, in Proc. 19th International Symposium on Computer Architecture, 268–277, 1992.

Dally, W. and Seitz, C.: Deadlock-free Message Routing in Multiprocessor Interconnection Networks, in IEEE Transactions on Computers, 547–553, 1987.

Duato, J., Yalamanchili, S., and Ni, L.: Interconnection Networks – An Engineering Approach, Morgan Kaufman Publishers, San Francisco, USA, 2003.

Glass, C. and Ni, L.: The Turn Model for Adaptive Routing, in Proc. 19th International Symposium on Computer Architecture, 278–287, 1992.

McKeown, N.: Scheduling Algorithms for Input-queued Cell Switches, Ph.D. thesis, University of California, Berkeley, 1995.

Third Generation Partnership Project: 3GPP home page, www.3gpp.org.

Thul, M. J., Gilbert, F., Vogt, T., Kreiselmaier, G., and Wehn, N.: A Scalable System Architecture for High-Throughput Turbo-Decoders, in Proc. 2002 Workshop on Signal Processing Systems (SiPS'02), San Diego, California, USA, 152–158, 2002a.

Thul, M. J., Gilbert, F., and Wehn, N.: Optimized Concurrent Interleaving for High-Throughput Turbo-Decoding, in Proc. 9th IEEE International Conference on Electronics, Circuits and Systems (ICECS'02), Dubrovnik, Croatia, 1099–1102, 2002b.

Thul, M. J., Wehn, N., and Rao, L. P.: Enabling High-Speed Turbo-Decoding Through Concurrent Interleaving, in Proc. 2002 IEEE International Symposium on Circuits and Systems (ISCAS'02), Phoenix, Arizona, USA, 897–900, 2002c.

7

Resonant charging

C. Saas and J. A. Nossek

Munich University of Technology, Institute for Circuit Theory and Signal Processing, Arcisstr. 16, 80290 Munich, Germany

Abstract. It has been shown (Athas et al., 1994) that adiabatic switching can significantly reduce the dynamic power dissipation in an integrated circuit. Due to the overhead in the realization of adiabatic logic blocks (Saas et al., 2000) the best results are achieved when it is used only for charging dominant loads in an integrated circuit (Voss and Glessner, 2001). It has been demonstrated (Saas et al., 2001) that a multi stage driver is needed for minimal power dissipation. In this article a complete three stage driver including the generation of oscillating supply is described. To obtain a minimal power dissipation during synchronization the resonant frequency has to be constant. Therefore the waveforms for the logic states of the signal and the realization of a single stage differ from those presented in (Saas et al., 2001). In the H-SPICE simulations losses of the inductor are taken into account. This allows to estimate the power reduction that is achievable in a real system.

1 Introduction

Adiabatic switching (Athas et al., 1994) is a method to reduce the dynamic dissipation of a circuit by charging the capacitances with a time-variant source. The minimal energy dissipation is achieved for a constant charging current.

$$E_{diss} = \frac{RC_L}{T} C_L V_{dd}^2 \qquad (1)$$

It has been shown in (Saas et al., 2000) that the realization of general complex logic blocks based on pass transistor logic utilizing adiabatic switching leads to a considerable overhead for maintaining true adiabatic behavior.

Therefore, to avoid this overhead it is worthwhile to consider the energy dissipation associated with dominant load capacitances only, without the inclusion of complex logic. Keeping in mind that usually a large part of the power is dissipated in the I/O-cells of complex chips (Sakurai et al., 1997), a significant reduction of this dissipation can be expected. Since the driver is only a single cell, which can be controlled by standard CMOS logic, it can easily be included into a standard design flow.

The standard pad driver has to charge the pad as quickly as possible. In particular this time has to be shorter than the clock period T_{clk}. For this reason the $\frac{W}{L}$ of the driving inverters has to be adjusted to keep the channel resistance R_{ch} sufficiently low. The dissipated energy $E_{diss} = \frac{1}{2}C_L V_{dd}^2$ is independent of the channel resistance R_{ch}. Therefore the properties of the transistor do not influence the energy dissipated but only the switching speed of the driver.

To fulfill the speed requirements in most cases an inverter chain is used. The dissipated energy is increased by the additional gate capacitances of the driving inverters in a multiple stage CMOS driver.

The basic idea of an adiabatic driver is very simple. The capacitor is charged through a transmission gate by the phase Φ. The gate is controlled by standard CMOS gates. Both, p and n channel transistors are used to obtain a full voltage swing at the output. Of course the $\frac{W}{L}$ has to be reasonably large as the dissipated energy is dependent on the channel resistance R of the transmission gate (1). Despite the simplicity of the circuit there is a significant saving of energy. The biggest part of the remaining power loss is dissipated in the controlling CMOS gates which charge the large gate capacitances of the transmission gate non-adiabatically. To minimize the losses in the controlling gates, the adiabatic stages can be cascaded. By this, only transistors in the first stage are charged non-adiabatically.

2 The multistage driver

For the present examination a three stage design has been chosen. To find the optimum number of stages is an open problem.

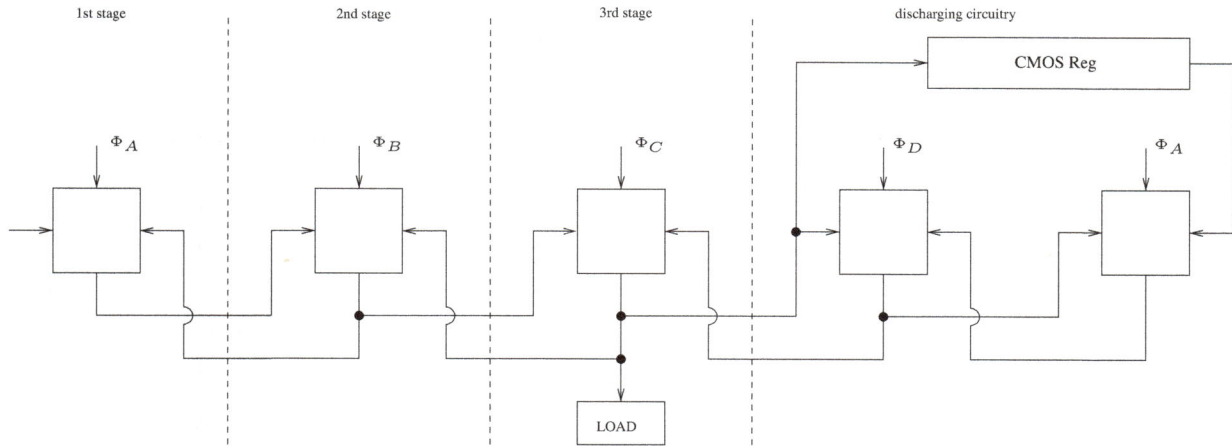

Fig. 1. Schematics for the multi stage driver.

A diagram for a three stage driver is shown in Fig. 1. About half of the energy is dissipated during the discharging process. Therefore, also the discharging has to be done adiabatically. For the discharging process the initial state in the controlling gate has to be present for the whole discharging process (Schlaffer, 2000). So, it is necessary to store this state. Of course, the discharging has to be done with appropriately sized transistors. For discharging the same cascade as for charging is obtained. For the last stage a CMOS register is used to store the state needed for adiabatic discharging. Only the minimal transistors in the register are discharged non-adiabatically, but this energy is negligible.

Four clock phases are used. Each stage is connected to a clock phase which is delayed by a quarter of the clock period to the phase of the predecessor (see Fig. 5). By this the output of the preceding stage is valid during the rising edge. Therefore, the output is well suited as a control signal for the charging of the succeeding stage. On the other hand, the succeeding stage is valid during the falling edge at the preceding stage. Therefore, its output can be used to discharge the output of the preceding stage adiabatically. For the charging of the first stage standard CMOS gates are used. They have to be valid for three quarters of the clock period. The output of a standard CMOS register is used to discharge the last stage. Its state has been set in one of the preceding stages.

Using this timing a three stage driver generates a latency of $\frac{3}{4}T_{clk}$.

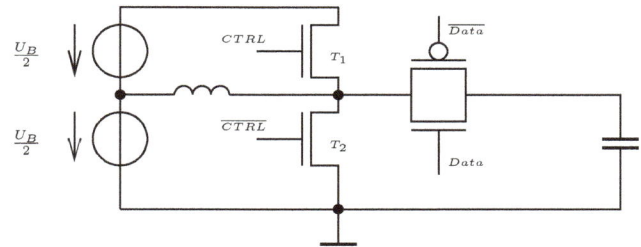

Fig. 2. Oscillator.

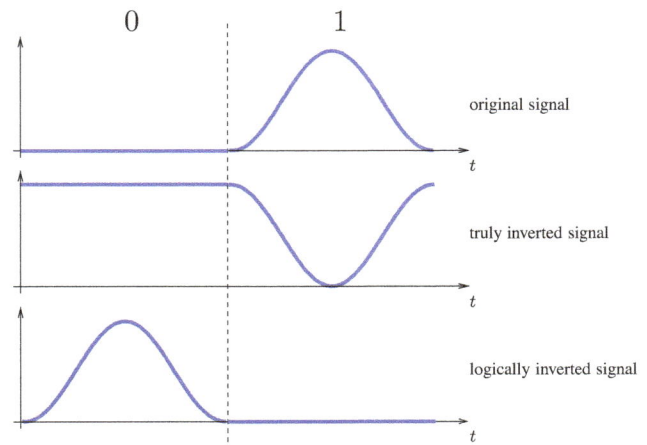

Fig. 3. Waveforms and inverted waveforms.

3   The oscillator

As the generation of ideal ramps is not possible with a high efficiency, they are approximated by a sinus. To generate the sinusoidal power supply, an oscillator is needed. The oscillator (Fig. 2) is a resonant LC oscillator working in class E operation. It is based on (Ziesler et al., 2001) and inherits some advantages when compared to the widely known blip circuit which is often proposed as the source for adiabatic cir-

cuits (Athas et al., 2000). It needs only a single inductor for each phase and generates a full sinusoidal voltage oscillation at the capacitor.

The transistors T_1 and T_2 are closed for a short period at the maximum, respectively the minimum, of the output voltage to compensate for losses during one cycle. In addition this keeps the oscillation synchronous to the CMOS clock. Although some work has been published (Ziesler et

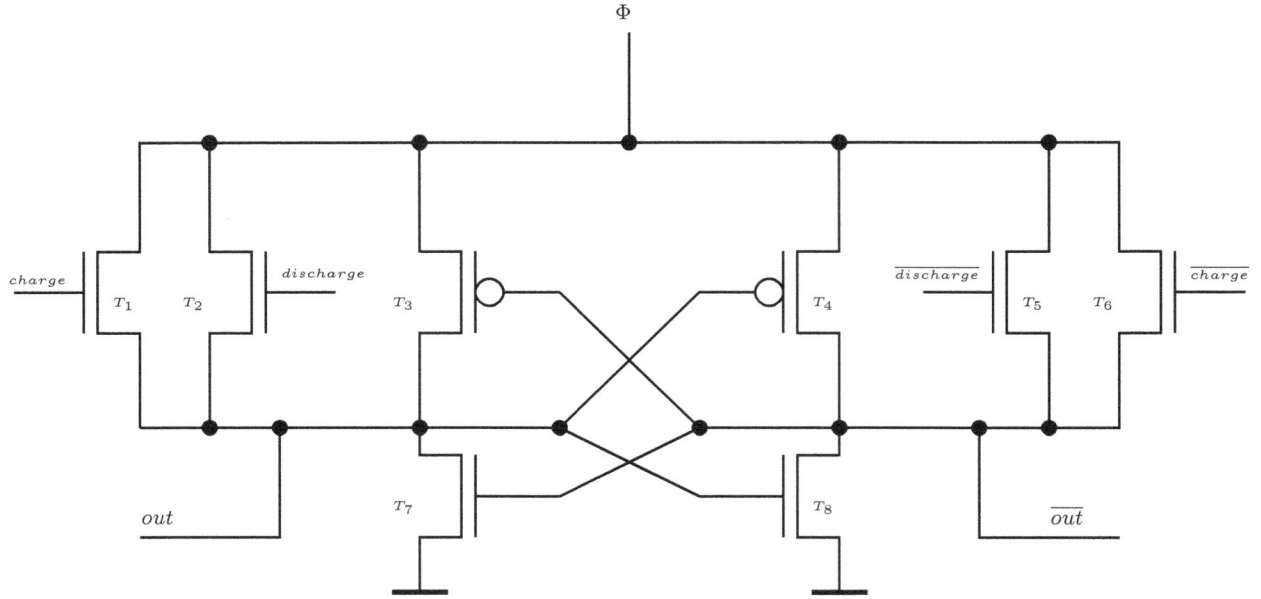

Fig. 4. Circuit of a single stage.

al., 2001), the controlling signals for the T_1 and T_2 are generated manually.

4   Different waveforms for the logic states

The idea of the adiabatic driver is to charge the output with the oscillating power supply in case of a logical 1 and to cut it off in case of a logical 0. A logic 1 will result in one period of the oscillation at the output. It starts and ends in the minimum of the sinus. The DC offset as well as the amplitude of the oscillation is $\frac{V_{dd}}{2}$.

As already mentioned above, dual rail encoding is required. There are two ways to invert the signal, which will be named "truly inverted" and "logically inverted". They are depicted in Fig. 3

If the original waveform is inverted, a signal which is at V_{dd} for a logical 0 and sinusoidal for a logical 1 is obtained. This signal would be the perfect one to control the p-MOS transistor of the transmission gate in the previous and the subsequent stage. The drawback of this solution is, that one can not combine the original and the inverted signal to a DC and a sinusoidal one just by switching between them. This is needed to present a constant load to the oscillator. Since the load capacitance is part of the resonator and therefore has a direct influence on the operating frequency it is mandatory to keep it constant for all logic states of the circuit. To achieve this, there has to be a constant sinusoidal oscillation on a constant number of load capacitances. Therefore, an "logically inverted" output is used. During a logic 1 the output will oscillate and the inverted one will stay at 0 Volt. For the logic 0 it is vice versa. If only a logically inverted signal is created at the output of each stage, there is no signal available which is suitable for controlling the p-channel transistors in the other

Table 1.

$\frac{1}{2} C U^2$	350.00 $\frac{pJ}{Bit}$
CMOS	351.52 $\frac{pJ}{Bit}$
adiabatic driver at 1 MHz	52.70 $\frac{pJ}{Bit}$

stages. Therefore, the individual stages of the driver have to be redesigned.

5   A single stage

It is well known, that a p-channel transistor is needed to obtain a full charging up to V_{dd}. If only a n-channel transistor is used, it is only conducting until V_{GS} is larger than V_{th}. Therefore, the source voltage can only reach $V_{dd} - V_{th}$. If one has a look at the waveform of a logical "1", one can see, that the p-channel transistor is only needed during a rather short period. The basic evaluation of the input of a single stage is done by the n-channel transistors. It seems to be a good idea to abandon the signal to control the p-channel transistors and use some internal signal which already reached its final level due to the n-channel transistors instead.

Such a solution is proposed in Fig. 4. It consists of 2 n-channel transistors per output signal (T_1, T_2 and T_5, T_6). One is used to charge the output, and the other one for discharging. The charging signal is "delayed" by $-\frac{T}{4}$ whereas the discharging signal is delayed by $\frac{T}{4}$.

There is only one p-channel transistor per output signal (T_3, T_4). If the output is logical 1 it has to be conducting

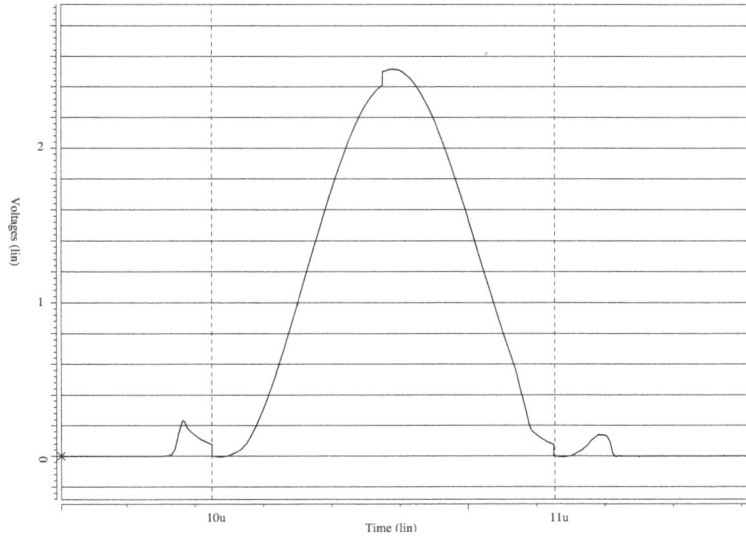

Fig. 5. Simulated output waveform.

during the maximum of Φ. On the other hand, it has to be non-conducting for $U_\Phi > V_{th}$ in the case of a logical 0.

This behavior can be achieved by controlling the p-channel transistor with the inverted output. The n-channel transistors T_7 and T_8 ensure the output signal to be 0 V for a logical "0". They are not present in stage 3, as the load capacitor is large enough to achieve a stable output.

6 Simulation results

The adiabatic multistage driver has been simulated using a $0.25\,\mu$m process and H-SPICE. As the inductor is intended to be an external one, a Q value of 100 seems to be realistic.

A driver for 4 off-chip connections has been simulated. Each of the connections represents a load of 28 pF. The adiabatic drivers are implemented to load one output at a time. Thus a group of 4 drivers represents a constant data independant load to the oscillators for all times.

The simulations results show that the circuit is working adiabatically. The driver dissipates much less energy than a conventional CMOS driver at useful operating frequencies. It has to be noted, that these results summarize the whole energy that is dissipated including the generation of the sinusoidal power supply.

7 Summary and outlook

In this paper a multi stage adiabatic driver has been presented. An oscillator has been chosen to generate the sinusoidal power supply. To ensure minimal losses during the synchronization to the CMOS clock the oscillator has to work on a constant load. The single stages of the driver have been designed to fit the requirements for constant load. To avoid the need for truly inverted signals the p-channel tran-

sistors which are needed to reach V_{dd} are realized as clamping devices. The simulation results show a pretty large potential for adiabatic drivers. Although a small latency of $\frac{T}{4} * Num.\ of\ stages$ is introduced the energy savings still make this concept interesting for a number of applications. Of course high speed applications are not the aim for adiabatic circuits, but the simulations have shown that reasonable operation is possible up to 100 MHz.

Further research has to be done on the modeling of the load and the other off-chip connections. Most probably the results can be improved by more advanced methods for the transistor sizing.

References

Athas, W., Svensson, L. J., Koller, J., Tzastzanis, N., and Chou, E.-C.: Low-power digital systems based on adiabatic-switching principles, IEEE Transactions on Very Large Scale Integration (VLSI) Systems, 2, 398–406, 1994.

Athas, W., Tzartzanis, N., Mao, W., Peterson, L., Lal, R., Chong, K., Moon, J.-S., Svenson, L., and Bolotski, M.: The design and implementation of a low-power clock-powered microprocessor, Journal of Solid State Circuits, 35, 1561–1570, 2000.

Saas, C., Schlaffer, A., and Nossek, J.: An adiabatic multiplier, PAT-MOS 2000, 276–284, 2000.

Saas, C., Schlaffer, A., and Nossek, J.: An adiabatic multi stage driver, ECCTD, 2001.

Sakurai, T., Kawaguchi, H., and Kuroda, T.: Low-power cmos design through v_{th} control and low-swing circuits, Institute of Industrial Science, Univ. of Tokyo, 1997.

Schlaffer, A.: Entwurf von adiabatischen Schaltungen, Ph.D. thesis, Munich University of Technology, 2000.

Voss, B. and Glesner, M.: A low-power sinusoidal clock, ISCAS, 2001.

Ziesler, C., Kim, S., and Papaefthymiou, M.: A resonant clock generator for single-phase adiabatic systems, ISLPED, 2001.

8

Test signal generation for analog circuits

B. Burdiek and W. Mathis

Institut für Theoretische Elektrotechnik und Hochfrequenztechnik, Universität Hannover, Appelstr. 9A, 30167 Hannover, Germany

Abstract. In this paper a new test signal generation approach for general analog circuits based on the variational calculus and modern control theory methods is presented. The computed transient test signals also called test stimuli are optimal with respect to the detection of a given fault set by means of a predefined merit functional representing a fault detection criterion. The test signal generation problem of finding optimal test stimuli detecting all faults form the fault set is formulated as an optimal control problem. The solution of the optimal control problem representing the test stimuli is computed using an optimization procedure. The optimization procedure is based on the necessary conditions for optimality like the maximum principle of Pontryagin and adjoint circuit equations.

1 Introduction

Advances in EDA technology have increased the size and complexity of integrated circuits. As a result the test costs have become a key part of the overall manufacturing costs. Although the area of the analog part of a mixed-signal IC is much smaller than the digital one, the test costs are dominated by the analog part because of its more complex specifications. For this reason tools and efficient techniques for the generation of specific tests, which have the ability to reduce the test time and thus the test costs, are needed.

In transient testing (Gomes and Chatterjee, 1999; Variyam et al., 1999; Burdiek, 2001), the circuit under test (CUT) is excited with a transient test stimulus and the circuit response is sampled at specified time points for fault detection. In this paper a new test signal generation method based on control theory techniques like Pontryagin's maximum principle is presented. It should be noted that optimal control theory methods such as the maximum principle are based on the variational calculus. The proposed test generation approach formulated as an optimal control problem generates optimum transient test stimuli for a general analog circuit. A Lagrangian merit functional required for the optimal control

problem serves as the fault detection criterion. The functional, which indirectly depends on the controls of the circuit representing the test stimuli, is based on the difference between the good test response and all faulty test responses. The solution of the control problem given by the optimal controls, is computed using an optimization procedure. The optimization procedure maximizes the merit functional with respect to the controls and thus enhances the fault detection capability. Since the procedure takes advantage from necessary optimality conditions such as the maximum principle, it does not need the gradient of the merit functional with respect to the controls. The solution of the optimal control problem must satisfy the necessary conditions for optimality like the maximum principle. Thus, all test stimuli, which do not satisfy Pontryagin's maximum principle, are not optimal and thus do not maximize the fault detection criterion. Therefore, they cannot be solutions of the optimal control problem.

In the next section the test generation problem is formulated as an optimal control problem. The optimization procedure and the necessary conditions for optimality, which have to be fulfilled by the optimal controls representing the test stimuli, are described in Sect. 3. Experimental results are presented in Sect. 4. A conclusion of the paper is given in Sect. 5.

2 Problem formulation

Let $\mathbf{F}(\dot{\mathbf{z}}(\mathbf{x}), \mathbf{x}, \mathbf{u}, t) = \mathbf{0}$ be the differential algebraic equations (DAE's) of the circuit under test (CUT). The DAE system, which models the circuit correctly, arises from the modified nodal analysis (MNA). The control vector $\mathbf{u}(t)$, with $\mathbf{u} \in U \subset R^p$, represents the controls of all independent voltage and current sources of the circuit. The state vector $\mathbf{x}$, with $\mathbf{x} \in R^n$, represents the node potentials $\mathbf{v}_n$ and the branch currents $\mathbf{i}_b$ of the modified nodal description. The vector function $\mathbf{z}(\mathbf{x})$ describes the charges of the voltage controlled capacitors and the fluxes of the current controlled inductors of the circuit. The variable vector of the adjoint network of the circuit (Director and Rohrer, 1969) is called the costate vector denoted by ψ.

$$\mathbf{x} = \begin{bmatrix} \mathbf{v}_n \\ \mathbf{i}_b \end{bmatrix} \quad \mathbf{u} = \begin{bmatrix} \mathbf{J} \\ \mathbf{E} \end{bmatrix} \quad \psi = \begin{bmatrix} \hat{\mathbf{v}}_n \\ \hat{\mathbf{i}}_b \end{bmatrix} \quad \mathbf{z} = \begin{bmatrix} \mathbf{q} \\ \phi \end{bmatrix} \tag{1}$$

The given fault list of the CUT containing k parametric and catastrophic faults is termed by the fault set $S_f = \{f_1, \cdots, f_k\}$. Throughout the paper good device is denoted with index g and the faulty devices with index f. Vectors and terms referring to the good and all faulty circuits are denoted with index a. All types of vectors used in this paper are shown in Eq. (1).

$$J_a(\mathbf{x}_a) = -\bar{J}_a(\mathbf{x}_a) \stackrel{!}{=} \min \tag{2}$$

$$\bar{J}_a(\mathbf{x}_a(t)) = \int_{t_0}^{t_f} \bar{f}_a\left(\mathbf{x}_g, \mathbf{x}_{f_1}, \cdots, \mathbf{x}_{f_k}\right) dt \tag{3}$$

$$m_j \leq u_j(t) \leq M_j, \quad j = 1, \ldots, p, \quad t \in [t_0, t_f] \tag{4}$$

$$\mathbf{F}_a(\dot{\mathbf{z}}_a, \mathbf{x}_a, \mathbf{u}, t) = \begin{bmatrix} \mathbf{F}_g(\dot{\mathbf{z}}_g, \mathbf{x}_g, \mathbf{u}, t) \\ \vdots \\ \mathbf{F}_{f_k}(\dot{\mathbf{z}}_{f_k}, \mathbf{x}_{f_k}, \mathbf{u}, t) \end{bmatrix} = \mathbf{0} \tag{5}$$

In Eqs. (2)–(5) the test generation problem is formulated as an optimal control problem. Without loss of generality we can describe the fault detection criterion $\bar{J}_a(\mathbf{x}_a(t))$ of the test generation problem by a Lagrangian merit functional. This is possible, since other types of functionals representing a fault detection criterion can be transformed into a Lagrangian functional. The argument of the merit functional is the state vector $\mathbf{x}_a$ containing the state vectors of the good and all faulty circuits $\mathbf{x}_a^t = (\mathbf{x}_g, \cdots, \mathbf{x}_{f_k})^t$. The functional $\bar{J}_a(\mathbf{x}_a(t))$, which only depends on the circuits states $\mathbf{x}_a$, is is based on the difference between the good test response and all faulty test responses. The functional $\bar{J}_a$ cannot depend on the controls $\mathbf{u}$, since only test response measurements can be used for fault detection. The optimal control problem defined in Eqs. (2)–(5) is to find a control vector $\mathbf{u}^*(t)$ form the set of admissible controls Eq. (4) which causes the DAE system Eq. (5) to follow an admissible trajectory $\mathbf{x}_a^*(t)$ that minimizes the merit functional $J_a(\mathbf{x}_a^*(t))$ in Eq. (2) and thus maximizes the fault detection criterion. Thus the solution of our problem is given by the optimal control vector $\mathbf{u}_a^*(t)$ and the optimal state vector $\mathbf{x}_a^*(t)$.

For a fixed $\mathbf{u}(t)$ the computation of the solution of the DAE system in Eq. (5) and the computation of the merit functional $J_a(\mathbf{x}_a(t))$ is performed by a fault simulation, simulating each fault of the fault set S_f sequentially. This can be done, since the DAE systems of the good and all faulty circuits are not coupled with each other.

3 Optimization procedure

In this section we describe the optimization procedure, which is used to solve the general test signal generation problem

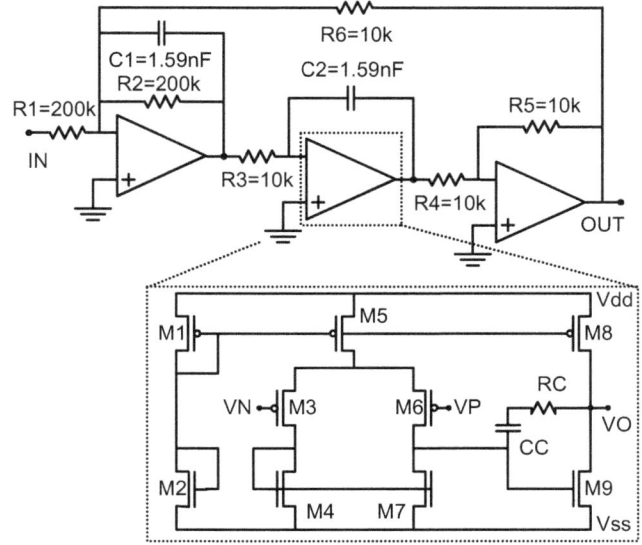

Fig. 1. Biquad filter (resonance frequency $f_r = 10\,\text{kHz}$).

defined in Sect. 2–5. We first formulate the necessary conditions for the optimal controls $\mathbf{u}$ required for the optimization process. To minimize the merit functional $J_a(\mathbf{x}_a)$ from Eq. (2) under the constrains given by Eq. (4) and Eq. (5), we form an augmented merit functional L_a including the constraints. Using the concept of Lagrange multipliers we include the DAE systems of the good and all faulty circuits into an augmented merit functional $L_a(\mathbf{x}_a, \psi_a, \mathbf{u}, t)$ Eq. (6). The Hamiltonian function H_a resulting from L_a is given by Eq. (7).

$$L_a = \int_{t_0}^{t_f} -H_a(\mathbf{x}_a, \psi_a, \mathbf{u}, t) dt \stackrel{!}{=} \min \tag{6}$$

$$H_a(\mathbf{x}_a, \psi_a, \mathbf{u}, t) = \bar{f}_a(\mathbf{x}_g, \mathbf{x}_{f_1}, \ldots, \mathbf{x}_{f_k}) +$$

$$\psi_g^t \mathbf{F}_g\left(\dot{\mathbf{z}}_g, \mathbf{x}_g, \mathbf{u}, t\right) + \sum_{i=1}^{k} \psi_{f_i}^t \mathbf{F}_{f_i}\left(\dot{\mathbf{z}}_{f_i}, \mathbf{x}_{f_i}, \mathbf{u}, t\right) \tag{7}$$

The costate vector ψ_a is composed of the costate vector ψ_g of the adjoint network of the good circuit and all costate vectors ψ_{f_i} of the faulty circuits. From the first variation of L_a we obtain the adjoint DAE system in Eq. (8) including the adjoint systems of the good and all faulty circuits.

$$\begin{bmatrix} \mathbf{S}_g & & \\ & 0 \ddots & \\ & & \mathbf{S}_{f_k} \end{bmatrix}^t \begin{bmatrix} \dot{\psi}_g \\ \vdots \\ \dot{\psi}_{f_k} \end{bmatrix} - \begin{bmatrix} \mathbf{G}_g & & \\ & 0 \ddots & \\ & & \mathbf{G}_{f_k} \end{bmatrix}^t \begin{bmatrix} \psi_g \\ \vdots \\ \psi_{f_k} \end{bmatrix} = \begin{bmatrix} \frac{\partial \bar{f}_a}{\partial \mathbf{x}_g} \\ \vdots \\ \frac{\partial \bar{f}_a}{\partial \mathbf{x}_{f_k}} \end{bmatrix} \tag{8}$$

The matrices S and G are abbreviations for time dependent Jacobian matrices and arise from the storage and resistive elements of the circuit.

$$\mathbf{S}(t) = \frac{\partial \mathbf{F}}{\partial \dot{\mathbf{z}}} \frac{\partial \mathbf{z}}{\partial \mathbf{x}}\bigg|_{\mathbf{x}(t)} \quad \mathbf{G}(t) = \frac{\partial \mathbf{F}}{\partial \mathbf{x}}\bigg|_{\mathbf{x}(t)} \tag{9}$$

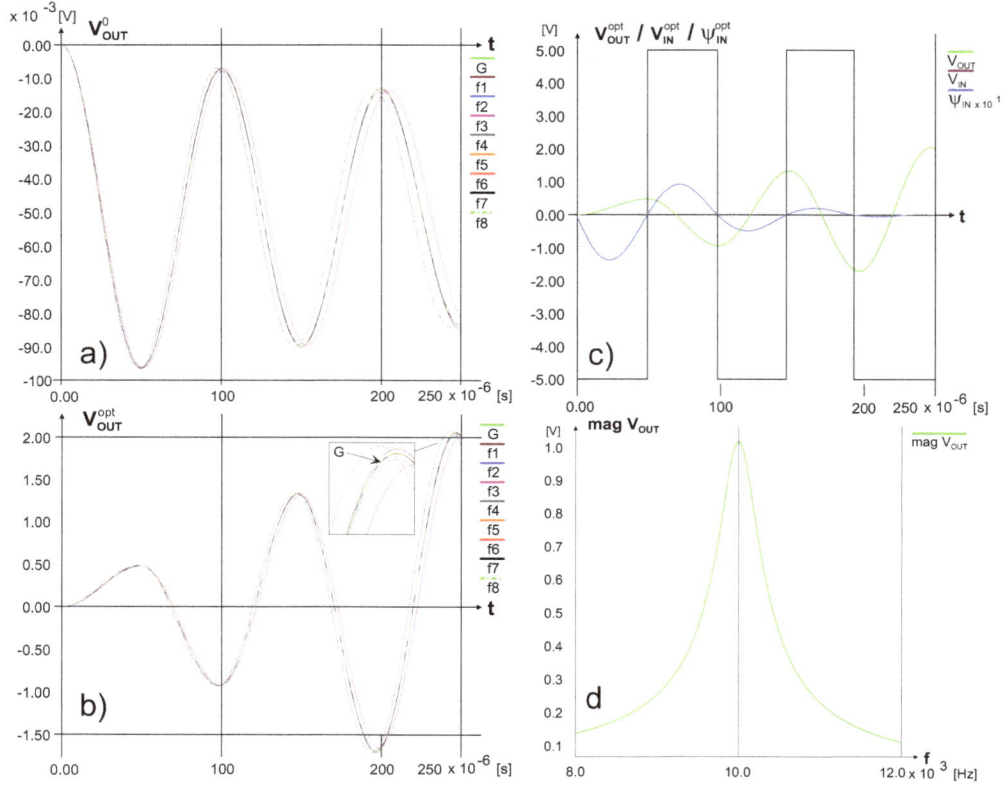

Fig. 2. (a) Test responses before optimization **(b)** Test responses after optimization **(c)** Test response V_{out} of the good device, optimal test stimulus V_{in} and its corresponding waveform $\psi_{s_{in}}$ **(d)** Frequency response of the biquad filter.

The classical methods of the variational calculus cannot be applied to problems with a closed set U as control region. For problems with bounded controls $\mathbf{u}$ a method known as Pontryagin's maximum principle (Pontrjagin et al., 1964) is used. In situations with constraints on the control variables the necessary condition for optimal control Eq. (10) is replaced with Eq. (11), which states the maximum principle.

$$\nabla_{\mathbf{u}} \, J_a = -\frac{\partial H_a}{\partial \mathbf{u}} = \mathbf{0} \tag{10}$$

$$\max_{\mathbf{u} \in U} \; H_a(\mathbf{x_a}(t), \psi_a(t), \mathbf{u}(t), t) = H_a(\mathbf{x_a}, \psi_a, \hat{\mathbf{u}}, t) \tag{11}$$

Suppose $\hat{\mathbf{u}}$ is an optimal control, which minimizes the merit functional of the control problem. Than Pontryagin's maximum principle Eq. (11) says, that $\hat{\mathbf{u}}$ and its corresponding pair $(\mathbf{x}_a, \psi_a)$ maximizes the Hamiltonian H_a for all $t \in [t_0, t_f]$ and for all admissible controls $\mathbf{u} \in U$. The maximum principle is a necessary condition for optimal control and is used as a calculation base for the optimal controls. In case of a free end point $\mathbf{x}(t_f)$ given with the test generation problem the optimal solution of the control problem $\mathbf{u}^*(t)$, $\mathbf{x}_a^*(t)$ must satisfy the system of the circuit equations (5) and their initial values $\mathbf{x}_a(t_0)$, the adjoint equations (8) and their final values $\psi_a(t_f) = \mathbf{0}$ and the maximum principle of Pontryagin. Since modified nodal equations are used for circuit description and these are linear in $\mathbf{u}$, the application of the

maximum principle lead to the qualitative form of optimal test signals (Burdiek, 2002), given by Eq. (12). Only test signals of this form are possible candidates for optimal controls.

$$u_j(t) = \begin{cases} M_j, & \mathbf{b}^t(\psi_g + \cdots + \psi_{f_k}) \geq 0 \\ m_j, & \mathbf{b}^t(\psi_g + \cdots + \psi_{f_k}) < 0 \end{cases}, \; 1 \leq j \leq p \tag{12}$$

Since the circuit equations (5) are coupled with the adjoint equations (8) an optimization procedure is needed for the calculation of the optimal controls. In the following the essential steps of the optimization procedure are explained. Starting with an initial control vector $\mathbf{u}^0(t)$ the solution $\mathbf{x}_a^0(t)$ of the DAE system in Eq. (5) is computed in the first step. The evaluation of the merit functional J_a^0 is performed during the simulation of Eq. (5). Using the state vector $\mathbf{x}_a^0(t)$ for the computation of the matrices S and G the solution vector $\psi_a^0(t)$ of the adjoint DAE system Eq. (8) is calculated in the next step. In the last step Pontriyagin's maximum principle is applied using Eq. (12) to obtain the control vector $\hat{\mathbf{u}}^0$. The next iterate $\mathbf{u}_a^1(t)$ is calculated with the aid of Eq. (13), whereby the stepsize α_l is optimal with $\alpha_l = 1$. The optimization procedure terminates when the minimal sequence $J_a^0 > J_a^1 > \dots$ aborts.

$$\mathbf{u}^{l+1} = (1 - \alpha_l)\mathbf{u}^l + \hat{\mathbf{u}}^l \quad l = 0, 1\dots \tag{13}$$

4 Experimental results

In this section the test generation method is applied to a bi-quad filter shown in Fig. 1. To demonstrate the approach the test generation procedure is applied to a small fault set of eight hard to detect parametric faults. For the fault detection criterion $\bar{J}_a$ we use the functional described by Eq. (14). The weighting factors w_{f_i} used to distinguish faults in $\bar{J}_a$ are determined from an initial fault simulation.

$$\bar{J}_a(\mathbf{x}_a(t)) = \sum_{i=1}^{k} w_{f_i} \phi_{f_i} \left(V_{out_g}, V_{out_{fi}} \right) \qquad (14)$$

$$\phi_{f_i} = \int_{t_0}^{t_f} \left| V_{out_g}(t) - V_{out_{fi}}(t) \right| dt, \qquad \phi_a = \sum_{i=1}^{k} \phi_{f_i} \qquad (15)$$

The unit step function $1(t)$ is used as initial test stimulus for the procedure denoted by u^0. The optimal test stimulus u^{opt} (V_{in}^{opt}) shown in Fig. 2c was generated within 2 iterations. The simulation results and the parameters of the test generation procedure are listed in Table 1. The test responses V_{out} before and after the optimization of the CUT are shown in Fig. 2a and Figure 2b. As one can see form the simulation results the generated test stimulus $u^{opt}(t)$ significantly enhances the fault detection capability. The switching points (SP's) of the test stimulus computed with Eq. (12) are given by the isolated zeros of the waveform $\psi_{s_{in}}$ shown in Fig. 2c. Obviously, it is good a strategy to test the filter in the near of its resonance frequency $f_r = 10\,\text{kHz}$, which is shown in Fig. 2d. For this reason suitable test signals for the biquad filter have to contain a first harmonic, which is approximately f_r. This is the case for our generated test signal.

After the test generation process a fault simulation of 150 faults was carried out to determine the performance of the generated test stimulus. This resulted in a fault coverage of 97 percent. The test generation approach proposed in this paper has been implemented in a C++ program named TORAD (Test Generator for Analog Devices). The simulator TORAD supports several circuit analyses, like transient analysis and transient sensitivity analysis. The last one includes the ability to simulate the transient behaviour of adjoint networks.

5 Conclusion

In this paper a new test signal generation approach based on modern control theory methods such as the maximum principle of Pontryagin was presented. The proposed method, which was formulated as an optimal control problem, generates optimum transient test signals for general analog circuits. An optimization procedure was used for the computation of the solution of the control problem. The procedure takes advantage from necessary optimality conditions such as the maximum principle, so that it does not need the gradient of the merit functional with respect to the controls. Since the procedure generates optimal test signals, it is best suited for the detection of hard to detect faults in analog circuits.

Table 1. Results and parameters of the optimization procedure

Results of the optimization procedure:			
Faults	w_{fi}	Φ_{fi}^0	Φ_{fi}^{opt}
f_1 : F(p,r2,resistance,5%)	5.56	5.898e-08	1.092e-06
f_2 : F(p,r2,resistance,-5%)	5.08	6.438e-08	1.194e-06
f_3 : F(p,c2,capacitance,5%)	0.26	1.193e-06	2.155e-05
f_4 : F(p,c2,capacitance,-5%)	0.25	1.238e-06	2.255e-05
f_5 : F(p,m:op2:1,l,5%)	276	1.116e-09	7.952e-08
f_6 : F(p,m:op2:1,l,-5%)	361	8.536e-10	7.836e-08
f_7 : F(p,m:op1:1,l,5%)	267	1.827e-09	7.982e-08
f_8 : F(p,m:op1:1,l,-5%)	350	1.476e-09	8.030e-08

Φ_a^0 = 2.5602346e-06	Φ_a^{opt}= 4.6711551e-05	Φ_a^{opt}/Φ_a^0 = 18.3

SPs of u^0 (t) = V_{in}^0(t) : { (0.00ms,1V) }

SPs of u^{opt}(t) = V_{in}^{opt}(t) : { (0.0us, -5V) (49.4us, 5V)
(98.1us, -5V) (146us, 5V)
(193us, -5V) (250us, 5V) }

Number of iterations I : 2

Parameters of the optimization procedure:	
Start point of time t_0 : 0.0us	End point of time t_f : 250us
Upper bound M_{in} of u_{in} : 5V	Lower bound m_{in} of u_{in} : -5V
Minimum time distance between SPs : 25us	

References

Gomes, A. V. and Chatterjee, A.: Minimal length diagnostic tests for analog circuits using test history, DATE-Conference 1999, pp. 189–194, 1999.

Variyam, P. N., Hou, J. and Chatterjee, A.: Efficient Test Generation for Transient Testing of Analog Circuits Using Partial Numerical Simulation, VLSI Test Symposium, pp. 214–219, 1999.

Burdiek, B.: Generation of Optimum Test Stimuli for Nonlinear Analog Circuits Using Nonlinear Programming and Time-Domain Sensitivities, DATE-Conference 2001, pp. 603–608, 2001.

Burdiek, B.: The Qualitative Form of Optimum Transient Test Signals for Analog Circuits Derived from Control Theory Methods, ISCAS-Conference 2002.

Pontrjagin, L. S., Boltyanskii, V. G., Gamkrelidze, R. V., and Mishchenko, E. F.: The Mathematical Theory of Optimal Processes, Pergamon, 1964.

Director, S. W. and Rohrer, R. A.: The Generalized Adjoint Network and Network Sensitivities, IEEE Trans. Circuit Theory, vol. CT-16, pp. 318–323, 1969.

9

Power estimation on functional level for programmable processors

M. Schneider, H. Blume, and T. G. Noll

Lehrstuhl für Allgemeine Elektrotechnik und Datenverarbeitungssysteme, RWTH Aachen, Schinkelstraße 2, 52062 Aachen, Germany

Abstract. In diesem Beitrag werden verschiedene Ansätze zur Verlustleistungsschätzung von programmierbaren Prozessoren vorgestellt und bezüglich ihrer Übertragbarkeit auf moderne Prozessor-Architekturen wie beispielsweise Very Long Instruction Word (VLIW)-Architekturen bewertet. Besonderes Augenmerk liegt hierbei auf dem Konzept der sogenannten Functional-Level Power Analysis (FLPA). Dieser Ansatz basiert auf der Einteilung der Prozessor-Architektur in funktionale Blöcke wie beispielsweise Processing-Unit, Clock-Netzwerk, interner Speicher und andere. Die Verlustleistungsaufnahme dieser Blöcke wird parameterabhängig durch arithmetische Modellfunktionen beschrieben. Durch automatisierte Analyse von Assemblercodes des zu schätzenden Systems mittels eines Parsers können die Eingangsparameter wie beispielsweise der erzielte Parallelitätsgrad oder die Art des Speicherzugriffs gewonnen werden. Dieser Ansatz wird am Beispiel zweier moderner digitaler Signalprozessoren durch eine Vielzahl von Basis-Algorithmen der digitalen Signalverarbeitung evaluiert. Die ermittelten Schätzwerte für die einzelnen Algorithmen werden dabei mit physikalisch gemessenen Werten verglichen. Es ergibt sich ein sehr kleiner maximaler Schätzfehler von 3%.

In this contribution different approaches for power estimation for programmable processors are presented and evaluated concerning their capability to be applied to modern digital signal processor architectures like e.g. Very Long Instruction Word (VLIW) -architectures. Special emphasis will be laid on the concept of so-called Functional-Level Power Analysis (FLPA). This approach is based on the separation of the processor architecture into functional blocks like e.g. processing unit, clock network, internal memory and others. The power consumption of these blocks is described by parameter dependent arithmetic model functions. By application of a parser based automized analysis of assembler codes of the systems to be estimated the input parameters of the
arithmetic functions like e.g. the achieved degree of parallelism or the kind and number of memory accesses can be computed. This approach is exemplarily demonstrated and evaluated applying two modern digital signal processors and a variety of basic algorithms of digital signal processing. The resulting estimation values for the inspected algorithms are compared to physically measured values. A resulting maximum estimation error of 3% is achieved.

1 Introduction

In the course of increasing complexity of digital signal processing applications, especially in the field of mobile applications, low power techniques are of crucial importance. Therefore, it is desirable to estimate the power consumption of a system at a very early stage in the design flow. By this means it is possible to predict whether a system will meet a certain power budget before it is physically implemented. Necessary changes in the system partitioning or the underlying architecture will then be much less time and money consuming, because no physical implementation of the system is required to determine its power dissipation.

Another important design criteria of modern electronic systems is the demand for flexibility, e.g. the ability to adapt a system to changing specifications or standards. This fact along with the continuous growth of their computational power makes programmable digital signal processor (DSP)-kernels a very attractive component for heterogeneous Systems-on-Chip.

Like any other architecture block the power consumption of a DSP (-kernel) depends on several factors like the switching activity of the input data, the clock frequency and of course the executed algorithm itself. Besides these dependencies there are many more DSP-specific influencing factors like the type and rate of memory accesses, the usage of specific architecture elements like DMA controllers or dedicated co-processors, different compiler optimization settings, pipeline stalls and cache misses but also different

Program cache		Register in the DSP core
MUL: 0111001110100110	Cycle 1 →	0111001110100110
ADD: 0010110111100011	Cycle 2 →	0010110111100011

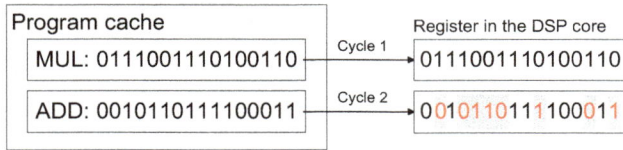

Fig. 1. Sequential execution of two different DSP instructions.

Program cache		Register in the DSP core
ADD: 0010110111100011	Cycle 1 →	0111001110100110
ADD: 0010110111100011	Cycle 2 →	0111001110100110

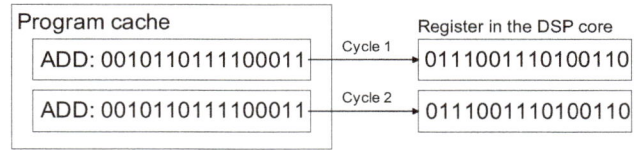

Fig. 2. Sequential execution of two identical DSP instructions.

programming styles or the choice of algorithmic alternatives which all strongly influence the power consumption of an algorithm that is executed on a DSP.

For this reason it is desirable to consider methodologies for power estimation that cover all significant influencing factors and provide a sufficient accuracy at moderate complexity. Such a methodology is presented in this paper and verified using several exemplary vehicles. The paper is organized as follows: Sect. 2 shortly reviews and discusses several existing power estimation techniques in terms of their portability to modern DSP architectures. The following section describes the so-called Functional-Level Power Analysis (FLPA) approach in detail. Section 4 lists some results concerning the application of the FLPA methodology for estimating the power of a variety of basic algorithms. A conclusion of the paper is given in Sect. 5.

2 Classical approaches for power estimation

One possible straight forward power estimation approach on DSPs is the so-called Physical-Level Power Analysis methodology. This approach is based on the analysis of the switching activity of all transistors of the DSP architecture. The requirement of this methodology is the availability of a description of the processor architecture on the transistor level, which is rarely given for modern DSPs. But the main disadvantage is the extremely high computational effort that makes approaches like this inapplicable for digital signal processors. Architectural-Level approaches like (Brooks et al., 2000) reduce this computational effort by modelling typical architecture elements like registers, functional units or load/store queues. These models are not based on physical measurements and require still exact knowledge of the processors architecture. Therefore, these two methodologies can be mainly found in the field of microprocessor development.

Another possibility for power estimation for DSPs is the so-called Instruction-Level Power Analysis (Tiwari et al., 1996). By means of physical measurements or low level simulations the energy consumption of each instruction out of the instruction set of a given processor is determined. By analysis of the assembler code of a program it is then possible to estimate the specific power consumption for this program performed on a certain processor. The advantage of this approach is the ability to cover a specific part of power consumption of DSPs: the so-called inter-instruction effects. In general, the energy consumption of a DSP instruction depends on the previously executed instructions, what can be explained by means of Figs. 1 and 2.

At a certain stage of a processors pipeline, instruction words are transferred from the program cache into a register in the DSP core for further processing. Figure 1 shows the situation that an ADD (addition) instruction word replaces a MUL (multiplication) instruction word in cycle 2. The numbers shaded with gray boxes show the bits in the register that switch their state in this case. In this example a Hamming distance (number of different bits of these two instruction words) of eight (H_d=8) is resulting. As can be seen in Fig. 2 the sequence of two identical instructions causes no switching activity (H_d=0). Effects like this occur in many stages of a processors pipeline and as a result of these effects the energy consumption of a DSP instruction obviously depends on the previously executed instruction (Marwedel, 2003). The Instruction-Level Power Analysis methodology allows to cover such inter-instruction effects by measuring the energy consumption of groups of DSP instructions, but that makes this approach very complex due to the huge number of possible combinations. The effort will even grow, if Very-Long-Instruction-Word (VLIW) architectures shall be modeled due to their increasing word length and their ability to issue several operations in parallel.

A more attractive approach for power estimation is the Functional-Level Power Analysis (FLPA) methodology. This methodology has been introduced in (Qu et al., 2000) and was first applied in (Senn et al., 2002) to a digital signal processor. Here, a refined extension of this methodology is presented in order to model complete DSP cores including the modeling of separate units like cache, internal RAM, EDMA and integrated co-processors, different types of memory accesses etc. The following section will demonstrate this methodology applying an exemplary vehicle – the TMS320C6416 DSP.

3 Functional-Level Power Analysis (FLPA)

The basic principle of the FLPA methodology is depicted in Fig. 3.

In a first step the DSP architecture is divided into functional blocks like fetch unit, processing unit, internal memory and others like the clocking system. By means of measurements it is possible to find an arithmetic function for each block that determines its power consumption in dependency of certain parameters. These parameters are for example the clock frequency, the degree of parallelism or the rate with

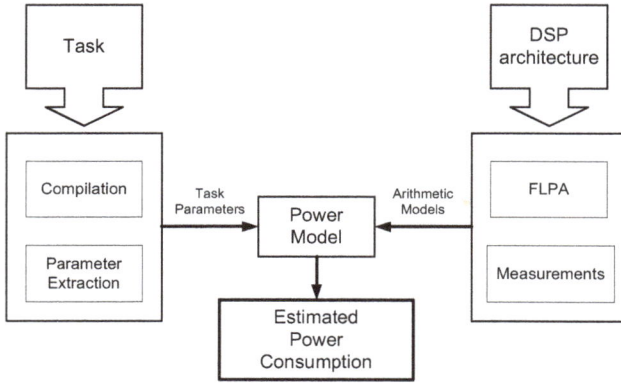

Fig. 3. The basic FLPA principle.

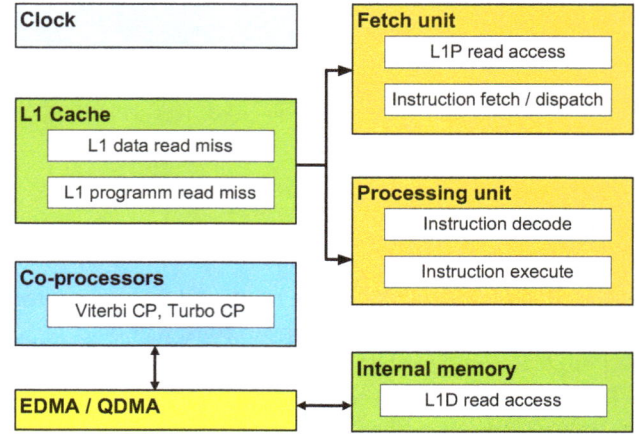

Fig. 4. The TMS320C6416 architecture.

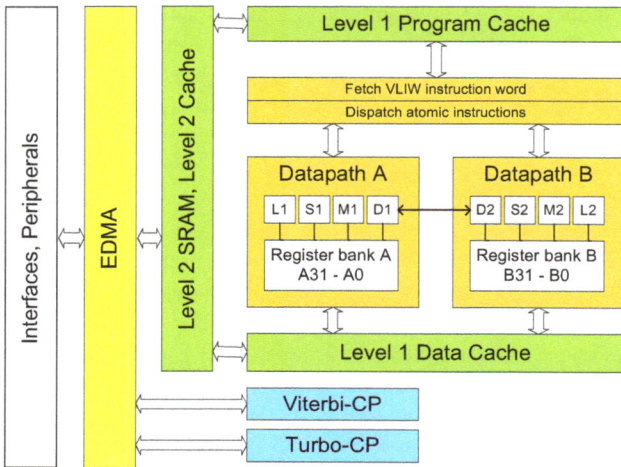

Fig. 5. Separation of the TMS320C6416 architecture into functional blocks.

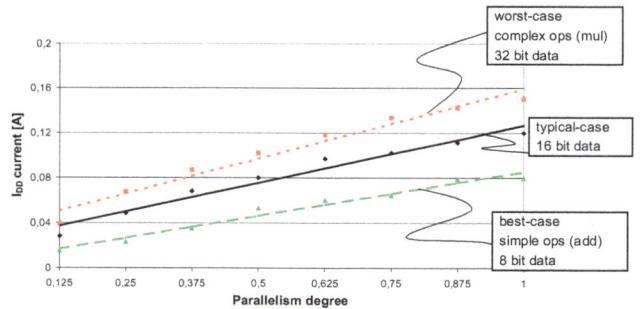

Fig. 6. Model function of the TMS320C6416 processing unit.

3.1 An exemplary vehicle: The TMS320C6416 DSP

The TMS320C6416 is a state-of-the-art VLIW DSP aiming for multimedia applications. Figure 4 depicts a block diagram of the DSP architecture.

It is based on a VLIW-architecture with two parallel data paths each including four issue-slots. Furthermore, this processor includes a couple of interfaces (ATM, PCI, etc.), an Enhanced DMA-controller (EDMA) and two dedicated co-processors (Viterbi and Turbo decoder co-processor). For this work the integrated software development environment Code Composer Studio (CCS) and the hardware test and evaluation board (TEB) including the C6416 have been utilized. For further details of this architecture see (TMS320C6416, SPRS164C documentation set).

This architecture can be divided into seven functional blocks as depicted in Fig. 5.

Arithmetic model functions describing the power consumption of a functional block can be found by means of measurements. Therefore, it is necessary to stimulate each block separately. This can be achieved by executing different parts of assembler code, that will be called scenarios according to (Senn et al., 2002).

A determination of a model function applying such scenarios will be described here considering the processing unit and the fetch unit as example.

which the internal memory is accessed. Most of these parameters can be automatically determined by a parser which analyzes the assembler file of a program code. The total power consumption is then given as the sum of the power consumption of each functional block:

$$P_{\text{total}} = \sum_i P_{\text{block } i}. \tag{1}$$

The left side of Fig. 3 depicts the process of extracting parameters from a program which implements a task. After compilation it is possible to extract the task parameters from the assembler code. Further parameters can be derived from a single execution of the program (e.g. the number of required clock cycles). These parameters are the input values for the previously determined arithmetic model functions. Thus, an estimation for the algorithms power consumption can be computed. This approach is applicable to all kinds of processor architectures. Further on, FLPA can be applied to a processor with moderate effort and no detailed knowledge of the processors architecture is necessary.

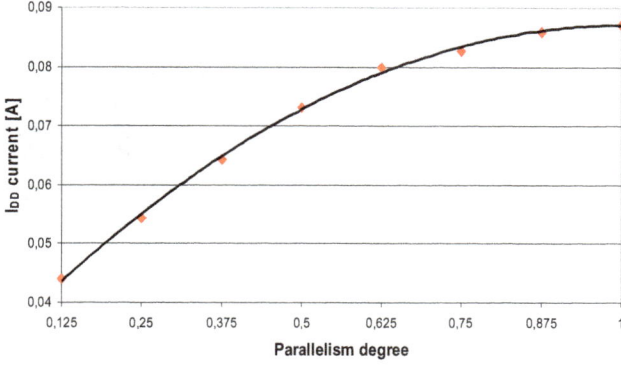

Fig. 7. Model function of the TMS320C6416 fetch unit.

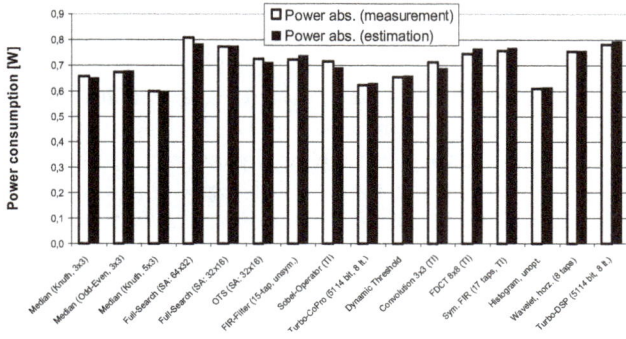

Fig. 8. FLPA power estimation results and measurements for the TMS320C6416 (absolute power consumption).

The power consumption of the processing unit has three significant parameters:

– the degree of parallelism α (percentage of parallel working functional units),

– the number of executed instructions,

– the type of input data.

The scenarios belonging to the processing unit vary these parameters separately.

In Fig. 6 the current drawn by the processing unit is depicted over the degree of parallelism. The applied test scenario includes a loop where within each loop iteration 1000 instructions are executed. The dotted line of Fig. 6 represents the worst-case power consumption of the processing unit, in which complex instructions (e.g. multiplications) with the maximum word length of the input data (32 bit) are executed. In contrast to that, the dashed line represents the best-case power consumption with simple instructions (e.g. additions) and a small word length of the input data (8 bit). The arithmetic function belonging to the straight line (typical-case: instruction mix, medium word length of the input data (16 bit)) is chosen as model function for the FLPA model of the TMS320C6416 processing unit and modeled by

$$P_{\text{processing unit}} = (1.02 \cdot 10^{-1} \cdot \alpha + 2.46 \cdot 10^{-2}) \cdot V_{\text{core}}$$
$$= I_{DD, \text{ processing unit}} \cdot V_{\text{core}}. \tag{2}$$

Table 1. Model functions of the functional blocks of the TMS320C6416 and belonging list of parameters.

functional block	block specific power consumption function
clock system	$P_{\text{clock system}} = (a \cdot F + b) \cdot V_{\text{Core}}$
fetch unit	$P_{\text{fetch unit}} = (c \cdot \alpha^2 + d \cdot \alpha + e) \cdot F \cdot (1\text{-}PSR) \cdot V_{\text{Core}}$
processing unit	$P_{\text{proc. unit}} = (f \cdot \alpha + g) \cdot F \cdot (1\text{-}PSR) \cdot V_{\text{Core}}$
internal memory	$P_{\text{internal memory}} = (h \cdot \beta + i \cdot \gamma) \cdot F \cdot (1\text{-}PSR) \cdot V_{\text{Core}}$
level-1 cache	$P_{\text{level-1 cache}} = (j \cdot \delta + k \cdot \varepsilon) \cdot F \cdot (1\text{-}PSR) \cdot V_{\text{Core}}$
EDMA/QDMA	$P_{\text{EDMA/QDMA}} = (m \cdot \zeta) \cdot F \cdot (1\text{-}PSR) \cdot V_{\text{Core}}$
co-processors (Turbo, Viterbi)	$P_{\text{copro}} = (n \cdot \eta + p \cdot \theta) \cdot F \cdot (1\text{-}PSR) \cdot V_{\text{Core}}$

parameter	description	parameter	description
α	parallelism degree	θ	Turbo co-processor activity rate
β	memory access rate (read)	η	Viterbi co-processor activity rate
γ	memory access rate (write)	V_{Core}	Core Voltage of the processor
δ	L1P cache miss rate	F	clock frequency
ε	L1D cache miss rate	PSR	pipeline stall rate
ζ	EDMA activity rate	$a,b,c,d,e,f,$ $g,h,i,j,k,m,$ n,p	coefficients for polynomials

Here, V_{Core} denotes the core-voltage of the processing unit and α the achieved degree of parallelism. The error of the estimated power consumption for algorithms with either extremely complex or extremely simple instructions will be examined in the next section.

The architecture of the fetch unit of the TMS320C6416 has the task to control the flow of VLIW instruction words to the DSP core and to dispatch the atomic instructions to the functional units. Though the architecture of the fetch unit of the TMS320C6416 is not known in detail it is possible to model this functional block. In a test scenario the only parameter having a strong impact on the power consumption of the fetch unit, the parallelism degree α, is varied and some working points are measured. Figure 7 depicts the current consumption drawn by the fetch unit.

According to the measured working points a polynomial function (here, a quadratic function) can be found which describes the power consumption of the fetch unit

$$P_{\text{fetch unit}} =$$
$$(-5.67 \cdot 10^{-2} \cdot \alpha^2 + 1.14 \cdot 10^{-1} \cdot \alpha + 3.02 \cdot 10^{-2}) \cdot V_{\text{core}}$$
$$= I_{DD, \text{ fetch unit}} \cdot V_{\text{core}}. \tag{3}$$

All the other FLPA blocks depicted in Fig. 5 can be modeled similarly. The complete FLPA power model of the TMS320C6416 including the complete list of required parameters is shown in Table 1.

4 Results

For the evaluation of the FLPA methodology the power consumption was measured as well as estimated for a variety of digital signal processing algorithms. The comparison of estimated and measured values shows a maximum error of 3%, as can be seen in Fig. 8. All algorithms which are marked

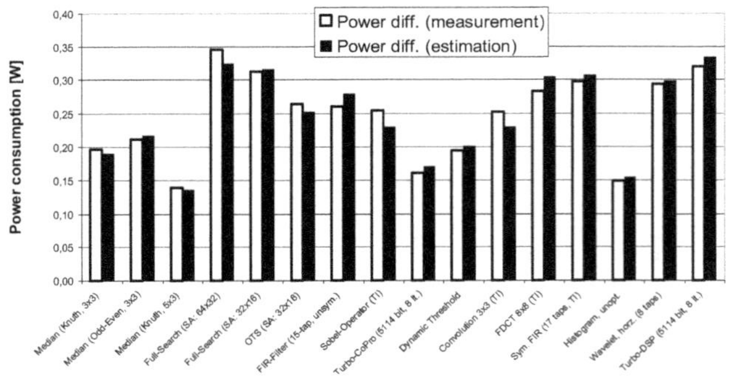

Fig. 9. FLPA power estimation results and measurements for the TMS320C6416 (differential power consumption).

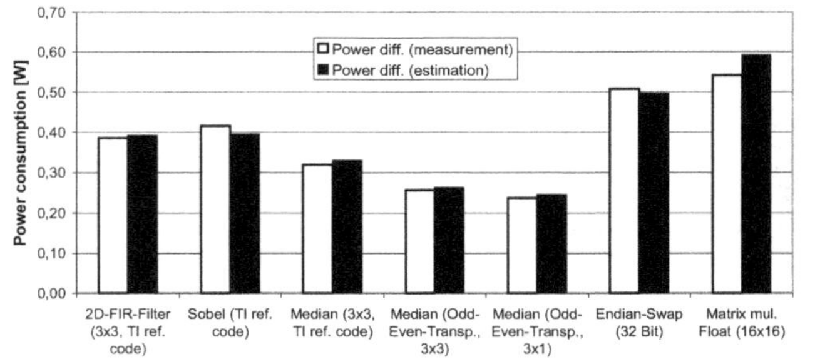

Fig. 10. FLPA power estimation results and measurements for the TMS320C6711 (differential power consumption).

with (TI) have been taken from the TI code library in order to apply the methodology also for DSP code which was optimized by the processor manufacturer himself.

Obviously, the part of the total power consumption according to the clock system is a constant offset for each algorithm which is performed on the processor. Therefore, for a fair comparison differential power consumption values (without the clock system) should also be regarded. The comparison depicted in Fig. 9 yields a maximum error of 10%. It should be noticed that according to the program to be performed on the processor the differential power consumption varies by more than 200 mW. This dynamics is much larger than the maximum estimation error of about ten to twenty mW.

The FLPA approach has also been applied to the C6711 processor which is a floating point processor providing no further co-processors. The C6711 FLPA model comprises seven model functions. Compared to the set of algorithms which have been taken as benchmarking set for the C6416 the benchmarking set for the C6711 also included dedicated floating point applications like floating point matrix multiplications. The maximum power consumption of the C6711 within the experiments amounted to 1.1 W and the dynamics concerning the power consumption of the different algorithms amounted to 350 mW. A comparison between the FLPA power estimation and physical measurements yields a maximum error of less than 5% for the absolute power consumption and less than 10% (40 mW) for the differential power consumption (see Fig. 10). Again this comparison proves that the FLPA methodology provides sufficient accuracy for a power estimation in an early stage of the design flow.

5 Conclusion

Different approaches for power estimation for programmable processors have been described and evaluated concerning their capability to be applied to modern digital signal processor (DSP) architectures like e.g. Very Long Instruction Word (VLIW)-architectures. The concept of so-called Functional-Level Power Analysis (FLPA) has been extended and refined and the belonging separation of the processor architecture

into functional blocks has been shown. The power consumption of these blocks has been described in terms of parameterized arithmetic model functions. A parser which allows to analyze automatically the assembler codes has been implemented. This parser yields the input parameters of the arithmetic functions like e.g. the achieved degree of parallelism or the kind and number of memory accesses. A demonstration and evaluation of this approach has been performed applying the DSPs TMS320C6416 and TMS320C6711 and a variety of basic algorithms of digital signal processing. Resulting estimation values for the inspected algorithms are compared to measured values. A resulting maximum estimation error of 3% for the absolute power consumption and 10% for the differential power consumption is achieved. The application of this methodology allows to evaluate efficiently different parameter settings of a programmable processor like different coding styles, compiler settings, algorithmic alternatives etc. concerning the resulting power consumption. Therefore, it is a valuable methodology for a system designer to explore the design space of programmable processors concerning the power aspect.

References

Brooks, D., Tiwari, V., and Martonosi, M.: Wattch: A Framework for Architectural-Level Power Analysis and Optimizations, Proceedings of the ISCA, 83–94, 2000.

Marwedel, P.: Fast, predictable and low-energy memory references through memory-architecture aware compilation, Proceedings of the DSP Design Workshop 2003, Dresden, 2003.

Qu, G., Kawabe, N., Usami, K., and Potkonjak, M.: Function Level Power Estimation Methodology for Microprocessors, Proc. of the Design Automation Conference 2000, 810–813, 2000.

Senn, E., Julien, N., Laurent, J., and Martin, E.: Power Consumption Estimation of a C Program for Data-Intensive Applications, Proc. of the PATMOS Conference 2002, 332–341, 2000.

Tiwari, V., Malik, S., and Wolfe, A.: Instruction Level Power Analysis and Optimization of Software, Journal of VLSI Signal Processing, 1–18, 1996.

TMS320C6416: Fixed Point Digital Signal Processor, Texas Instruments, SPRS164C, 2001.

TMS320C6711: datasheets, http://www.ti.com.

10

Different approaches of high speed data transmission standards

M. Ehlert

Micronas GmbH, Freiburg, Germany

Abstract. A number of standards addresses the problem of high-speed data transmission on serial or serial-parallel data lines. Serial-parallel data transmission means the transmitted information is distributed on parallel data lines. Even though several standards exist, there are only a few basic techniques used in most of these standards. This paper is giving an overview of these different basic techniques used in the physical layer of today's data transmission standards, for example DVI/HDMI, USB2.0, Infiniband, SFI5, etc. [1–9]. The main focus lies on the approaches used for physical signaling, line coding and information synchronization in serial and serial-parallel systems. In addition, currently discussed techniques to improve data transmission in the future will be presented.

1 Introduction

High-speed data communication standards originally have addressed fiber optics systems for telecommunication and datacom. Nowadays, the increasing operation speed of modern processors, computer busses as well as low cost data interfaces e.g. USB2.0 have reached a high-speed data communication of several hundreds of megabits data rate per second and gigabits per second which are about to come to mass markets soon.

Therefore, in addition to fiber optics systems, backplane and bus architectures are addressed by high-speed data transmission standards. Since these applications usually do not need to cover long distances, copper cables are used as interconnection media for speeds up to several gigabit per second to reduce system cost which is the key to the mass market. Today, distances of 20–30 m can be implemented with such cables which are sufficient for wide areas of applications.

Despite the different applications like fiber optics, copper cable, or backplane data transfer, the various synchronization techniques used in the different standards are still quite close to each other. Differences mostly originate from the various coding schemes used to encode the data and to transmit special characters used in the synchronization process. This paper will address the most often-used design techniques in today's standards. The interfaces are described starting with the electrical signal form as the lowest physical level and going up to line coding and frame processing as the highest level of data transmission.

First, an example of an electrical data path and the electrical signals used for data transmission is explained in Sect. 2. Furthermore, timing budget issues are discussed here. In Sect. 3, phase detection techniques used for bit synchronization are explained. Then, in Sect. 4, data coding schemes are discussed using the examples of 8B10B and TMDS signaling. Word synchronization and speed matching techniques are presented in Sect. 5. A summary and conclusions are given in Sect. 6.

2 Physical data transmission

Electrical data paths suffer from damping and dispersion and, in the case of parallel data transmission, from skew between different signal lines. Moreover, connectors add considerable distortion to the signals due to their usually bad high frequency characteristics which leads to reflections. An example of a data path for backplane data transmission is given in Fig. 1.

For backplane transmission, two ICs placed on different PCBs are connected via two connectors to a backplane or main board. With FR4 PCB material, skew and damping are about 250 fs/mm and 12 mdB/mm, respectively. For example, in the Infiniband standard [6], 20" of transmission length is allowed. This leads to a damping of about 6.1 dB and a skew of 130 ps, respectively, only for the signal lines. Since Infiniband allows an eye closure of 2/3, at 3.125 Gbit/s there is only 106 ps of eye opening left which is smaller than the total skew. In addition, there is the influence of the connectors which usually is even worse than that of the signal

Table 1. Electrical signal specification ranges.

Parameter	Encoder	Decoder
Eye Opening	60–80%	30–50%
Amplitude	250–1000 mV	150–400 mV

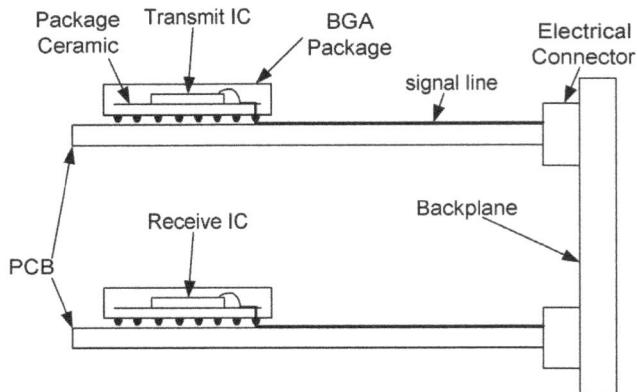

Fig. 1. Datapath for backplane data transmission.

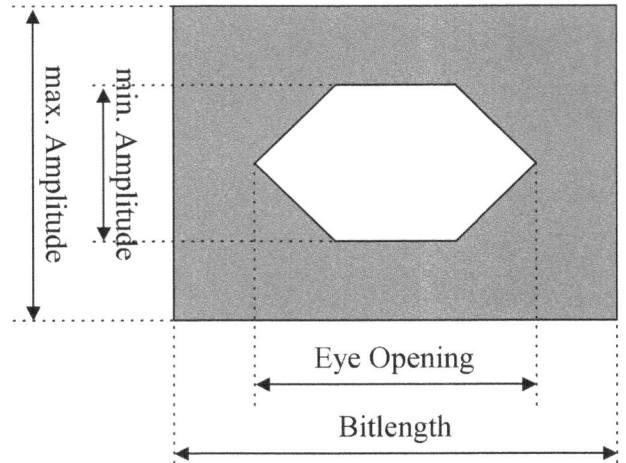

Fig. 2. Definition of eye diagram.

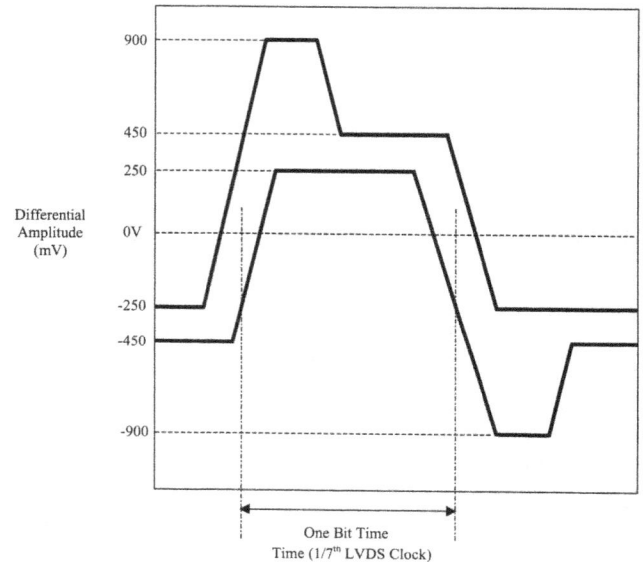

Fig. 3. OpenLDI specification for optional pre-amphasis.

lines. Therefore, retiming of several parallel data lines with only one common clock phase is not possible. Techniques to achieve retiming under such conditions will be discussed in Sect. 5.

The particular minimum eye opening and amplitude of the various standards specify the electrical signals (see Fig. 2). The ranges for these values are given in Table 1. Typically, with these values, a bit error rate (BER) of 1e12 must be achieved. So far, only DVI/HDMI accept a higher BER of 1e9 since single bit errors are less important in video applications.

In addition to the specification of a simple eye diagram, some standards like OpenLDI [22] specify waveforms with pre-amphasis (Fig. 3) as an option to reduce the effects of band limitation of the data path. This, however, is not a must, yet.

Moreover, it is discussed to include reconstruction filters at the receiver to widen up the eye opening again for the retiming. These filtering options are interesting for three reasons. One is to upgrade older equipment for higher data transmission rates without a need to change the transmission cables or PCBs, second is to reduce system cost allowing low cost connectors and PCB material, and finally to enhance the transmission distance.

The most important difference in the specification of the electrical signal between the various standards is the minimum eye opening allowed at the receiver. The retiming clock at the receiver needs to retime the data within this residual eye opening. All internal error sources of the receiver must be subtracted from this opening. Error sources are e.g. clock jitter of the retiming clock, phase error of the clock alignment, setup and hold time of the retiming Flip-Flops as

well as influences from the receiver package. All these error sources must be accounted for in an overall timing budget, which yields the final real eye opening usable for the retiming operation.

3 Bit synchronization

The next level of data transmission following the physical signal is the recovering of single bits from the electrical signal. This process is called bit synchronization. The most common technique to establish bit synchronization is to use a phase locked loop (PLL) [10–15]. This can either be a clock-and-data recovery (CDR, see Fig. 4) or, in case a correct clock is available already, a delay-locked-loop (DLL).

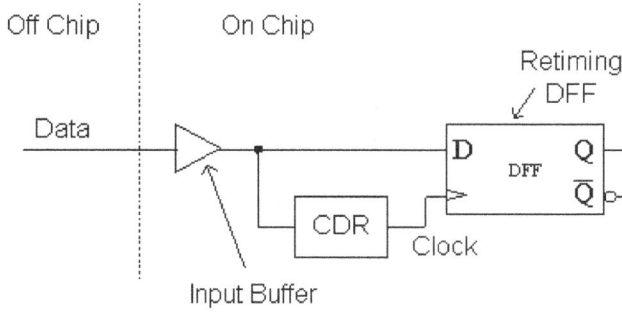

Fig. 4. Bit synchronization using clock-and-data recovery.

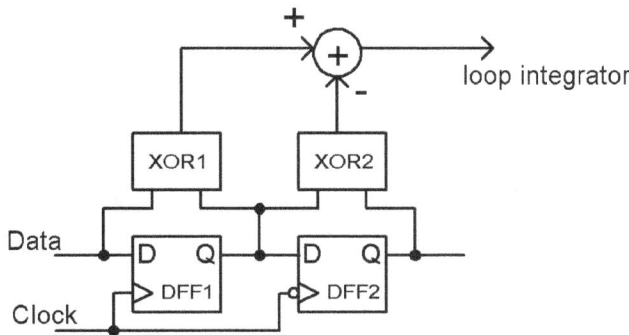

Fig. 5. Principle of phase detector proposed by Hooge [16].

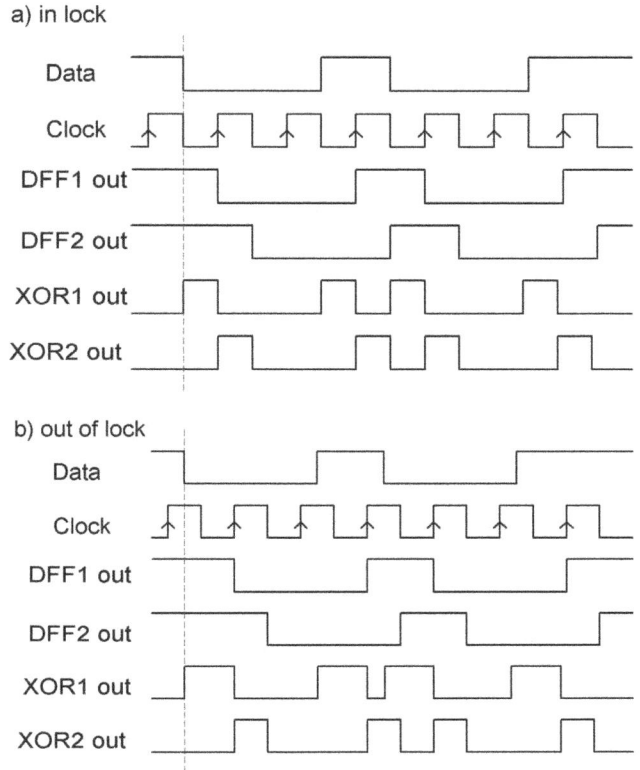

Fig. 6. Waveforms of Hooge phase detector (**a**) in lock (**b**) out of lock.

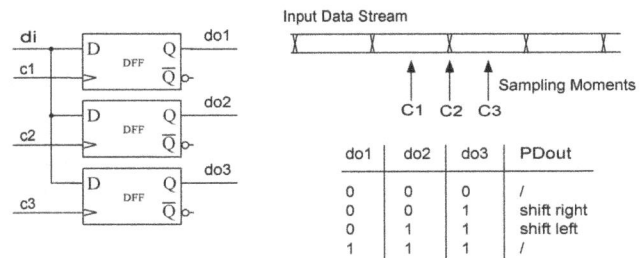

Fig. 7. Example of bang-bang phase detector with timing and truth table.

Since this technique allows high phase accuracy, it is used for small eye openings at the receiver. In contrast, a majority vote is used if lower phase accuracy and/or if a very short lock in time is needed (as e.g. in USB2.0 with 50% eye opening and phase lock within 12 bit).

The evaluation of the phase error in a CDR can be done either analog or digital.

3.1 Analog phase detection

Analog phase detection is the traditional way of phase detection [16–18]. In analog phase detectors, an output voltage is generated which is proportional to the phase error. This is mostly done with a structure based on the principle proposed by Hooge (Fig. 5, [16]).

In the Hooge phase detector, two flip-flops are sampling the incoming data. The clock for both retiming operations is shifted by 180°. An XOR gate evaluates the input and output of each flip-flop. In lock, the output pulse trains of both XOR gates are identical but shifted in time (Fig. 6a).

If the data signal is not sampled in the bit middle than the output of the first XOR of the Hooge phase detector changes its duty cycle depending on whether the clock is late or early. The second XOR gate stays unaffected thus working as a reference. Therefore, the phase error can be evaluated by the difference between the output pulses of the two XOR.

The problem of analog phase detectors is that the pulses generated for low phase errors are very narrow. This requires fast evaluation circuits to process the small pulses. Otherwise, the loop dynamics can be degraded due to dead zones. Therefore, it has become more common lately to use digital or bang-bang phase detectors for high speed applications.

3.2 Digital phase detection

In contrast to analog phase detectors, digital phase detectors only make the decision if the clock phase is early or late compared to the data (e.g. [19, 20]). The output is independent of the amount of phase error. In Fig. 7, an example of a bang-bang phase detector is given, together with a timing diagram and a truth table.

In this example, the incoming data is sampled at three times, two succeeding bits (phases C1 and C3) and the

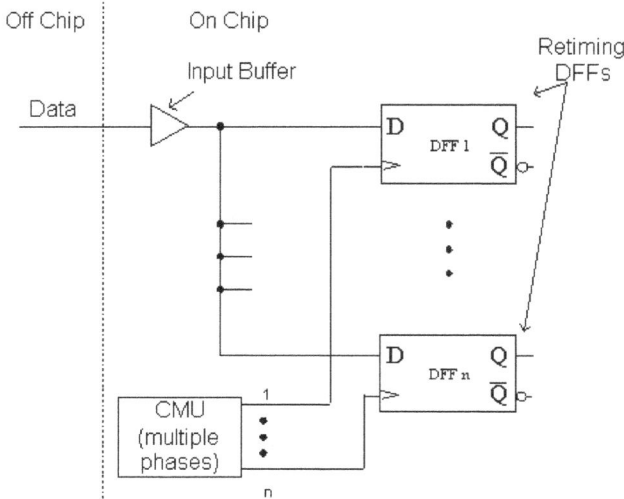

Fig. 8. Example for majority vote system, incoming data is retimed with n phases.

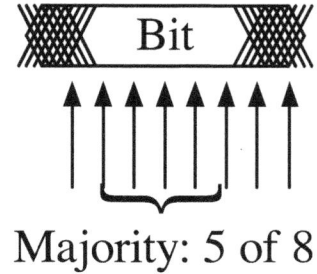

Fig. 9. Majority vote: more than half the sampling phases inside the eye opening.

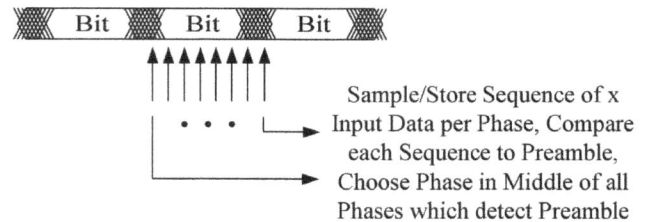

Fig. 10. Majority vote with preamble or training sequence.

transmission between those bits (phase C2). If C2 is identical to C1, the clock has been too fast. If C2 is identical to C3 the clock has been too slow. If all phases yield the same output, no transition has occurred and no decision could be made. In lock, C2 theoretically samples directly in the bit middle and therefore its output is not defined.

Other bang-bang structures sample the internal clock with the data signal which also yields the needed phase information. The advantage of bang-bang detectors, which are evaluating the data, is the perfect matching of phase evaluation and retiming operation since both actions are done at the same time and with the same flip-flops.

The general drawbacks of these kinds of phase detectors are an increased noise at the control node of the VCO and a more difficult handling. This is because a PLL using a digital phase detector is not a linear system anymore.

3.3   Majority vote phase detection (oversampling)

In a majority vote or oversampling phase detection system, the incoming data is retimed several times with different equally spaced clock phases (see Fig. 8).

The correct phase is then found either by a majority decision or by comparison with a test data sequence during an initial training interval. For a majority decision, the eye opening must be wide enough to have more than half of the sampling phases inside the data eye (Fig. 9). Therefore, this is only possible for eye openings wider than 50%.

More common for the usage of higher frequencies is the comparison of the sampled data to a preamble or training sequence during startup or certain intermediate training intervals (Fig. 10). For example, USB2.0 has an initial sequence of a minimum of 12 bit of 01 pattern followed by a single 1 which can be searched for in the different sampled data streams of the phases.

3.4   Chip border crossing

A specialty of data transmission is the communication of two chips sitting adjacent to each other on the same board or even in the same package. This is the case for serializer-deserializer (serdes) applications where it is not yet possible to implement the whole serdes function within one IC and InP or GaAs ICs are needed. This special problem is not directly addressed by any standard yet but has been discussed for parallel data transmission with data rates higher than 10 Gbit/s/channel. At those speeds, the eye opening is so small that even the skew introduced by the packages can be too high to retime the data on the receiving IC without deskewing.

Since the amount of circuitry on the upstream side IC with InP and GaAs is limited due to technology reasons, the idea is to implement a DLL across the chip borders to do the deskewing. To achieve this, on the upstream side only the phase detector must be implemented. Then, during a training period delay elements are trained on the downstream side to equalize the skew between the different data lines. As a result, the amount off circuitry on the upstream side is greatly reduced (see Fig. 11).

Common to all the bit synchronization techniques is the need for transitions in the input signal. If there are no bit changes than no phase information can be derived. Therefore, a minimum number of transitions must be specified in order to guarantee correct operation of the synchronization. Since this usually cannot be guaranteed by the data itself, line coding is used to establish a minimum number of data signal transitions.

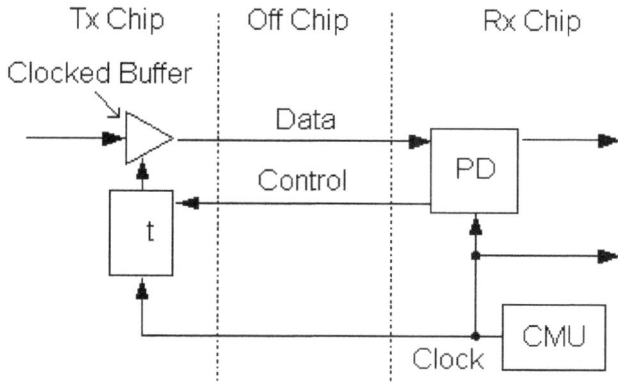

Fig. 11. Block diagram for deskewing of data lines across chip borders, only a phase detector (PD) is placed on the receiver (RX), the deskewing is done on the transmitter (TX).

4   Line coding

The next level of data recovery following the bit synchronization is the extraction of the information from the bit stream. To achieve this, the word boundaries must be found from the data stream.

In package oriented transmission standards, this can simply be done by defining a known start pattern. For example, in USB2.0, a minimum sequence of 12 succeeding low-high pattern followed by a single low is used. After this preamble, the transmitted bits contain real data information. Therefore, the start of the word boundary is known after the initial synchronization of the link.

In standards where a constant data stream is applied the word boundary must be found without initial knowledge of the position of a special bit. For standards that allow the distribution of the transmitted information on parallel data lines (serial-parallel data transmission), also the problem of skew compensation is addressed. In that case, deskewing information must be added to the data stream. Only if the skew on the bit lines do not exceed one bit (this includes the bit synchronization!) [15] this is not necessary.

All these tasks are achieved by encoding of the transmitted data. The biggest differences between the various standards result from theses kinds of line coding that are used. There are several more reasons to do line coding in general. Some of these are:

1. Provide a 50% mark ratio for DC balance (equal numbers of "0" and "1" on average).

2. Allow frame border detection through synchronization patterns or known sequences.

3. Transmission error detection/correction.

4. Supply special control patterns.

5. Achieve a certain spectrum for the transmitted data.

Table 2. TMDS coding table for 20 bit alternating sequence, bits are transmitted LSB first.

8 bit data word (MSB...LSB)	10 bit transmit word (MSB...LSB)
0 0 0 0 0 1 0 1	1 1 1 1 1 1 1 1 0 0
0 0 0 0 0 0 0 0	1 1 1 1 1 1 1 1 1 1
1 1 1 1 1 0 1 1	0 0 0 0 0 0 0 0 1 1
1 1 1 1 1 1 1 0	0 0 0 0 0 0 0 0 0 0

Table 3. TMDS coding table for maximum run length of 22 high bits, bits are transmitted LSB first.

8 bit data word (MSB...LSB)	10 bit transmit word (MSB...LSB)
0 0 0 0 0 0 0 0	0 1 0 0 0 0 0 0 0 0
0 0 0 0 0 1 0 1	1 1 1 1 1 1 1 1 0 0
0 0 0 0 0 0 0 0	1 1 1 1 1 1 1 1 1 1
0 0 0 1 0 0 0 1	0 1 0 0 0 0 1 1 1 1

As an example, the coding schemes 8B10B and TMDS (Transmission Minimized Differential Signal) will be chosen for comparison.

Both codes transfer an 8 bit data word into a 10 bit transmit word. While for 8B10B the coding is done with a complex logic function [21], TMDS only needs a simple logic to generate the 10 bit transmit words [2, 9].

The 8B10B code provides a 50% mark ratio with a maximum run length of only 5 bits. This maximum run length is only achieved within the 12 special characters. Therefore, the special characters can easily be detected by this 5 bit sequence. They are used to recover the 10 bit word boundary and to transfer control information.

In the case of serial-parallel data transmission, special characters furthermore are used for deskewing. The information in each data line is framed by special characters. The receiver detects the special characters in each channel and compensates the skew by alignment of those characters. This is possible as long as the system skew does not reach half the length of the total data package including the framing.

The short run length of the 8B10B code helps in the design of the bit synchronization circuit at the receiver. This is because the jitter generated by those blocks depends on the number of transitions in the data stream.

For TMDS on the other side, there is no certain run length given in the specification. With some patience, one can find an input pattern that yields an alternating sequence of 20 highs and lows or a maximum run length of 22 bits (see Tables 2 and 3). Furthermore, fast bit changes after a long sequence (e.g. $21\times$ high, followed by one low and one high) can be found. This makes the design of the bit synchronization circuit more difficult.

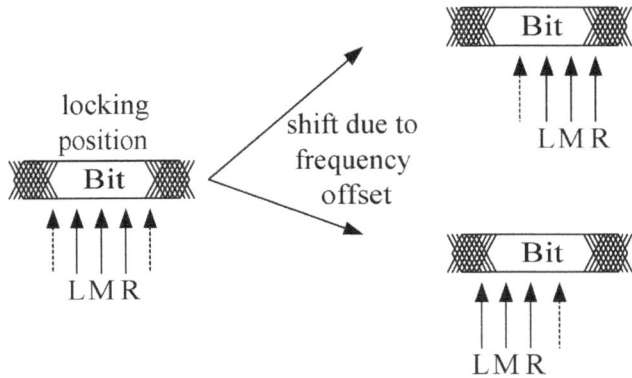

Fig. 12. Phase tracking for oversampling and majority vote synchronization systems.

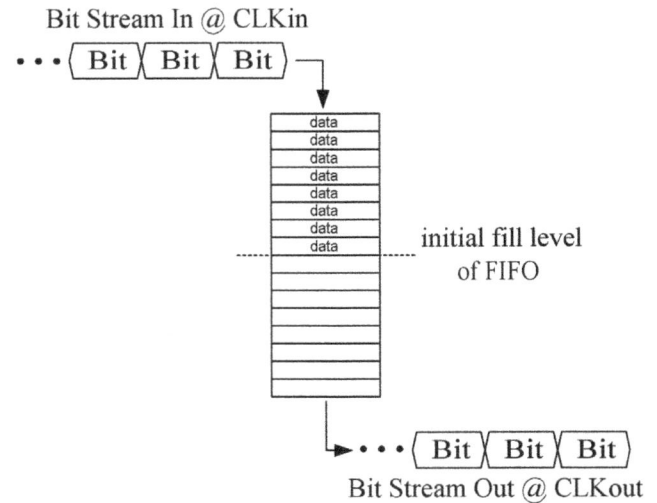

Fig. 13. Block diagram of simple elastic buffer.

As 8B10B, TMDS offers special transmit characters meant to transmit control information. The difficulty is that the four available patterns do not offer a unique sequence in the data stream, as does 8B10B. In contrast, control patterns must always be sent and detected in a sequence. This is because it is not possible to achieve a data pattern that matches two subsequent special characters from sending coded data patterns so the frame borders can be detected by at least two of these patterns in sequence.

One advantage of TMDS patterns is that electro-magnetic interference is reduced due to the reduced number of transitions.

Both codes have a 25% overhead for transmission data.

As can be seen from this evaluation, both codes have optimized different things. Despite looking close on the first side, they have some major implementation differences. While 8B10B encoding and decoding takes considerably more circuitry to implement, TMDS is more difficult to deal with for the bit synchronization at the receiver.

In that way, every other code, e.g. 64/66, scrambling etc. is optimizing different things but in general, the main tasks of line coding given above apply to all of them.

5   Matching transmission speeds

If transmitter and receiver are far apart as in most applications, they do not have the same physical clock source. Therefore, the clock speeds at the transmitter domain and at the receiver domain do not perfectly match. Usually, clock differences of up to 1000 ppm are allowed.

This difference in the clock speeds leads to two problems. First, for oversampling and majority vote systems, which run at a local clock, the bit synchronization circuit at the receiver must be tuning all the time. This is necessary to keep the optimal sampling point for the incoming data for those systems and to prevent data loss. The tuning is done by constant evaluation of more than the optimal sampling phase found in the initial start up. If e.g. three adjacent phases are evaluated, no phase shift is necessary as long as all phases yield the same sampling result. If for example the evaluation of the

left of three phases yields a different result than the other two phases, the former right phase will become the new middle phase, an additional phase is added at the right side of the sampling group, the former middle phase becomes the new left phase and the former left phase is not used for evaluation in the next time step anymore. This process is illustrated in Fig. 12.

For analog or digital phase detectors in a CDR, reference clock differences at transmitter and receiver are not a problem. Here, the CDR loop always runs at the clock speed of the transmitter due to the tuning of the local VCO.

The second problem related to the clock mismatches at transmitter and receiver is to loose or add data at the receiver. This is because the clock differences add up and eventually exceed one bit length.

In package-oriented standards like USB2.0, a simple elastic buffer can be used to compensate for these clock speed differences. An elastic buffer is a FIFO buffer where the write and the read operations are done at different clock speeds. In the case of USB2.0, not more than 12 bit mismatch can occur during the transmission of one package. Therefore, if an elastic buffer with a depth of 24 bits is filled up with 12 bits, no data loss or addition can happen. This is illustrated in Fig. 13.

For standards using constant data streams, the elastic buffer is more complicated. For example, in Infiniband, there are special areas defined in the data stream which do not contain real data. The elastic buffer detects these areas by the framing of special characters. Then, the elastic buffer adds or subtracts words from or to these areas respectively depending on the current fill level of the buffer. Due to this process, differences in the clock speeds are compensated. In Fig. 14, this is shown in principle. The shaded words are the fill pattern. In the Infiniband standard for example, with the maximum allowed clock offset of 200 ppm, a 10 bit word must be filled in or extracted from the data stream approximately every 16 us.

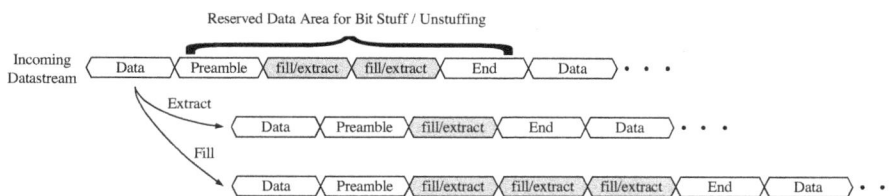

Fig. 14. Principle of clock speed matching process for constant data stream systems.

6 Summary and conclusions

It has been shown that for the physical layer of today's high-speed standards, there are not many different design techniques used to recover the data from the data stream. In general, two different techniques, oversampling and bang-bang based CDR loops, are used. In addition, techniques to enhance the transmission distance (or to reduce system cost) like pre-amphasis and reconstruction filters are about to be included in future implementations.

The main difference between standards today results from the various kinds of data codes used to encode/decode the data and to add synchronization information. These various coding techniques are the main source of differences in the implementation of transmitters and receivers.

Acknowledgement. The author would like to thank J. Höhn and A. Weggel for helpful discussions.

References

[1] Electrical Characteristics of Low Voltage Differential Signaling (LVDS) Interface Circuits, Revision A, ANSI/TIA/EIA-644-A Electrical characteristics standard.

[2] DVI standard, www.ddwg.org.

[3] USB standard, www.usb.org.

[4] FireWire, www.apple.com/firewire.

[5] Optical Internetworking Forum (OIF), Physical Link Layer (PLL) Working Group, SFIX, www.oiforum.com.

[6] Infiniband standard, www.infinibandta.org/home.

[7] Intel Developer Network for PCI Express Architecture www.intel.com/technology/pciexpress/devnet.

[8] Gigabit Ethernet Alliance (XAUI), www.10gea.org/Tech-whitepapers.htm.

[9] HDMI standard, www.hdmi.org.

[10] Fukaishi, M., Nakamura, K., Sato, M., Tsutsui, Y., and Yotsuyanagi, M.: A 4.25-Gb/s CMOS Fiber Channel Transceiver with Asynchronous Tree-Type Demultiplexer and Frequency Conversion Architecture, IEEE Journal of Solid-State Circuits, 33, 12, 1998.

[11] Sakamoto, T., Tanaka, N., and Ando, Y.: Low-latency Skew-compensation Circuits for Parallel Optical Interconnections, ECTC 1999.

[12] Wadhwa, R., Aggarwal, A., Edwards, J., Ehlert, M., Höhn, J., Miao, G., Lakshmikumar, K., and Khoury, J.: A Low Power 0.13 μm CMOS OC-48 SONET and XAUI Compliant SERDES, CICC, 2003.

[13] Schmale, I., Heinemann, M., Drögemüller, K., Kuhl, D., Blank, J., Ehlert, M., Kraeker, T., Höhn, J., Klix, D., Plickert, V., Melchior, L., Hildebrandt, P., Leininger, L., Dröge, E., Kropp, J.-R., Wolf, H.-D., Wipiejewski, T., and Johnson, R.: High Speed 12×2.5 GBit/s parallel optical links (PAROLI) for increased transmission length, European Conference on Optical Communication (ECOC), 4–7 September, München, 2000.

[14] Drögemüller, K., Kuhl, D., Blank, J., Ehlert, M., Kraeker, T., Höhn, J., Klix, D., Plickert, V., Melchior, L., Schmale, I., Hildebrandt, P., Heinemann, M., Schiefelbein, F. P., Leininger, L., Wolf, H.-D., Wipiejewski, T., and Ebberg, A.: Current Progress of Advanced High Speed Parallel Optical Links for Computer Clusters and Switching Systems, 50th Electronic Components & Technology Conference (ECTC), 21–24 May Las Vegas, Nevada, USA, 2000.

[15] Kuhl, D., Drögemüller, K., Blank, J., Ehlert, M., Kraeker, T., Höhn, J., Klix, D., Plickert, V., Melchior, L., Hildebrandt, P., Heinemann, M., Beier, A., Leininger, L., Wolf, H.-D., Wipiejewski, T., and Engel, R.: PAROLI, A Parallel Optical Link with 15GBit/s Throughput in a 12 Channel Wide Interconnection, 6th Parallel Interconnect (PI), 17–19 October, Anchorage, Alaska, USA, 1999.

[16] Hooge, Jr., C. R.: A Self Correcting Clock Recovery Circuit, Journal of Lightwave Technology, LT-3, 6, December, 1985.

[17] Lee, T. H. and Bulzacchelli, J. F.: A 155-MHz Clock Recovery Delay- and Phase-Locked Loop, IEEE Journal of Solid-State Circuits, 27, 12, December, 1992.

[18] Savoj, J. and Razavi, B.: A 10-Gb/s CMOS Clock and Data Recovery Circuit, 2000 Symposium on VLSI Circuits Digest of Technical Papers, 2000.

[19] Pottbäcker, A., Langmann, U., and Schreiber, H.-U.: A Si Bipolar Phase and Frequency Detector IC for Clock Extraction up to 8 Gb/s, IEEE Journal of Solid-State Circuits, 27, 12, December, 1992.

[20] Reinhold, M., Dorschky, C., Rose, E. et al.: A Fully Integrated 40-Gb/s Clock and Data Recovery IC With 1:4 DEMUX in SiGe Technology, IEEE Journal of Solid-State Circuits, 36, 12, December, 2001.

[21] Widmer, A. X. and Franaszek, P. A.: A DC-Balanced, Partitioned-Block, 8B/10B Transmission Code, IBM J. Res. Develop., 27, 5, September, 1983.

[22] Open LDI standard, v0.95.

11

Incremental refinement of a multi-user-detection algorithm

M. Vollmer and J. Götze

University of Dortmund, Information Processing Lab, Germany

Abstract. Multi-user detection is a technique proposed for mobile radio systems based on the CDMA principle, such as the upcoming UMTS. While offering an elegant solution to problems such as intra-cell interference, it demands very significant computational resources.

In this paper, we present a high-level approach for reducing the required resources for performing multi-user detection in a 3GPP TDD multi-user system. This approach is based on a displacement representation of the parameters that describe the transmission system, and a generalized Schur algorithm that works on this representation. The Schur algorithm naturally leads to a highly parallel hardware implementation using CORDIC cells. It is shown that this hardware architecture can also be used to compute the initial displacement representation.

It is very beneficial to introduce incremental refinement structures into the solution process, both at the algorithmic level and in the individual cells of the hardware architecture. We detail these approximations and present simulation results that confirm their effectiveness.

1 Introduction

The third generation of mobile radio systems, UMTS, has a very high demand for signal processing hardware, both in the base station as well as in the mobile terminal. This is due to the amount of computations needed to combat the negative effects in a CDMA system. One particularly elegant approach is the employment of multi-user detection at the receiver (Verdu, 1998). With this technique, the data symbols of one user are estimated by using knowledge of all other users that are transmitting at the same time but on different codes.

While the basic formulation of a linear multi-user detector is quite simple, the resulting raw amount of computing power is quite simple, the resulting raw amount of computing power required for its real-time realization is formidable (Haardt and Mohr, 2000). However, for the specific case of the TDD mode of UMTS (Haardt et al., 2000) with its burst-structured transmission and relatively few simultaneously active users, there exist algorithms and implementation strategies that can significantly reduce the computational requirements (Vollmer et al., 2001).

In this paper, we concentrate on the Schur algorithm (Schur, 1917, 1986). It leads to a fine-grained, parallel formulation of its computations which should prove beneficial for a dedicated hardware implementation.

One key characteristic of the Schur algorithm is that it exploits the inherent mathematical structure of the specific problem, such as the periodicity of the spreading code and the (assumed) time-invariance of the radio channel during one burst. Another important insight is that limited inter-symbol-interference of the system allows for far reaching approximations. The approximations can be chosen such that they directly lead to less consumed resources (area, clock frequency, power).

We will shortly present the mathematical formulation of our specific linear multi-user detector in Sect. 2. In Sect. 3, we explain its displacement representation and the corresponding Schur algorithm, together with approximations that work on the algorithmic level. Section 4 shows the parallel and pipelined formulation of this algorithm, using approximating hyperbolic CORDIC cells for complex valued signals. Simulation results are shown in Sect. 5.

2 Linear multi-user detection in a burst-structured system

As explained in, for example (Vollmer et al., 2001), the transmission in a burst-structured CDMA system can be modeled by the matrix equation

$$x = \mathbf{T}d + n.$$

In this equation, d represents the collection of the data symbols sent by K users, where each user contributes N sym-

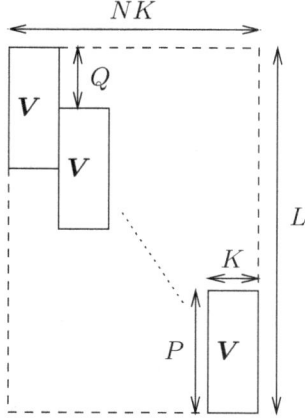

Fig. 1. The structure of **T**. It is build from copies of the unstructured matrix **V**, resulting in a Block-Toeplitz-Structure for **T**. K: number of active CDMA codes, Q: spreading factor, P: $Q + W - 1$ with W the channel length.

bols; **T** models the (assumed) linear and time-invariant transmission; n contains noise; and x contains the received samples. The task of the receiver is to compute an estimate $\hat{d}$ of d from the knowledge of x and **T**. This can be done by using the pseudo-inverse $\mathbf{T}^+$ of **T** (Lewis and Odell, 1971) and computing

$$\hat{d} = \mathbf{T}^+ x = (\mathbf{T}^H \mathbf{T})^{-1} \mathbf{T}^H x. \tag{1}$$

Computing $\mathbf{T}^+$ in general is expensive. However, the structure of **T** as shown in Fig. 1 allows the application of specialized methods that can reduce the computational requirements significantly (Vollmer et al., 2001).

3 Displacement representation and Schur algorithm

Instead of computing $\hat{d}$ via Eq. (1), one can also compute the QR factorization of **T** such that $\mathbf{T} = \mathbf{QR}$ and

$$\hat{d} = \mathbf{R}^{-1} \mathbf{Q}^H x. \tag{2}$$

The matrix $\mathbf{Q} \in \mathbb{C}^{L \times NK}$ satisfies $\mathbf{Q}^H \mathbf{Q} = \mathbf{I}$ and $\mathbf{R} \in \mathbb{C}^{KN \times KN}$ is upper triangular. The matrix **R** is also known as the Cholesky factor of $\mathbf{T}^H \mathbf{T}$.

The Schur algorithm computes **R** and $z = \mathbf{Q}^H x$ while exploiting the Block-Toeplitz structure of **T**. It starts by computing a *displacement representation* for **T** and x and then continues to gradually transform them into **R** and z by applying local unitary and hyperbolic transformations (Chun et al., 1987; Kailath and Chun, 1994).

The displacement representation for **T** can be written down as

$$\mathbf{T}^H \mathbf{T} = \mathbf{R}^H \mathbf{R} = \mathbf{U}_A^H \mathbf{U}_A - \mathbf{U}_B^H \mathbf{U}_B, \tag{3}$$

where $\mathbf{U}_A$ and $\mathbf{U}_B$ are as depicted in Fig. 2. As can be seen, $\mathbf{U}_A$ is constructed from copies of the matrix **A**, while $\mathbf{U}_B$ is constructed from copies of **B**. Moreover, **B** can be found in

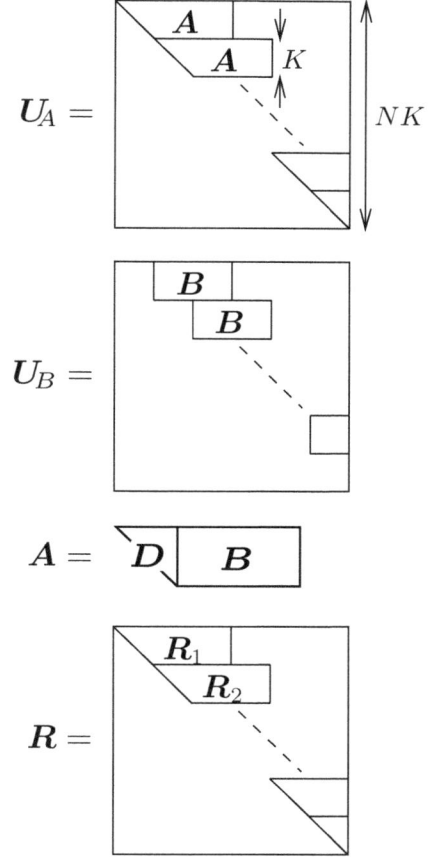

Fig. 2. The structure of $\mathbf{U}_A$, $\mathbf{U}_B$ and **A**, **B**. Both $\mathbf{U}_A$ and $\mathbf{U}_B$ inherit the Block-Toeplitz and the band-structure of **T**.

A: the two matrices differ only in the upper triangular part, which is zero in **B** and the non-zero matrix **D** in **A**.

The matrices **A** and **B** are called the generators of the displacement representation and need to be computed before the Schur algorithm can start to compute **R**. By using the matrix partitionings of Fig. 2, one can show that $\mathbf{A} = \mathbf{R}_1$. However, $\mathbf{R}_1$ can also be found in a partial QR-Decomposition of **T** that only annihilates the first block-row of **T**, as shown in Fig. 3. This partial transformation can be computed using a parallel processor array and fortunately, we will find that we can use the same processor array to compute the rest of the Schur algorithm.

As explained in (Vollmer et al., 1999), z can be computed by the Schur algorithm from the right hand side generator y that is chosen such that $\mathbf{T}^H x = \mathbf{U}_D^H y$ where $\mathbf{U}_D = \mathbf{U}_A - \mathbf{U}_B$ is a block-diagonal, Block-Toeplitz matrix with **D** on its diagonal. Just as **A** can be found in the first rows of **GT**, it can be shown that the first K elements of **Gx** contain the first elements of y. The next K elements of y can be found in the first elements of $\mathbf{Gx}'$ where x' is a shifted version of x with $x'_i = x_{i+Q}$. Figure 4 shows graphically how this can be extended to all parts of y.

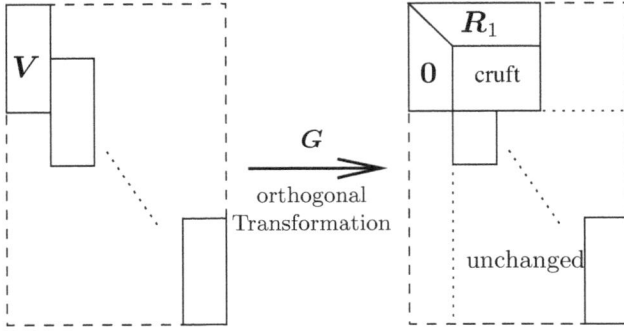

Fig. 3. Partial QR-Decomposition of **T**. After annihilating the indicated part with a orthogonal transformation, the first rows of the result **GT** contain $\mathbf{R}_1$.

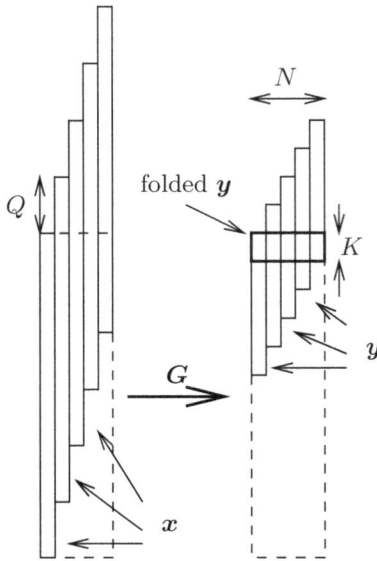

Fig. 4. Transforming x piecewise into y, using **G** from Figure 3

Equation (3) can be rewritten as

$$\mathbf{T}^H\mathbf{T} = \begin{bmatrix} \mathbf{U}_A \\ \mathbf{U}_B \end{bmatrix}^H \begin{bmatrix} \mathbf{I} & \\ & -\mathbf{I} \end{bmatrix} \begin{bmatrix} \mathbf{U}_A \\ \mathbf{U}_B \end{bmatrix}.$$

The Schur algorithm then completes the QR-Decomposition by applying a transformation Θ with the properties:

$$\Theta \begin{bmatrix} \mathbf{U}_A \\ \mathbf{U}_B \end{bmatrix} = \begin{bmatrix} \mathbf{R} \\ \mathbf{0} \end{bmatrix} \quad \text{and} \quad \Theta^H \begin{bmatrix} \mathbf{I} & \\ & -\mathbf{I} \end{bmatrix} \Theta = \begin{bmatrix} \mathbf{I} & \\ & -\mathbf{I} \end{bmatrix}.$$

It can be shown that the result of applying Θ to two stacked y contains z:

$$\Theta \begin{bmatrix} y \\ y \end{bmatrix} = \begin{bmatrix} z \\ * \end{bmatrix} \quad \text{since} \quad \mathbf{T}^H x = \begin{bmatrix} \mathbf{U}_A \\ \mathbf{U}_B \end{bmatrix} \begin{bmatrix} \mathbf{I} & \\ & -\mathbf{I} \end{bmatrix} \begin{bmatrix} y \\ y \end{bmatrix}. \quad (4)$$

Analogous to the more well-known QR decomposition with unitary rotations (Golub and van Loan, 1996), the transformation Θ itself is composed of individual elementary

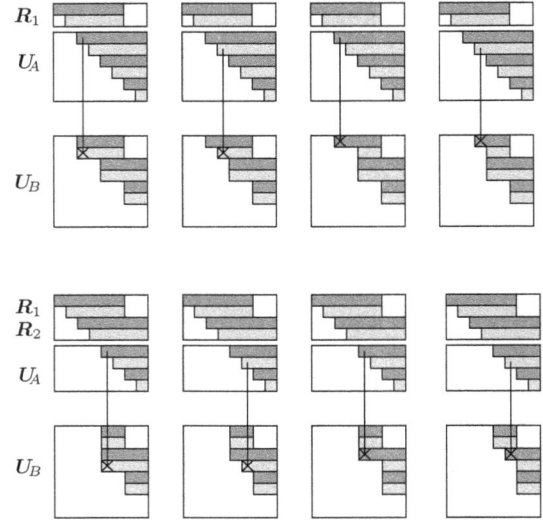

Fig. 5. Eliminating two blocks in $\mathbf{U}_B$ for $K = 2$, $N = 4$.

transformations that each eliminate one element in $\mathbf{U}_B$. Such an elementary transformation is either a unitary or a hyperbolic rotation, depending on whether it affects only the $\mathbf{U}_B$-part of **x**, or both the $\mathbf{U}_A$- and $\mathbf{U}_B$-part.

An elementary hyperbolic rotation $\mathbf{H} \in \mathbb{C}^{2 \times 2}$ is defined by the conditions

$$\mathbf{H}\begin{bmatrix} a \\ b \end{bmatrix} = \begin{bmatrix} r \\ 0 \end{bmatrix} \quad \text{and} \quad \mathbf{H}^H\begin{bmatrix} 1 & 0 \\ 0 & -1 \end{bmatrix}\mathbf{H} = \begin{bmatrix} 1 & 0 \\ 0 & -1 \end{bmatrix}.$$

Similar to a unitary rotation, it can be computed as

$$\mathbf{H} = \begin{bmatrix} \cosh(\phi) & -\sinh(\phi) \\ -\sinh(\phi) & \cosh(\phi) \end{bmatrix}, \quad \phi = \tanh^{-1}\left(\frac{b}{a}\right)$$

$$\text{or} \quad \mathbf{H} = \begin{bmatrix} c^* & s^* \\ s & c \end{bmatrix}, \quad \begin{matrix} c = a/r, \ s = -b/r, \\ r = \sqrt{a^*a - b^*b} \end{matrix}$$

The first variant is only valid for real valued a and b, while the second variant is also valid for complex values. Hyperbolic rotations are only defined for $|b| < |a|$.

The task of Θ is to transform $\mathbf{U}_B$ to zero. Figure 5 shows a possible sequence of elementary hyperbolic transformations to eliminate two blocks. It should be easy to see how to eliminate the remaining blocks.

This complicated way of arriving at **R** does not seem to gain anything compared to a more straightforward QR decomposition of **T**. The trick is to observe that $\mathbf{U}_A$ and $\mathbf{U}_B$ inherit the Block-Toeplitz structure of $\mathbf{T}^H\mathbf{T}$ and that this structure is preserved to a large degree while $\mathbf{U}_B$ is transformed to zero and $\mathbf{U}_A$ to **R**. For example, the second block in Fig. 5 can be eliminated with the same sequence of rotations as the first block block, only placed differently. Therefore, we don't need to explicitly carry out these computations and can just copy their result from their previous applications. Applying this to Fig. 5 allows us to skip the second batch of transformations completely.

Looking more closely at the process, as for example done in (Vollmer et al., 1999), we can see that $\mathbf{U}_B$ remains Block-

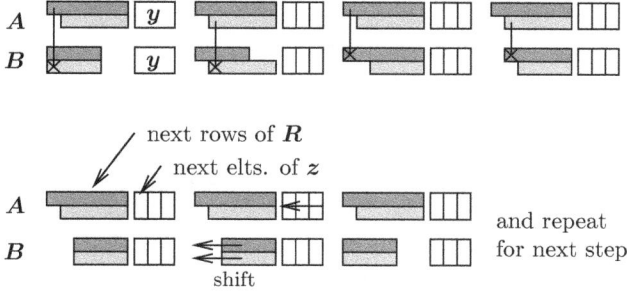

Fig. 6. Eliminating one block in **B** while exploiting the structure.

Toeplitz throughout, and it therefore suffices to store only the first K rows, that is, **B**. Additionally, $\mathbf{U}_A$ can be partitioned into two parts: an upper one which is not Block-Toeplitz, and a lower one, which is. The border between these two parts moves downwards during the transformation and it can be seen that the elements in the upper part will not be touched again. This is depicted in Fig. 5 by separating the upper rows of $\mathbf{U}_A$ and labeling them $\mathbf{R}_1$ and $\mathbf{R}_2$, respectively. Thus, the upper part of the gradually transformed $\mathbf{U}_A$ contains more and more rows of the desired result **R**. Figure 6 shows how to exploit these insights by only working with non-redundant data.

The generator for the right hand side y needs to be transformed in the same way as $\mathbf{U}_A$ and $\mathbf{U}_B$. Since the actual computation will only apply the non-redundant parts of $\boldsymbol{\Theta}$, but y is in principle affected by all parts, this requires a small twist. The redundant parts of $\boldsymbol{\Theta}$ can be resurrected by transforming the folded version of y (indicated in Fig. 4) alongside with **A** and **B**. As shown above in (4), two copies of y enter the transformation and thus two copies of the folded y are present in the final algorithm: one next to **A**, and one next to **B**; see Fig. 6.

It should be noted that is is of course possible (and straightforward) to exploit the band structure that is present in **T** (and thus **A** and **B**) by avoiding operations that are known to process only zeros.

In addition to exploiting the structure of **T**, we can also introduce approximations into the solution process. The limited inter-symbol-interference of the system leads to the fact that **B** converges to zero quite rapidly and thus the transformations can be stopped early. In other words, the later transformations will find **B** to be already quite close to zero, and can be omitted entirely. The remaining rows of **R** are produced by continuing to shift **A** and y and copying them into the appropriate parts of **R** and z.

4 Parallel processor array for the Schur algorithm

The partial QR-Decomposition for computing the generators can be carried out on a partial QR array, as depicted in Fig. 7. The relevant part of **T** is fed into the array and when the partial decomposition has finished, the feedback registers contain the indicated elements of **A** and y. One step of the Schur

algorithm (Fig. 6) can then be computed on the same array by replacing the orthogonal rotations with hyperbolic ones. After the K rows of **B** have been put through this array, their new contents can be retrieved at the outputs. The registers will then contain new rows for **R**. This process is repeated until **B** is sufficiently close to zero. At the same time, the generator y for the right hand side is transformed and after each step, new elements for z are produced.

A hyperbolic rotation for complex values can be build from three real-valued rotations: two orthogonal ones and one hyperbolic. The orthogonal rotations are used to extinguish the imaginary parts of **A** and **B**, the hyperbolic rotation eliminates the remaining real part of **B**. This can be reduced to just one orthogonal rotation and one hyperbolic one when care is taken to produce only real-valued elements on the diagonal of **A**. Figure 8 shows this graphically.

The complete array can then be composed from these complex-valued, hyperbolic vector and rotation cells as shown in Fig. 7 for $K = 2$ and $N = 2$.

The real-valued cells in Fig. 8 can be implemented using CORDIC devices (Volder, 1959; Hu, 1992). Each CORDIC operation consists of a number of micro rotations (Götze and Hekstra, 1995) that are chosen to be practical for hardware implementations.

A CORDIC device implements the complete transformation **H** by approximating it by a sequence of micro rotations of the form

$$\mathbf{M}_i = K_i \begin{bmatrix} 1 & -\mu_i 2^{-s(i)} \\ -\mu_i 2^{-s(i)} & 1 \end{bmatrix},$$
$$\mu_i = \pm 1, \qquad K_i = (1 - 2^{-2s(i)})^{-\frac{1}{2}}$$

such that $\mathbf{H} \approx \prod_{i=0}^{n-1} \mathbf{M}_{n-i}$. The parameter μ_i determines the direction of rotation and i determines the angle. The function $s(i)$ specifies the shift sequence and can, for hyperbolic rotations, be taken as (Walther, 1971)

$$s(i) = \{1, 1, 2, 3, 4, 4, 5, \ldots, 12, 13, 13, 14, \\ \ldots, 39, 40, 40, 41, 42, \ldots\}.$$

The hardware operations corresponding to $\mathbf{M}_i$ (except for the multiplication by K_i) are additions and shifts. Figure 9 shows a schematic for implementing a micro rotation. The scaling factor K_i is independent of μ_i and can be accumulated over the sequence of micro rotations.

The number of micro rotations n performed per elementary rotation is an indicator of the amount of hardware resources required to implement the CORDIC device, and of the accuracy attained. The goal is to keep this number as low as possible while still achieving useful results from the multi-user detector.

5 Simulation results

To verify the effectiveness of the described approximation method, simulations of a linear multi-user detector in a CDMA system as specified in TS 25.102 (3GPP, 2002) were

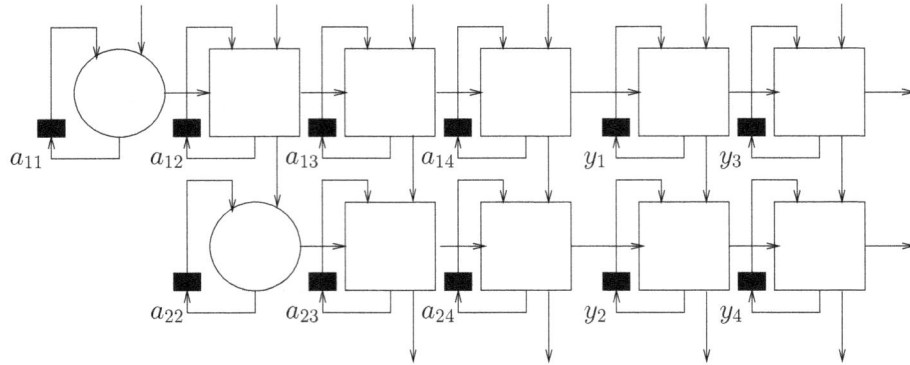

Fig. 7. Systolic array for Schur algorithm.

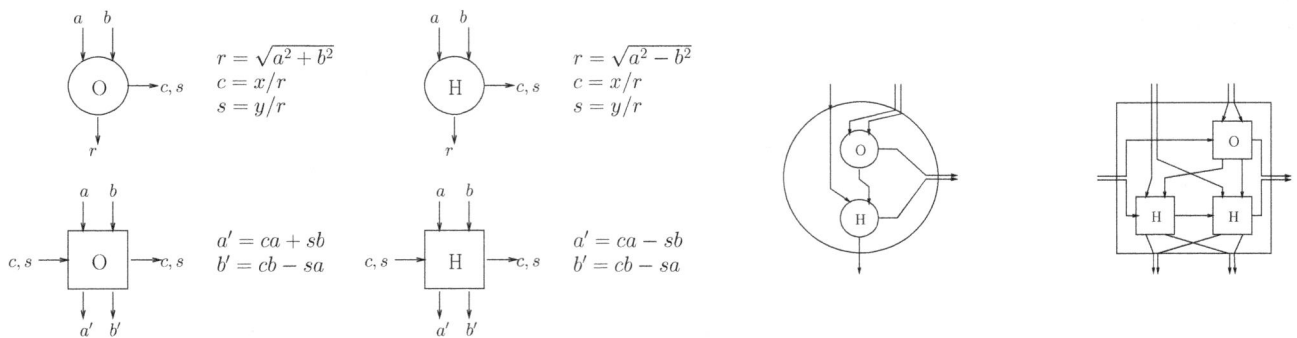

Fig. 8. Real-valued orthogonal and hyperbolic cells and how to build complex hyperbolic cells from them.

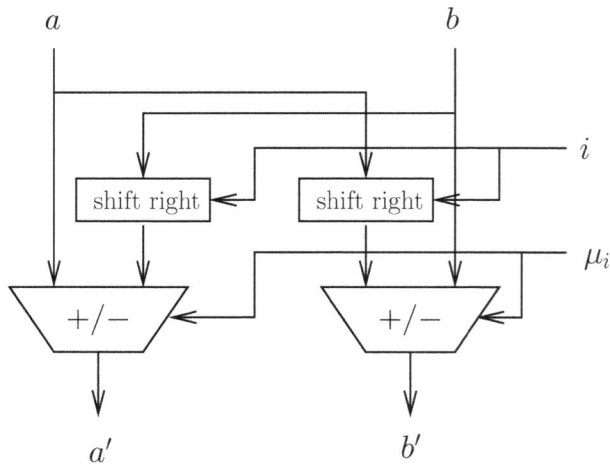

Fig. 9. Hardware for computing an unscaled micro rotation.

Table 1. Simulation parameters

spreading factor	Q	8
number of active codes	K	6
symbols per half-burst	N	122
channel length in chips	W	57
midamble length in chips		512
mobile speed in km/h		120
relative tap delays in ns		$0, 260, 521, 781$
corresp. rel. power in dB		$0, -3, -6, -9$
simulated bits per code		10^6

was varied. The quality of the resulting receiver was assessed by measuring the bit error ratio for different signal to noise ratios. The CORDIC cells in the receiver were simulated with IEEE double precision floating point numbers.

Each instantiation of the array received the exact same input data. That is, although the simulation used random numbers to model the data source and the time variance of the radio channel, each BER measurement used the same sequence of random numbers. As can be seen in Fig. 10, the bit error ratios for a user depend heavily on the concrete realization of the radio channel and it is thus important to keep all parameters of a simulation constant (including the random numbers), except for the parameter that is being examined.

Figure 11 shows results for a decreasing number of com-

performed. The scenario was chosen according to Annex A.2.1: uplink service with 12.2 kbs using the 3.84 Mcps TDD option. The channel was modeled according to Annex B.2.1, case 3. The resulting parameters are summarized in Table 1. The channel estimation was assumed to be perfect.

The described processor array was used as the receiver in this system. The number of Schur steps and the number of micro rotations per hyperbolic/orthogonal transformation

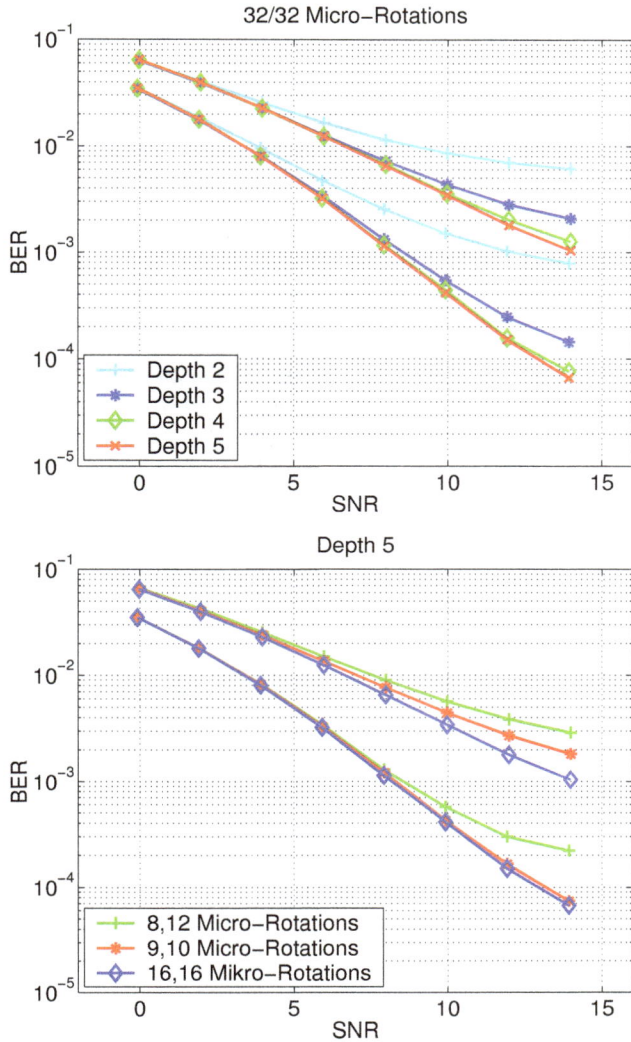

Fig. 11. Simulation results.

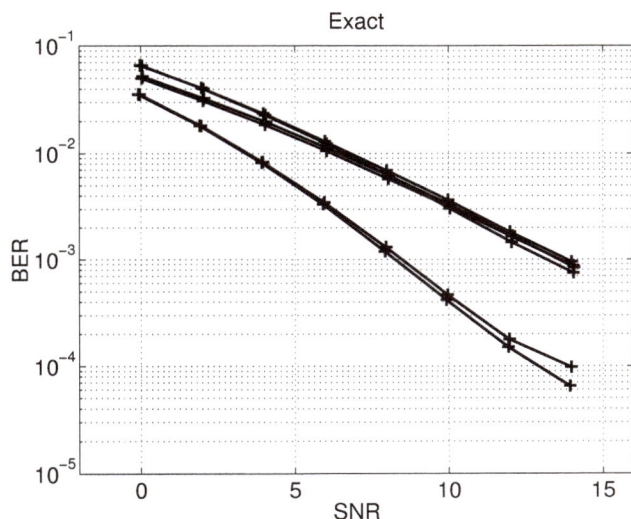

Fig. 10. BER without approximations for each user.

puted rows in **R**, for two selected users. The legend "depth=5" indicates that $5K$ rows have been computed before stopping the transformation, for example. It can be seen that already $5K$ rows might suffice (out of $122K$) for this simulation scenario. Also, results for a decreasing number of micro rotations when $5K$ rows of **R** are computed are shown. The legend "n,m Micro-Rotations" means that n micro-rotations are used for the orthogonal cells of Fig. 8 and m are used for the hyperbolic ones. It can be seen that 9 orthogonal and 10 hyperbolic micro-rotations suffice to nearly attain the performance of double precision floating point.

6 Summary

We have presented a hardware oriented, systolic architecture for implementing a complex valued, linear multi-user detector for a burst structured system described by a Block-Toeplitz structured system matrix. The architecture is able to exploit this inherent structure of the matrix.

The presented algorithmic modifications directly lead to less power consumption. As always, a parallel and pipelined implementation is the first step in reducing the power/time consumption (Mehra et al., 1996). Refinement structures (Nawab et al., 1997) are then introduced on different levels. First the original algorithm is modified by re-interpreting it as an iterative method and introducing a "depth" parameter that controls the number of iterations. This is justified by the observation that the algorithm converges quickly towards a steady state due to the band structure of the system matrix.

At the architectural level, the complex rotations (realized by two or three real CORDIC devices) are approximated by reducing their number of micro rotations.

Future work will compare the switching activity of the realizations at different incremental refinement steps in order to explicitly show the possible reduction of power consumption by the presented methodology.

References

3GPP: TS 25.102 UE Radio Transmission and Reception (TDD), 2002.

Chun, J., Kailath, T., and Lev-Ari, H.: Fast Parallel Algorithms for QR and Triangular Factorization, SIAM J. Sci. Stat. Comput., 8, 1987.

Golub, G. and van Loan, C.: Matrix Computations, The John Hopkins University Press, third edn., 1996.

Götze, J. and Hekstra, G.: An algorithm and architecture based on orthonormal μ–rotations for computing the symmetric evd, INTEGRATION – The VLSI Journal (Special Issue on Algorithms and Parallel VLSI Architectures), 20, 21–39, 1995.

Haardt, M. and Mohr, W.: The complete solution for third generation mobile communications: Two modes on air – One winning strategy, in Proc. IEEE Int. Conference on Third Generation Wireless Communications, San Francisco, USA, 2000.

Haardt, M., Klein, A., Koehn, R., Oestreich, S., Purat, M., Sommer, V., and Ulrich, T.: The TD-CDMA based UTRA TDD mode,

IEEE J. Select. Areas Commun., 18, 1375–1386, special issue on "Wideband CDMA", 2000.

Hu, Y. H.: CORDIC-Based VLSI Architectures for Digital Signal Processing, IEEE Signal Processing Magazine, 1992.

Kailath, T. and Chun, J.: Generalized Displacement Structure for Block-Toeplitz, Toeplitz-Block, and Toeplitz-Derived Matrices, SIAM J. Matrix Anal. Appl, 15, 114–128, 1994.

Lewis, T. and Odell, P.: Estimation in Linear Models, Prentice-Hall, Inc. Englewood Cliffs, New Jersey, 1971.

Mehra, R., Lidsky, D., Abnous, A., Landman, P., and Rabaey, J.: Algorithm and architectural level methodologies for low power, Low Power Design Methodologies, Kluwer Academic Publishers, 1996.

Nawab, S., Oppenheim, A., Chandrakasan, A., Winograd, J., and Ludwig, J.: Approximate signal processing, J. of VLSI Sig. Proc. Syst., 15, 177–200, 1997.

Schur, I.: Über Potenzreihen, die im Inneren des Einheitskreises beschränkt sind. I., J. für reine und angewandte Mathematik, 147, 205–232, 1917.

Schur, I.: On power series which are bounded in the interior of the unit circle. i., in: Operator Theory: Advances and Applications, (Ed) Gohberg, I., pp. 31–59, Birkhäuser Verlag, 1986.

Verdu, S.: Multiuser Detection, Cambridge University Press, 1998.

Volder, J. E.: The (CORDIC) Trigonmetric Computing Technique, IRE Transactions on Electronic Computers, EC, 330–334, 1959.

Vollmer, M., Haardt, M., and Götze, J.: Schur algorithms for Joint Detection in TD-CDMA based mobile radio systems, Annals of Telecommunications (special issue on multi user detection), 54, 365–378, 1999.

Vollmer, M., Haardt, M., and Götze, J.: Comparative Study of Joint-Detection Techniques for TD-CDMA Based Mobile Radio Systems, IEEE J. Select. Areas Commun., 19, 1461–1475, 2001.

Walther, J.: A unified algorithm for elementary functions, Spring Joint Computer Conf., pp. 379–385, 1971.

12

Numerical modelling of nonlinear electromechanical coupling of an atomic force microscope with finite element method

J. Freitag and W. Mathis

Institut für Theoretische Elektrotechnik, Appelstraße 9A, 30167 Hannover, Germany

Abstract. In this contribution, an atomic force microscope is modelled and in this context, a non-linear coupled 3-D-boundary value problem is solved numerically using the finite element method. The coupling of this system is done by using the Maxwell stress tensor. In general, an iterative weak coupling is used, where the two physical problems are solved separately. However, this method does not necessarily guarantee convergence of the nonlinear computation. Hence, this contribution shows the possibility of solving the multiphysical problem by a strong coupling, which is also referred to as monolithic approach. The electrostatic field and the mechanical displacements are calculated simultaneously by solving only one system of equation. Since the Maxwell stress tensor depends nonlinearly on the potential, the solution is solved iteratively by the Newton method.

1 Motivation

Due to the tremendous development of science in the field of nanotechnology over the last years, microsystems can increasingly be used on many levels. Compared to macrosystems, the advantage lies in cost savings (less consumption of materials, parallel production) and efficiency (low energy and power requirements allow autonomous systems). Applications for these microsystems are primarily medical and security systems, life sciences and logistics. This process makes the use of high resolution instrumentation and its optimisation indispensable to produce these systems. In this context, scanning probe microscopy (SPM) has been developed to be the most appropriate procedure. Here a sample is scanned pointwise using a probe. The resulting measurements are assembled and used to finally produce an image of the object to be measured in terms of the considered interactions. Since atomically small structures are measured, the process is highly error liable due to various interferences, which makes the interpretation of the results difficult. The

aim of this work is to simulate the measurement process to reflect the impact of such measurement errors and eliminate them from the measurements data.

2 Scanning Probe Microscopy

The existing types of SPMs differ in the kind of interaction (electric, magnetic, etc.) between the tip and the sample. This contribution is considered with the atomic force microscope (AFM). The AFM was developed in 1986 and measures atomic forces on a nanometer scale. With a resolution from 0.1 to 10 nm and a scanning velocity from 0.5 to 10 lines per second, it could last 20 min until a digital picture emerges. The AFM consists of a cantilever spring with a sharp tip on the end, which is ideally conical and has a diameter of one atom (Fig. 1). During the process of measuring, the tip navigates above the sample, leading to the potential difference. The cantilever deflects due to the adhesive interactions. This change on the position of the cantilever is tracked by a laserbeam, which points at a position sensitive photodetektor. This signal is transformed and results finally in a picture that shows the sample surface.

3 Approach

The first step of the approach, is to model the AFM with the preprocessor GiD (including geometry, meshing, assign materials and boundary conditions).

The physical process is idealised by partial differential equations, whereas the coupling of the mechanical and the electrostatical field is performed by the Maxwell stress tensor. To solve the partial differential equations, the Finite Element Method (FEM) is used (Reddy, 1993). In this context, the elements definitions and the solving of the system of equations are programmed in Matlab. In former studies the electrostatic and mechanical equations were considered to be weak coupled, which means that the mechanical and electrostatical forces are calculated separately (Helmich, 2007; Bala, 2008; Greiff, 2009). In this case convergence of the

Correspondence to: J. Freitag
(freitag@tet.uni-hannover.de)

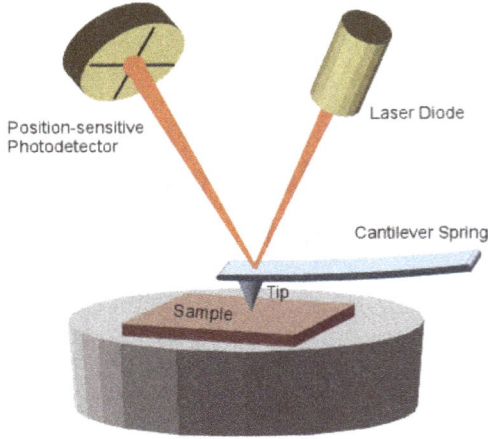

Fig. 1. Principle of an atomic force microscope (Atomic Force Microscope, 2009).

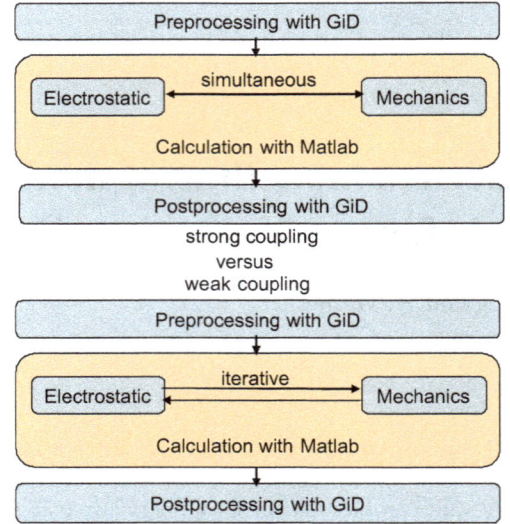

Fig. 2. Process of strong coupling versus weak coupling.

solution is not guaranteed. In this contribution, the partial differential equations are solved by a strong coupled formulation (Fig. 2). If the linearisation of the nonlinear differential equations is consistent, the starting values of the degrees of freedom are near to the solution values and the convergence is guaranteed. At last the postprocessing is done with GiD.

4 Computational methods

The equation describing the electrostatical part (Kuepfmueller, 2006) is given by the reduced form of the Maxwell equation

$$\mathrm{div}(\boldsymbol{D}) = 0, \tag{1}$$

because of the non-existence of free charges. $\boldsymbol{D}$ denotes the electric flux density. The constitutive equations are given by

$$\boldsymbol{D} = \epsilon_0 \, \boldsymbol{E} \tag{2}$$

and

$$\boldsymbol{E} = -\mathrm{grad}(\varphi) \tag{3}$$

with the electrical permittivity ϵ_0, the electrostatic field $\boldsymbol{E}$ and the electrostatic potential φ. The mechanical problem is described by the balance of linear momentum

$$\mathrm{div}(\boldsymbol{\sigma}) + \rho \boldsymbol{b} = \rho \dot{\boldsymbol{v}}, \tag{4}$$

however no inertia forces $\rho \boldsymbol{b}$ or volume forces $\rho \dot{\boldsymbol{v}}$ are considered (Wriggers, 2001). Therefore the strong formulation of the balance of linear monumentum is written as

$$\mathrm{div}(\boldsymbol{\sigma}) = \boldsymbol{0}. \tag{5}$$

$\boldsymbol{\sigma}$ denotes a stress tensor that is given by

$$\boldsymbol{\sigma} = \boldsymbol{\sigma}^M + \boldsymbol{\sigma}^C, \tag{6}$$

with the Cauchy stress tensor $\boldsymbol{\sigma}^C$ and the Maxwell stress tensor $\boldsymbol{\sigma}^M$. The Cauchy stress tensor is computed by a linear relation to strains, which is referred to as Hooke's law

$$\boldsymbol{\sigma}^C = \mathbb{C} \, \boldsymbol{\epsilon}. \tag{7}$$

$\mathbb{C}$ denotes the elastic constitutive tensor. The strain tensor

$$\boldsymbol{\epsilon} = \mathrm{grad}^s(\boldsymbol{u}) = \frac{1}{2}(\mathrm{grad}(\boldsymbol{u}) + \mathrm{grad}^t(\boldsymbol{u})), \tag{8}$$

is assumed by the symmetrical part of the displacement gradient. The Maxwell stress tensor $\boldsymbol{\sigma}^M$ is given by

$$\sigma_{ij}^M = \epsilon_0 \left(E_i E_j - \frac{1}{2} E_l E_l \delta_{ij} \right). \tag{9}$$

δ_{ij} denotes the Kroneker delta with

$$\begin{aligned} \delta_{ij} &= 1, \text{ for } i = j \\ \delta_{ij} &= 0, \text{ for } i \neq j. \end{aligned} \tag{10}$$

4.1 Variational formulations

In the following, the weak forms of the mechanical and the electrostatical part of the system are presented. For the mechanical part the strong form (5) is multiplied with the test function $\boldsymbol{\eta}_u$, and then integrated over the calculation domain Ω

$$\int_\Omega \mathrm{div}(\boldsymbol{\sigma}^C) \cdot \boldsymbol{\eta}_u \, d\Omega = 0. \tag{11}$$

Using partial integration and the Gauss' theorem we get

$$\int_\Gamma (\boldsymbol{\sigma}^C \, \boldsymbol{n}) \cdot \boldsymbol{\eta}_u \, d\Gamma - \int_\Omega \boldsymbol{\sigma} \, \mathrm{grad}^s(\boldsymbol{\eta}_u) \, d\Omega = 0, \tag{12}$$

whereas on the boundary Γ only the Maxwell stress tensor and in the region Ω only the Cauchy stress tensor acts,

$$\int_\Gamma (\boldsymbol{\sigma}^M \, \boldsymbol{n}) \cdot \boldsymbol{\eta_u} \, d\Gamma - \int_\Omega \boldsymbol{\sigma} \, \text{grad}^s(\boldsymbol{\eta_u}) \, d\Omega = 0. \tag{13}$$

Substituting Eqs. (7) and (8), the result for the weak form of the mechanical problem is given by

$$R_{\text{u}} = \int_\Gamma (\boldsymbol{\sigma}^M \, \boldsymbol{n}) \cdot \boldsymbol{\eta_u} \, d\Gamma - \int_\Omega \text{grad}^s(\boldsymbol{u}) \, \mathbb{C} \, \text{grad}^s(\boldsymbol{\eta_u}) \, d\Omega = 0. \tag{14}$$

The strong form of the electrostatical part combines Eqs. (1), (2) and (3) to

$$\text{div}(-\epsilon_0 \, \text{grad}(\varphi)) = 0. \tag{15}$$

Equivalent to the derivation of the mechanical weak form, the electrostatical weak form is described by

$$R_\varphi = \int_\Omega \text{grad}(\varphi) \, \epsilon_0 \, \text{grad}(\eta_\varphi) \, d\Omega = 0. \tag{16}$$

4.2 Linearisation

The degrees of freedom of the coupled 3-D problem are firstly the three components of the displacement vector u_x, u_y and u_z and secondly the potential φ. Since the weak form of the balance of linear momentum depends nonlinearly on the E-field, a direct solution of the multiphysical problem is not possible. This is justified by the use of the Maxwell stress tensor. Thus, the system is solved iteratively by the Newton algorithm. This leads to the coupled system of equations

$$\begin{bmatrix} \frac{\partial R_{\text{u}}}{\partial u} & \frac{d R_{\text{u}}}{d\varphi} \\ \frac{\partial R_\varphi}{\partial u} & \frac{d R_\varphi}{d\varphi} \end{bmatrix} \begin{bmatrix} \Delta \boldsymbol{u} \\ \Delta \varphi \end{bmatrix} = \begin{bmatrix} -R_{\text{u}} \\ -R_\varphi \end{bmatrix}. \tag{17}$$

The derivatives of R_φ and R_{u} can be written as follows:

$$\frac{\partial R_{\text{u}}}{\partial u} \cdot \Delta \boldsymbol{u} \mathrel{\hat{=}} -\int_\Omega \text{grad}^s(\Delta \boldsymbol{u}) \, \mathbb{C} \, \text{grad}^s(\boldsymbol{\eta_u}) \, d\Omega, \tag{18}$$

$$\frac{d R_\varphi}{d\varphi} \Delta \varphi \mathrel{\hat{=}} \int_\Omega \text{grad}(\Delta \varphi) \, \epsilon_0 \, \text{grad}(\eta_\varphi) \, d\Omega \tag{19}$$

and

$$\frac{d R_{\text{u}}}{d\varphi} \Delta \varphi \mathrel{\hat{=}} \int_\Gamma \left(\left\{ \frac{d\boldsymbol{\sigma}^M}{d\varphi} \Delta \varphi \right\} \boldsymbol{n} \right) \cdot \boldsymbol{\eta_u} \, d\Gamma. \tag{20}$$

As a first approach in this contribution we define that the electrostatic field is independent of the idealised displacement variable. This leads to

$$\frac{\partial R_\varphi}{\partial u} = \boldsymbol{0}. \tag{21}$$

The results of the derivatives, that are included in the stiffness matrix in Eq. (17) are described here in a simplified form. While the term on the left side is present in discretized form, the term on the right side is determined in an analytical form. Equation (20) is solved by the derivative of the Maxwell stress tensor with respect to the electrostatic potential φ. Using the chain rule we get

$$\frac{d\sigma_{ij}^M}{d\varphi} \Delta \varphi \mathrel{\hat{=}} \underbrace{\frac{\partial \sigma_{ij}^M}{\partial E_k}}_{A_{ijk}} \text{grad}(\Delta \varphi)_k, \tag{22}$$

where A_{ijk} is a third order tensor, that can be solved by partial derivation of σ_{ij}^M with respect to E_k. The results of A_{ijk} are

$$A_{ij1} = \begin{bmatrix} E_x & E_y & E_z \\ E_y & -E_x & 0 \\ E_z & 0 & -E_x \end{bmatrix}, \tag{23}$$

$$A_{ij2} = \begin{bmatrix} -E_y & E_x & 0 \\ E_x & E_y & E_z \\ 0 & E_z & -E_y \end{bmatrix} \tag{24}$$

and

$$A_{ij3} = \begin{bmatrix} -E_z & 0 & E_x \\ 0 & -E_z & E_y \\ E_x & E_y & E_z \end{bmatrix}. \tag{25}$$

With implementation of the linear shape functions, the concept of isoparametric elements and Gaussian quadrature, the system now can be solve.

5 AFM-modell

For this computation a cantilever with a length of $115\,\mu\text{m}$ and height and width of $1\,\mu\text{m}$ was modelled. The length of the conical tip is $0.5\,\mu\text{m}$. The young's modulus E is 300 GPa (silicon nitride Si_3N_4) and the Poisson ratio ν is set to 0.24 (Rombach, 2009). For the finite element mesh, tetrahedral elements where used with overall 3245 nodes and 147 575 elements. On the left hand side, the cantilever is fixed in all directions. In this area, a potential of 10 V is given and on the bottom (to model a sample) a potential of 0 V. In the first step of this project, a sample with a flat surface is selected for the verification of the formulations. The coupling is computed on the boundaries of the cantilever.

6 Computational results

Figure 4 shows the potential distribution in a slice through the three-dimensional calculation model. Here, the increased

Fig. 3. Displacement of an AFM in [μm].

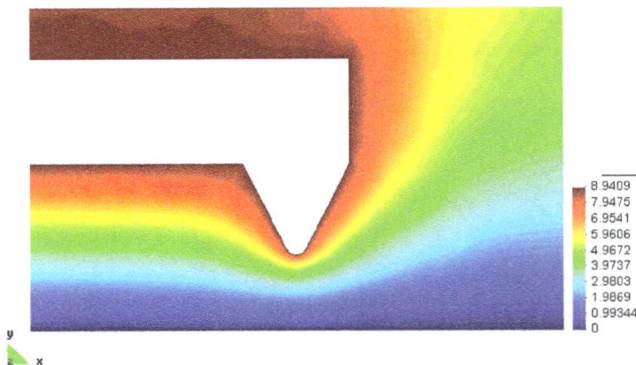

Fig. 4. Potential distribution of an AFM in [V].

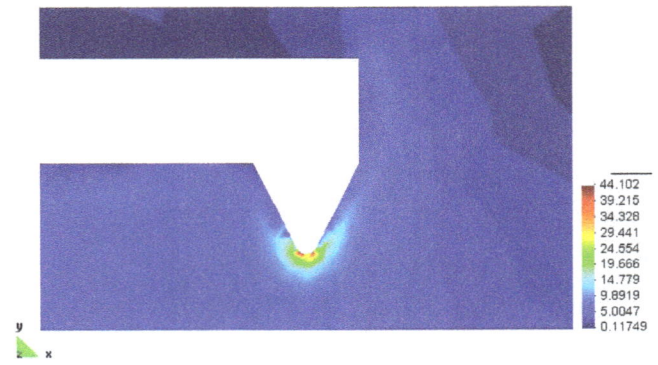

Fig. 5. Electrostatic field of an AFM in [V/μm].

gradient is clearly visible at the tip of the cantilever. This is shown significantly by the representation of the electrostatic field in Fig. 5. The combination of the mechanical loads and the electrostatic field by using the Maxwell stress tensor leads to relatively large forces, that act on the tip of the cantilever. The resulting deflection of the cantilever is shown in Fig. 3. On the other boundaries that link the Cantilever with the surrounding medium, the electrostatic field tends against the value 0, and thus the mechanical loads become negligible. To minimize the computing time, it can be considered to restrict the calculation field to an area around the tip, since mechanical loads are negligible outside this area.

7 Conclusions

In this contribution the electromechanical coupling of a working AFM was solved by using FEM. As a first approach, the electrostatic field was adopted as independent of the displacement. Therefore, it is necessary to extend this procedure. Furthermore, the presented results of strong coupling should be compared with those of weak coupling.

Acknowledgements. This work is funded by German Research Foundation (DFG) via a Ph.D. scholarship in the Research Training Group 615.

References

Atomic Force Microscope: http://www3.physik.uni-greifswald.de/method/afm/eafm.htm, 2009.

Bala, U. B.: Hybrid Numerical Modelling and Simulation of Electrostatic Force Microscope, Ph.D. thesis, Leibniz Universitaet Hannover, 2008.

Greiff, M.: Ein Konzept zur Modellierung und Simulation mikroelektronischer Systeme mit Anwendung auf ein elektrostatisches Kraftmikroskop, Ph.D. thesis, Leibniz Universitaet Hannover, 2009.

Helmich, T.: Elektromechanisch gekoppelte Kontaktmodellierung auf Mikroebene, Ph.D. thesis, Leibniz Universitaet Hannover, 2007.

Kuepfmueller, K. and Mathis, W. R. A.: Theoretische Elektrotechnik – Eine Einfuehrung, Springer, 17th edn., 2006.

Reddy, J. N.: An Introduction to the Finite Element Method, 2nd edn., 1993.

Rombach, M. and Hollstein, T.: Untersuchungen zum mechanischen Verhalten von Siliciumnitrid in einem Kugel-Platte-Kontakt, Materialwissenschaft und Werkstofftechnik, 26, 276–282, 2009.

Wriggers, P.: Nichtlineare Finite-Element-Methoden, Springer, 2001.

13

TLM modeling and system identification of optimized antenna structures

N. Fichtner[1]**, U. Siart**[1]**, Y. Kuznetsov**[2]**, A. Baev**[2]**, and P. Russer**[1]

[1]Institute for High Frequency Engineering, Technische Universität München, Arcisstr. 21, 80333 München, Germany
[2]Moscow Aviation Institute (State Technical University) Volokolamskoe sh. 4, A-80, GSP-3, Moscow, 125993, Russia

Abstract. The transmission line matrix (TLM) method in conjunction with the genetic algorithm (GA) is presented for the bandwidth optimization of a low profile patch antenna. The optimization routine is supplemented by a system identification (SI) procedure. By the SI the model parameters of the structure are estimated which is used for a reduction of the total TLM simulation time. The SI utilizes a new stability criterion of the physical poles for the parameter extraction.

1 Introduction

The modeling of the electromagnetic behavior and the design optimization is an important issue in the development of new systems and services. Fullwave techniques operating in time-domain (TD) as e.g. TLM and FDTD are advantageous for structures where the system response at several frequencies is needed. An efficient global optimization method is available by the genetic algorithm (GA) (Johnson and Rahmat-Samii, 1997, 1999). The GA is typically used with a methods of moments whereas here it is adopted to the transmission line matrix (TLM) method (Fichtner et al., 2007). Any GA optimization requires a repeated computation of the electromagnetic structure under consideration. In order to lower the computational burden a system identification (SI) methods is applied. On the one side the SI allows a reduction of the total TLM simulation time by estimation of the future signal response. On the other side the SI results in the extraction of the model order and the model's parameters allow for the determination of a lumped element network. The network oriented modeling of passive microwave circuits and antennas provides a compact representation of the electromagnetic

Correspondence to: N. Fichtner
(fichtner@tum.de)

structures and requires low computational effort and memory capacity.

2 TLM modeling

In this paper we investigate methods to reduce the simulation time of the TLM simulation through the model parameter estimation from significant parts of the impulse response using SI methods. As soon as the model parameters can be estimated with sufficient stability, a network oriented equivalent model may be established and the TLM simulation can be truncated at this point. In high-Q resonant structures with long impulse responses this method can save typically up to 90% of the total TLM simulation time (Kuznetsov et al., 2005). But even in low-Q structures a reduction of 50% is achievable. Therefore, a SI for the modeling of broadband structures is also advantageous.

The simulation results of the TLM electromagnetic modeling package YATSIM (Yat, 2007) have been used as a basis for the SI procedure. YATSIM provides an open source software package for time-domain electromagnetic full-wave simulation based on the TLM method. Our benchmark example is a low–profile microstrip fed patch antenna that has been optimized for maximum impedance bandwidth using a genetic algorithm (GA). The description of the GA optimization approach is outside the scope of this paper, see e.g. Johnson and Rahmat-Samii (1997, 1999) and Fichtner et al. (2007). In the following we compare the SI results for the initial (non-optimized) antenna to the results for the GA optimized antenna. The CAD antenna model of the antenna is shown in Fig. 1. The antenna is designed to have its first resonance mode TM_{100} at 6 GHz and the second resonance mode TM_{010} at 12 GHz (for patch antenna in the xy-plane). The green bridge-like objects in Fig. 1 do not correspond to physical objects but indicate the locations of probes for E- and H-fields, where the time domain impulse response has

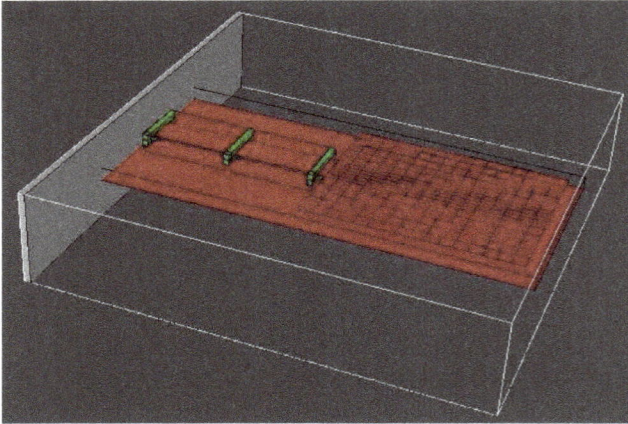

Fig. 1. CAD rendered view of the simulated and optimized patch antenna.

been computed with 10 000 time steps with the increment $\Delta t = 8.3 \times 10^{-13}$ s.

3 System identification

The SI method applied in this contribution is based on a system model deduced from the singularity expansion method (SEM) (Baum et al., 1991; Heyman and Felsen, 1985). Therein a scattering process in TD is subdivided into the entire part and the pole part. The entire part is responsible for the high frequency processes inside the EM structure concerning reflections, delays and other transformations of the initial wave front. The pole part describes the low frequency processes given by the quasi static exchange of the electric and magnetic energy stored at different locations in the microwave circuit. For a given input signal $a(t)$ we compute the response signal $b(t)$ from the convolution

$$b(t) = a(t) \times \left[h_{\text{POLE}}(t) + h_{\text{ENT}}(t) \right] \tag{1}$$

where

$$h_{\text{POLE}}(t) = \sum_{i=1}^{P} C_i e^{-p_i t} \tag{2}$$

$$h_{\text{ENT}}(t) = \sum_{k=1}^{K} D_K a(t - T_k) \tag{3}$$

are the impulse responses for the pole and entire parts, respectively. The estimation of D_k, T_k and K in Eq. (3) is not a part of this contribution because the early time part has short duration and will not be approximated by a system model. The residues C_i and the poles p_i in Eq. (2) characterize the late time response of the scattered signal and are subject to the SI.

In conventional parameter estimation methods such as the Matrix Pencil method (MPM) published by Hua and Sarkar

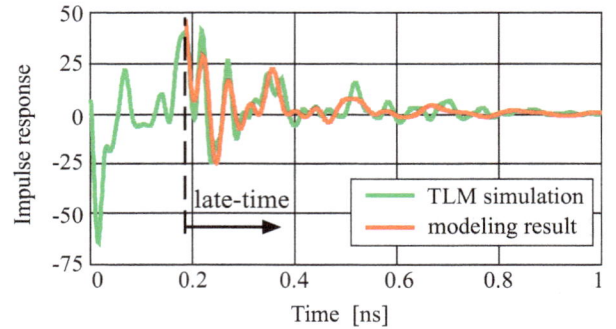

(a) Non–optimized antenna

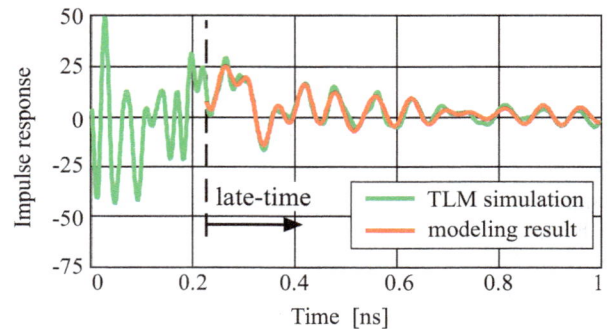

(b) Optimized antenna

Fig. 2. Patch antenna impulse response: TLM simulated and reconstructed by system identification.

(1990) the extracted poles can be categorized into physical and unphysical poles. The first are structure dependent whereas the latter only increase the quality of the approximation. From the SEM we know that the late time response of a back-scattered signal is given by structure dependent resonances. Thus a parameter estimation with a constant observation time window at different times in the late time part should result in constant pole positions for the physical poles. In order to classify the poles we applied the MPM for a constant observation window at different times in the response signal $b(t)$. Two effects were uncovered: Firstly one part of the obtained poles have positions in the complex plane which are independent from the actual observation time. Secondly there exists a minimal time border below which none of the poles has a constant position. The poles independent of the observation times can be considered as physical poles and the remaining poles as unphysical (numerical) poles.

The number of physical poles P_0 can also be obtained from the proposed SI method. We consider the case with $P > P_0$ and compute the pole locations at two different observation times. By this way we obtain two sets of poles $\mathbf{P}_1 = \{p_1^{(1)}, \ldots, p_P^{(1)}\}$ and $\mathbf{P}_2 = \{p_1^{(2)}, \ldots, p_P^{(2)}\}$. Next we

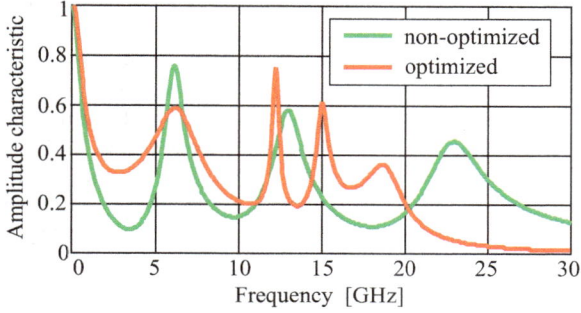

Fig. 3. Modeled patch antenna amplitude spectral response.

compute the Euclidian metric of the two pole sets from

$$\Delta P = ||\mathbf{P}_1 - \mathbf{P}_2|| = \sqrt{\sum_{n=1}^{P} \left| p_n^{(1)} - p_n^{(2)} \right|^2}. \qquad (4)$$

Note that in the metric (4) for a pole in set 1 $p_n^{(1)}$ the closest pole of set 2 must be chosen in order to minimize ΔP. However, the physical poles have almost constant position, i.e. $p_n^{(1)} \approx p_n^{(2)}$, whereas the unphysical poles deviate considerably and therefore increase ΔP. The metric ΔP mainly depends on the number of unphysical poles presented in the expansion (2). Choosing a smaller pole number P decreases the metric because some of the unphysical poles will vanish. This works down to $P=P_0$. For smaller vales $P<P_0$ the physical poles will start to change their position for different observation times which increases the metric ΔP. Thus the number of physical poles P_0 is found by varying P and looking for the minimum value of ΔP.

4 Modeling results

The impulse responses of the non-optimized (initial) antenna and the bandwidth optimized patch antenna described in Sect. 2 have been computed by the proposed SI method. The results are shown in Fig. 2. Figure 3 shows the corresponding amplitude characteristics. It can be seen that the optimized antenna has additional resonant frequencies. These provide a larger impedance bandwidth compared to the initial antenna with two resonances. It has to be mentioned that the additional resonances may considerably influence the group delay properties. Therefore the impedance bandwidth criterion used here applies well to wideband antennas for several narrowband services. However, antennas optimized in that way may not be suitable for ultrawide band pulse transmission.

The estimated model parameters can be used to establish network oriented equivalent models of the microwave structures by means of network synthesis. As an example the antenna input impedance derived from the all-pole model is shown in Fig. 4.

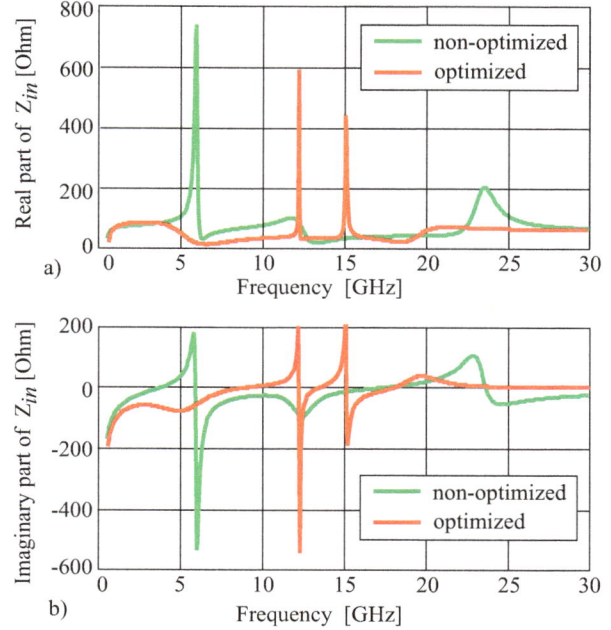

Fig. 4. Modeled patch antenna input impedance.

5 Conclusions

The time domain simulation of EM structures is very efficient for broadband structures but may require long simulation time for low loss resonant components. System identification can help to establish network oriented models of the structures from parts of the impulse response thus avoiding the need to simulate the total duration of the scattered signal. A combination of time domain electromagnetic modeling and system identification has therefore the potential to reduce computation time considerably. In this paper system identification has been applied to a bandwidth optimized patch antenna and to a band filter structure with good agreement between full time TLM response and SI based estimation.

Acknowledgements. The authors are indebted to the Deutscher Akademischer Austauschdienst (DAAD) for supporting the scientific exchange between the Moscow Aviation Institute and the Technische Universität München.

References

Baum, C., Rothwell, E., Chen, K.-M., and Nyquist, D.: The Singularity Expansion Method and Its Application to Target Identification, Proc. IEEE, 79, 1481–1492, 1991.

Fichtner, N., Siart, U., and Russer, P.: Antenna Bandwidth Optimization Using Transmission Line Matrix Modelling and Genetic Algorithms, in: ISSSE Proc. Int. Symp. on Signals, Systems and Electronics, Montreal, Quebec, 2007.

Heyman, E. and Felsen, L.: A Wavefront Interpretation of the Singularity Expansion Method, IEEE Trans. Antennas Propagat., 33, 706–718, 1985.

Hua, Y. and Sarkar, T.: Matrix pencil method for estimating parameters of exponetially damped/undamped sinusoids in noise, IEEE Trans. Acoustics, Speech and Signal Processing, 38, 814–824, 1990.

Johnson, J. and Rahmat-Samii, Y.: Genetic Algorithms in engineering electromagnetics, IEEE Trans. Antennas Propagat. Magazine, 39, 7–25, 1997.

Johnson, J. and Rahmat-Samii, Y.: Genetic Algorithms and Method of Moments (GA/MOM) for the Design of Integrated Antennas, IEEE Trans. Antennas Propagat., 47, 1606–1614, 1999.

Kuznetsov, Y., Baev, A., Shevgunov, T., Zedler, M., and Russer, P.: Transfer Function Representation of Passive Electromagnetic Structures, in: Proc. MTT-S Int. Microwave Symp., 2005.

Kuznetsov, Y., Baev, A., Shevgunov, T., Lorenz, P., and Russer, P.: Application of the Stability Criterion to the Passive Electromagnetic Structures Modeling, in: Proc. 36th European Microwave Conf., 2006.

YATPAC Homepage: http://www.yatpac.org, 2007.

14

Position finding using simple Doppler sensors

S. Schelkshorn and J. Detlefsen

Technische Universität München
Lehrstuhl für Hochfrequenztechnik
Fachgebiet Hochfrequente Felder und Schaltungen
80333 München, Germany

Abstract. An increasing number of modern applications and services is based on the knowledge of the users actual position. Depending on the application a rough position estimate is sufficient, e. g. services in cellular networks that use the information about the users actual cell. Other applications, e. g. navigation systems use the GPS-System for accurate position finding. Beyond these outdoor applications a growing number of indoor applications requires position information. The previously mentioned methods for position finding (mobile cell, GPS) are not usable for these indoor applications.

Within this paper we will present a system that relies on the simultaneous measurement of doppler signals at four different positions to obtain position and velocity of an unknown object. It is therefore suiteable for indoor usage, extendig already existing wireless infrastructure.

1 Introduction

Usually Doppler sensors can only provide relative distance information and therefore normally are not used for position finding purposes. The system presented here relies on the simultaneous Doppler measurement of four sensors at different positions. The four Doppler signals are evaluated to obtain position and velocity of a single moving target by iteratively solving a nonlinear system of equations.

Solving the nonlinear system of equations may be done in different ways. First simulation results concerning the reliability and accuracy of the procedure of position finding using the well known Newtons Method as well as results obtained by using three similar but enhanced algorithms will be discussed. A simple demonstration system to verify the simulation results will be presented.

Correspondence to: S. Schelkshorn
(simon.schelkshorn@tum.de)

2 Basic principle

In the case of a single doppler sensor at a fixed position one can only measure the velocity, i.e. the change in distance, of an moving object relative to the fixed sensor. This relative velocity is simply the projection of the velocity vector on the directional vector from the sensor to the object. Even if the objects position is known the velocity vector can not be determined as an infinite number of velocity vectors has the same projection (Fig. 1a shows two possibilities).

Adding a second doppler sensor at another fixed position to this system it is possible to combine the two relative velocities and the knowledge of the two directional vectors to obtain the correct velocity vector of the moving object (Fig. 1b).

If on the other hand the position of the object is unknown but the velocity vector is known, any number of object positions result in the same measured relative velocity at a single sensor. Adding again a second sensor now the unknown position can be found (see Fig. 2).

Combining the two preceeding examples one ends up with the case where neither the objects position nor its velocity is known. Both properties can be calculated from the relative distance information acquired by four doppler sensors.

3 Solving the system of nonlinear equations

Each of the four sensors measures the change in distance $\delta d^{i,j}$ from its location x_B^i with $i = 1 \dots 4$ to the object x_P^j.

$$\delta d^{i,j} = \left| x_P^j(kT) + v_P^j(kT) - x_B^i \right| - \left| x_P^j(kT) - x_B^i \right| \quad (1)$$

The measured values of all four sensors can be used to setup a system of equations for the unknowns. This system is a nonlinear system of equations.

A well known method for solving a system of nonlinear equations is the so-called Newtons Method (Werner, 1992). This method is based on the iterative solution of a linearized

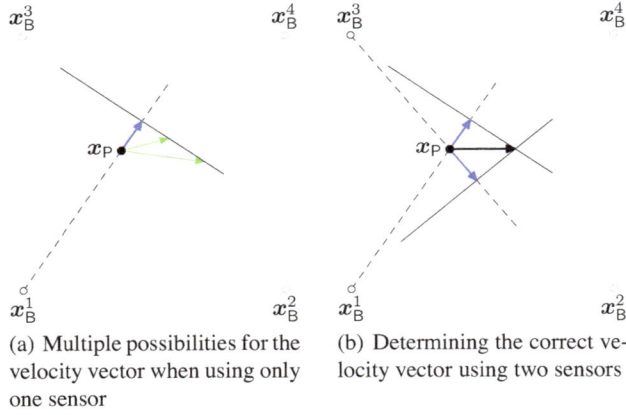

(a) Multiple possibilities for the velocity vector when using only one sensor

(b) Determining the correct velocity vector using two sensors

Fig. 1. Velocity measurement using one or two doppler sensors, position x_P is known.

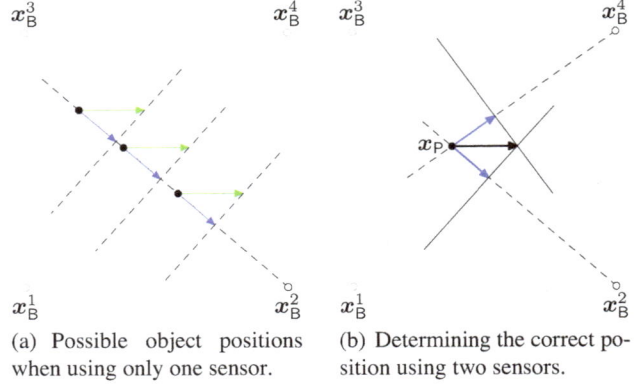

(a) Possible object positions when using only one sensor.

(b) Determining the correct position using two sensors.

Fig. 2. Position finding using one or two doppler sensors, velocity v_P is known.

version of the system of equations. Linearization is done at assumed values b_0^ℓ for the unknowns. The linearized change in distance can be written as follows

$$\delta d^{i,j}(\Delta) = \delta d_0^{i,j,\ell} + a^i \cdot \delta x_0^\ell + b^i \cdot \delta y_0^\ell + c^i \cdot \delta v_{x0}^\ell + d^i \cdot \delta v_{y0}^\ell \quad (2)$$

where $a^i \dots d^i$ are the according taylor coefficients. Combining all four sensors using this linearized expression above yields

$$\begin{bmatrix} a^1 \ b^1 \ c^1 \ d^1 \\ a^2 \ b^2 \ c^2 \ d^2 \\ a^3 \ b^3 \ c^3 \ d^3 \\ a^4 \ b^4 \ c^4 \ d^4 \end{bmatrix} \cdot \begin{bmatrix} \delta x_0^\ell \\ \delta y_0^\ell \\ \delta v_{x0}^\ell \\ \delta v_{y0}^\ell \end{bmatrix} = \begin{bmatrix} \delta d^{1,j}(\Delta) \\ \delta d^{2,j}(\Delta) \\ \delta d^{3,j}(\Delta) \\ \delta d^{4,j}(\Delta) \end{bmatrix} - \begin{bmatrix} \delta d_0^{1,j,\ell} \\ \delta d_0^{2,j,\ell} \\ \delta d_0^{3,j,\ell} \\ \delta d_0^{4,j,\ell} \end{bmatrix} \quad (3)$$

Equation (3) can also be written in a short form.

$$S_0^\ell \cdot \delta b_0^\ell = \left(\delta d^{i,j}(\Delta) - \delta d_0^{i,j,\ell} \right) \quad (4)$$

δb_0^ℓ represents the deviation of the unknowns from the assumed values b_0^ℓ. It is calculated by inverting S_0^ℓ.

$$\delta b_{opt}^\ell = \left[S_0^{\ell T} S_0^\ell \right]^{-1} S_0^{\ell T} \cdot \left(\delta d^{i,j}(\Delta) - \delta d_0^{i,j,\ell} \right) \quad (5)$$

For the next iteration b_0^ℓ is improved by δb_0^ℓ.

$$b_0^{\ell+1} = b_0^\ell + \delta b_0^\ell \quad (6)$$

This procedure is continued until convergence is reached.

When solving a system of nonlinear equations by using its linear couterpart one has to choose an initial value for the unkowns. If this guess lies close enough to the solution the iterative solution process will converge to the correct values (Hettwer and Benning, 2001). If the initial guess is an unsuitable one the iterative solution may diverge or in the worst case converge to a wrong value. In the case of divergence one can simply choose a different set of initial values and try again but in the case of convergence to a wrong result it is

impossible to detect. So choosing the right initial value for the linearization turns out to be the key issue in our problem.

In our case we have no information on the properties of the unknown object, so choosing an appropriate initial value for the linearization is difficult. Therefore it would be very helpful if the number of possible initial values that achieve convergence can be increased.

In the following four different methods for solving our system of equations are to be compared. The first as well as the simplest is the standard Newtons Method. The three additional methods are all based on it.

The second method is called Newton Method with additional attenuation (Hettwer and Benning, 2001). It differs from the standard version only in the way how the improved values are computed from the initial ones. Here δb_0^ℓ is multiplied by an attenuation factor before it is added to b_0^ℓ.

$$b_0^{\ell+1} = b_0^\ell + \lambda^\ell \delta b_0^\ell \quad (7)$$

The attenuation factor $\lambda^\ell \leq 1$ is chosen individually for each iteration. The selection is based on a residual that is a measure for the difference between the linearized and the nonlinear system of equations.

Both methods mentioned so far utilize the measured change in distance during one timestep T. To take advantage of the measurement at two consecutive timesteps, a second timestep can be incorporated in Eq. (3) as four additional equations thus leading to an overdetermined system of equations. This extension can be done for both methods, the standard Newtons Method and the one with additional attenuation (Schelkshorn, 2006).

To be able to compare the performance of the four mentioned methods a simulation was carried out where the object was placed at several different positions and the initial value for the unknowns used for the linearization was kept constant. Figure 3 shows a comparison of the results as a histogram. The four methods are designated as follows:

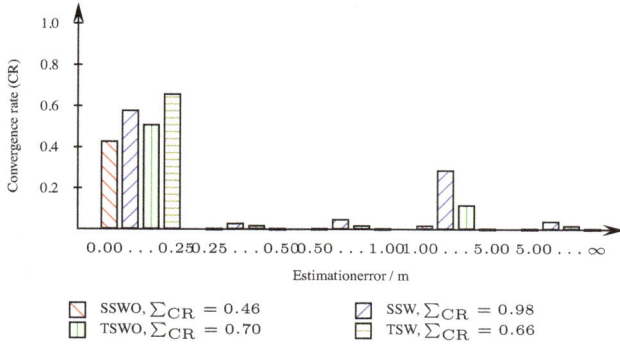

Fig. 3. Convergence rate and error with different algorithms.

- Single step without attenuation (SSWO): standard Newtons Method

- Single step with attenuation (SSW): Newtons Method with additional attenuation

- Two step without attenuation (TWSO): Newtons Method without attenuation, combined for two timesteps

- Two step with attenuation (TSW): Newtons Method with attenuation, combined for two timesteps

As it is possible to detect whether the solution converges or not, but wrong results can not be detected, one should use the ratio of the number of correct results versus the number of wrong results as a figure of merit.

One can see that introducing the attenuation yields a higher convergence rate but also a higher number of wrong results. Adding the information from the second timestep also increases the convergence rate compared to the standard Newtons Method with a lower number of wrong results compared to the attenuated version. Finally combining both, a second timestep and the attenuation turns out to be the best alternative, as there are almost no wrong results.

4 Multi target environments

Independent of the method of solving the resulting system of equations the whole principle is based on doppler measurement. This doppler measurement can easily be implemented in existing RF-infrastructure. As the system of equations relys on the combination of four doppler sensors one has to assure that each sensor observes only one doppler signal. That means in this simple setup only single target scenarios can be handled.

To deal with multi target scenarios it is necessary to separate the occuring doppler signals before further processing. After separation the described system of equations can be solved for each set of doppler signals.

Table 1. 4-Ch. Radar, System parameters.

Parameter	Value
Number of channels	4
Center frequency	2.45 GHz
Radar modes	CW, FSK, SFCW
Max. bandwidth	600 MHz
Sweeptime (SFCW)	5 ms
Max. sampling frequency	25 kHz
Output power	15 dBm

Fig. 4. 4-Ch. CW/FSK/SFCW-Radar @ 2.45 GHz.

5 Demonstration system

To verify the simulation results obtained so far, a demonstration system at 2.45 GHz is setup at the moment. This demonstration system consists of four identical channels with a central control unit based on an FPGA-Evaluation board. Each channel mainly consists of a PLL for signal generation and an I/Q-mixer in the receiver section. The I/Q-channels then are A/D-converted and the resulting data is transferred via LAN to a PC for processing. By continuously reprogramming the four PLLs it is also possible to generate modulated signals. So far FSK- and SFCW-modulation is considered additionally to the simple CW operation.

All relevant system parameters are summarized in Table 1. A photo of the actual design is shown in Fig. 4.

6 Conclusions

The presented setup is an easy way of position finding especially for indoor applications where other methods (e.g. GPS) won't work. Due to the simple approach of doppler measurement it can easily be implemented in already existing RF-infrastructure. In addition to the position information of the unkown object also its velocity is obtained. A notably advantage of this approach is that no active participation of the unknown object is required.

As this approach relys on the measurement of doppler signals it is only applicable in enviroments with moving targets.

References

Hettwer, J. and Benning, W.: Erweiterung des Konvergenzbereichs bei nichtlinearen Ausgleichungsaufgaben, Allgemeine Vermessungs-Nachrichten, 254–259, 2001.

Schelkshorn, S.: Konvergenzuntersuchung: Lösungsverfahren für Vier-Sensor-System, Tech. rep., TU München, Lehrstuhl für Hochfrequenztechnik (HFS), 2006.

Werner, J.: Numerische Mathematik, Bd. 1, Lineare und nichtlineare Gleichungssysteme, Interpolation, numerische Integration, Vieweg, 1992.

15

Bistatic extension for coherent MMW-ISAR-Imaging of objects and humans

S. Bertl, A. Dallinger, and J. Detlefsen

Technische Universität München, Fachgebiet Hochfrequente Felder und Schaltungen, Arcisstr. 21, 80333 München, Germany

Abstract. We present a bistatic extension of a broadband monostatic FMCW Radar working in the Millimetre-Wave (MMW) region and its bistatic imaging properties used for imaging purposes. Due to the different perspective of a bistatic setup compared to a monostatic one, additional information can be obtained.

A wide bandwidth of approx. 10 GHz is used for the task of high resolution imaging as it could be used for the detection of threats at a person's body in security-sensitive environments. Since MMWs propagate easily through common clothing, it is feasible to image objects like concealed weapons worn under the clothing. MMW-Imaging of humans is one possibility to enhance the capabilities of nowadays security checkpoints, e.g. at airports.

1 Introduction

For detection of concealed objects on the body of a person, different perspectives have to be applied in order to avoid shadowing effects. Therefore a setup with two sensors, one above and one below the person, is used as shown in Fig. 1. Both sensors are equipped with two receiving channels slightly displaced from each other. This allows the evaluation of the interferometric phase in order to obtain a three-dimensional image. This aspect is covered in Bertl et al. (2007) and is not considered in the following. The original setup is a purely monostatic one. The two sensors are moved on a circular trajectory all around the person. The image is reconstructed by means of the Synthetic Aperture Radar (SAR) principle.

By combining the two sensors coherently, both the monostatic signal and the signal of the other sensor can be pro-

Correspondence to: S. Bertl
(sebastian.bertl@tum.de)

cessed. In the following the properties of this bistatic signals and the resulting images will be discussed. Also a realisation of a bistatic extension will be presented. Since the resolution capabilities of a bistatic setup differ from that of the monostatic setup, additional information can be obtained by such an extension.

2 Bistatic imaging properties

2.1 Reconstruction algorithm

In order to reconstruct calibrated bistatic signals an adapted backprojection algorithm can be used. The location of the reconstructed point r_o and aperture coordinates r_a are given by

$$r_{o,a} = (x, y, z)_{o,a} = (\rho_{o,a} \cdot \cos \varphi_{o,a}, \rho_{o,a} \cdot \sin \varphi_{o,a}, z_{o,a}). \quad (1)$$

The distance R between the aperture position of the transmitter, the object position and the aperture position of the receiver then becomes $R = |r_{a,Tx} - r_o| + |r_{a,Rx} - r_o|$. The backprojected bistatic signal can be written as

$$h(r_o) = \int_{\varphi_{a,1}}^{\varphi_{a,2}} u(R, \varphi_a') \cdot \exp(jk_c R) d\varphi_a'. \quad (2)$$

2.2 Bistatic resolution

For the case of a point scatterer in the horizontal plane at the bisection of the two sensors, limits for the resolution will be derived in the following. This approach can be applied to arbitrary positions.

For the monostatic case the resolution in range becomes

$$\Delta r_{ms} = \frac{2\pi}{\Delta k_{ms}} = \frac{c_0}{2B}. \quad (3)$$

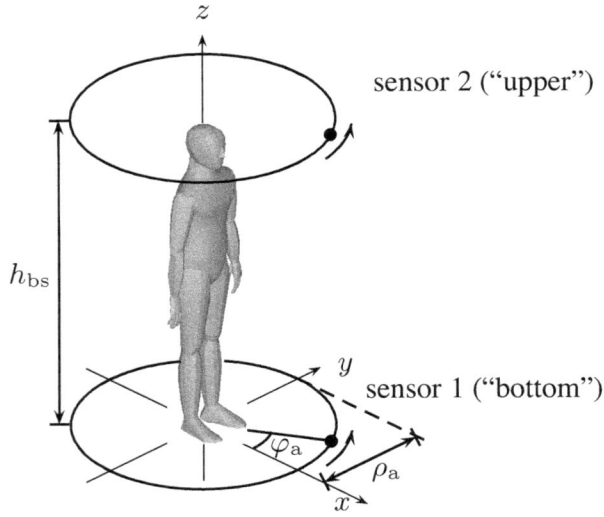

Fig. 1: Two sensors moving on a circular trajectory all around a person. The distance between sensor 1 and 2 is h_{bs}.

(a) mono- and bistatic k-vectors

(b) bistatic k-space

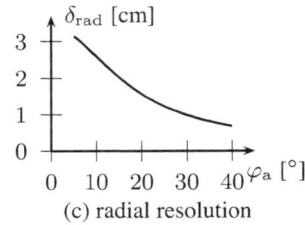

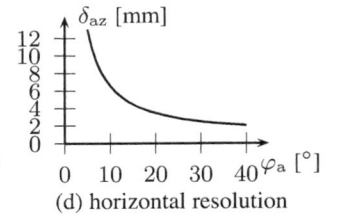

(c) radial resolution

(d) horizontal resolution

Fig. 2: Bistatic resolution; Used parameters in (c) and (d): $f = 90 \ldots 100\,\text{GHz}$, $\rho_{\text{a}} = 0.6\,\text{m}$, $h_{\text{bs}} = 1.62\,\text{m}$, $\rho_0 = 0.2\,\text{m}$, $z_0 = 0.81\,\text{m}$.

where

$$\Delta k_{\text{ms}} = (k_{\text{Tx,max}} + k_{\text{Rx,max}}) - (k_{\text{Tx,min}} + k_{\text{Rx,min}}) \tag{4}$$

$$= \frac{4\pi}{c_0} \cdot (f_{\max} - f_{\min}) = \frac{4\pi}{c_0} \cdot B. \tag{5}$$

and $f \in [f_{\min}, f_{\max}]$. These considerations can be applied to the bistatic case as well. In contrast to the monostatic case, the k-vector has to be decomposed into in a k_x- and a k_z-component, depending on the position of the point under consideration for the Tx- and Rx-vector. For a point between the two sensors at a position $(x, y, z) = (\rho_0, 0, 0.5 \cdot h_{\text{bs}})$ this results in

$$\boldsymbol{k}_{\text{Tx, bs}} = (k_x, k_z) = (\sin \alpha, \cos \alpha) \cdot k_0 \tag{6}$$

$$\boldsymbol{k}_{\text{Rx, bs}} = (k_x, k_z) = (\sin \alpha, -\cos \alpha) \cdot k_0 \tag{7}$$

as shown in Fig. 2a. The resulting bistatic k-vector then becomes

$$\boldsymbol{k}_{\text{bs}} = \boldsymbol{k}_{\text{Tx, bs}} + \boldsymbol{k}_{\text{Rx, bs}} = (2 \sin \alpha, 0) \cdot k_0. \tag{8}$$

By modulating the frequency by $\Delta f = f_{\max} - f_{\min}$ we obtain

$$\Delta k_{\text{bs}} = 2 \sin \alpha \frac{2\pi}{c_0} (f_{\max} - f_{\min}) = \sin \alpha \cdot \Delta k_{\text{ms}}. \tag{9}$$

Taking the SAR-processing step into account, the radial bistatic resolution can be improved considerably. The relevant angle of observation φ_{asp} (see Fig. 2b) can be written in terms of the object location (ρ_0) and the aperture coordinates ($\rho_{\text{a}}, \varphi_{\text{a}}$) as

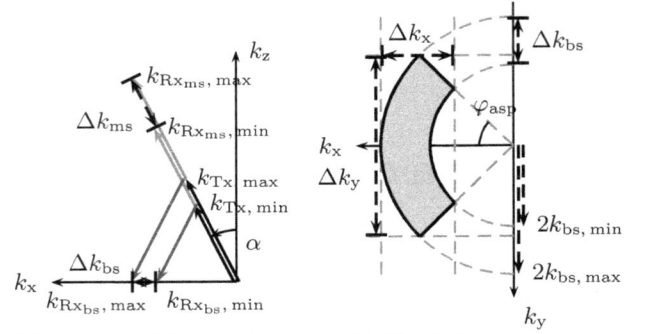

$$\tan \varphi_{\text{asp}} = \frac{\sin \varphi_a \cdot \rho_a}{\cos \varphi_a \cdot \rho_a - \rho_0}, \tag{10}$$

where φ_a is the aperture angle defined in Eq. (1). To determine the radial resolution capability, δ_{x} respectively δ_{rad}, the width of the occupied k-space along this direction has to be taken into account. For the maximum and minimum we obtain

$$2k_{\text{max, bs}} = \frac{4\pi f_{\max}}{c_0} \cdot \sin \alpha \tag{11}$$

$$2k_{\text{min, bs}} = \frac{4\pi f_{\min}}{c_0} \cdot \cos \varphi_{\text{asp}} \cdot \sin \alpha. \tag{12}$$

The radial resolution therefore becomes

$$\delta_x \ (= \delta_{\text{rad}}) = \frac{2\pi}{\Delta k_{\text{bs,x}}} = \frac{2\pi}{2k_{\text{max, bs}} - 2k_{\text{min, bs}}} \tag{13}$$

$$= \frac{c_0}{2(f_{\max} - f_{\min} \cdot \cos \varphi_{\text{asp}}) \cdot \sin \alpha}. \tag{14}$$

For the resolution along azimuth the same considerations lead to a resolution δ_{az}

$$\delta_y \ (= \delta_{\text{az}}) = \frac{c_0}{4 f_{\max} \cdot \sin \alpha \cdot \sin \varphi_{\text{asp}}}. \tag{15}$$

With Eq. (10) the resolutions δ_{rad} and δ_{az} can be written as a function of the aperture angle φ_{a}. A plot of the two expressions for the resolution can be seen in Fig. 2c and d.

3 Implementation of a bistatic measurement setup

The proposed system setup uses one common voltage controlled oscillator (VCO) to combine the two originally monostatic sensors coherently. According to Fig. 3 one of the sensor signals is delayed in time by using a coaxial delay line. In addition to the monostatic signal, both sensors also receive the bistatic signal generated by the other.

The monostatic and the bistatic signals can be separated in the time domain because of a suitable choice of the length of the delay line. By adequate windowing the monostatic part can be extracted, and the further processing of the monostatic signal is done by common SAR algorithms as presented in Dallinger et al. (2006).

The SAR processing of the bistatic signal is done according to Eq. (2).

3.1 Sensor description

The FMCW-radar principle is used for each sensor. A schematic of the two sensors is given in Fig. 3a. A Voltage Controlled Oscillator (VCO) generates a frequency-modulated signal. The generated signal is multiplied to obtain an output-signal in the MMW-domain. A part of the Tx-signal is transmitted by the Tx-antenna, the other part is used as the local oscillator (LO) signal. Since the homodyne detection principle is used, one MMW-source is sufficient. The received signal (Rx) is fed to the mixer via a Low Noise Amplifier (LNA). The Rx-signal is shifted to baseband. Instead of an I/Q demodulator the analytic signal is calculated using the Hilbert transform. The frequency-sweep is done in 1–2 ms. In conjunction with a circularly moving sensor a person can be measured within approx. 20 s over a full rotation of 360°. A more detailed description can be found in Detlefsen et al. (2006).

3.2 Bistatic setup

With the combined setup, given in Fig. 3a, mono- and bistatic measurements can be performed simultaneously.

Instead of two separately controlled signal sources only the source from (in this case) the bottom sensor is used. This signal is sent to the upper sensor via a coaxial delay line and is fed to the multiplier's input.

This means, that the same signal is transmitted by the bottom sensor first and then with a delay of $\Delta\tau_{12}$ by the upper sensor as well.

3.3 Spectra of the difference frequencies

The location of the resulting spectra observed at the two sensors is given in Fig. 3b, c. Black lines are related to Tx|Rx-signals generated by the bottom sensor, gray lines belong to the upper sensor. At the bottom sensor the monostatic part is received first. The delay τ_{ms} of this measurement is proportional to the distance between transmitter Tx_1, the scat-

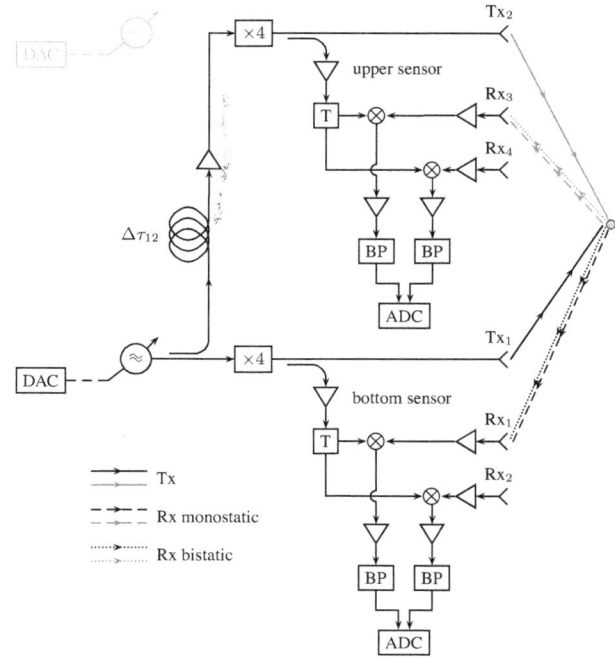

(a) System setup

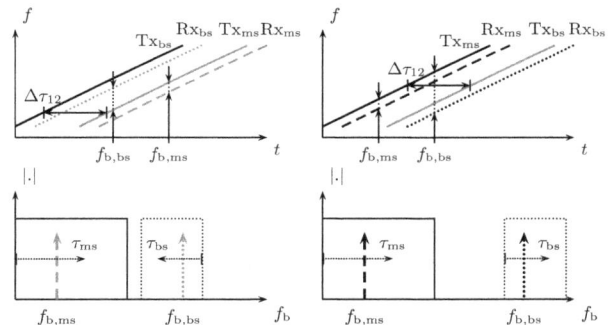

(b) Frequencyspectrum upper sensor (Tx2, Rx3|4)

(c) Frequencyspectrum bottom sensor (Tx1, Rx1|2)

Fig. 3: Bistatic setup with two coherent (interferometric) Sensors for acquisition of mono- and bistatic measurement data.

tering center and the considered receiver $Rx_{1|2}$. Here no changes compared to a purely monostatic measurement occur. The signal delayed by $\Delta\tau_{12}$, and transmitted from the upper sensor (Tx_2), is received directly by the bottom sensor or scattered at an object at first, which results in an additional travel time. After down-conversion using the Tx-signal (LO-signal) of the bottom sensor, the minimal beat-frequency for the bistatic part is obtained for the direct path. Scatterers at greater distances with respect to this direct path are represented by a larger beat frequency.

Considering the upper sensor, the chirp signal delayed by $\Delta\tau_{12}$ is radiated and used as LO-signal for the upper sensor.

(a) Photo (side view)

(b) Horizontal cut at $z_o = 0.81$ m (in dB)

(c) monostatic (side view)

(d) bistatic (side view)

(e) incoherent mono- and bistatic combination

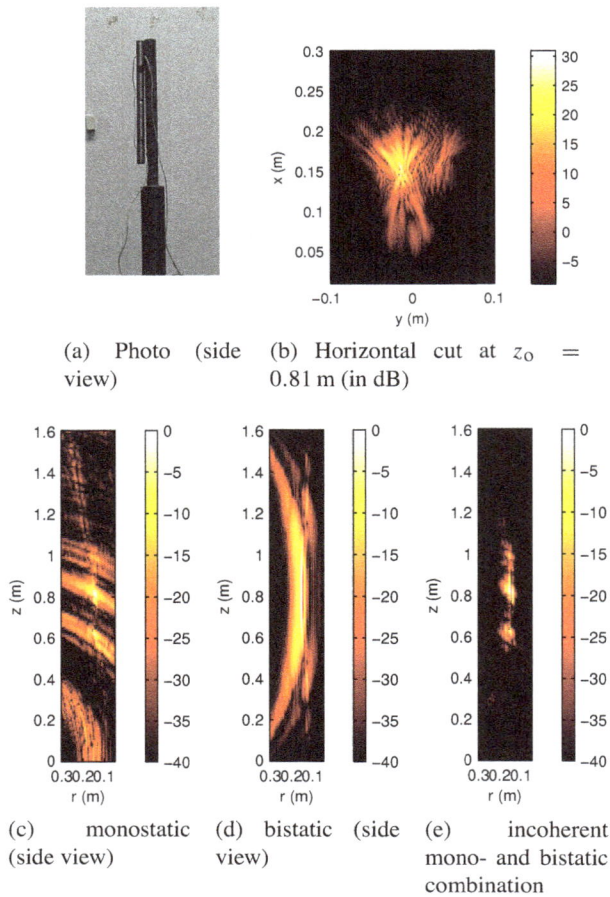

Fig. 4: Metallic rod mounted on a PVC-rod; aperture segment used for reconstruction $\varphi_a = \pm 40°$; Bandwidth $B = 10$ GHz; $f = 90 \ldots 101$ GHz

Again for the monostatic signals at the upper sensor, no changes arise compared with a purely monostatic measurement.

For the bistatic signal the highest possible difference frequency with respect to the delayed reference signal is obtained by the direct crosstalk between sensor 1 and 2. All responses from objects with a non-zero distance to this direct path will be mapped to a lower difference frequency. The spectrum of the bistatic component of the upper sensor is mirror-inverted, i.e. longer distances "Tx_1 – scatterer – $Rx_{3|4}$" are mapped to smaller difference frequencies as can be seen in Fig. 3c.

3.4 Determination of the delay-line length

In order to avoid overlapping of the spectra of the different signal components, several system parameters must be adjusted thoroughly. The maximum difference frequency occurs at the bistatic signal component of the bottom sensor.

The upper limit for the beat frequency can either result from the bandpass filter (denoted by "BP" in Fig. 3) or the available sampling rate. In this implementation the bandpass filter with a cut-off frequency of 300 kHz is the limiting factor compared to the sampling rate of 1 MSample/s. The lowest difference frequency appears for bistatic signals at the upper sensor. This frequency must not lie in the interval that is reserved for the monostatic signal components. The electrical length of the delay line was chosen to be 10.4 m, which assures that the mono- and bistatic signals do not interfere and that the ambiguity range and the BP's cut-off frequency are not exceeded.

4 Reconstruction results

For testing purposes a vertically placed metallic rod with an approximate height of 20 cm and a diameter of 1 cm was used. This isolated object was used to test the properties of the bistatic images. The mono- and bistatic reconstruction results and their non-coherent combination can be seen in Fig. 4. The theoretically determined values for the horizontal resolution are met. The resolution in z-direction is, as expected, poor.

5 Conclusions

We presented an extension of two monostatic sensors up to a fully mono- and bistatic sensor network. In addition to the monostatic signals, the information of the bistatic signals can be evaluated too. The data processing and the quality of the reconstructed monostatic images is not affected. As a nice side-effect only one signal generator (VCO) is needed with this extension instead of two.

Calibration and reconstruction algorithms capable to handle bistatic data were developed. First reconstruction results for isolated objects show that the resolution which is achieved, is according to the theoretical values for this setup.

References

Bertl, S., Dallinger, A., and Detlefsen, J.: Broadband circular interferometric millimetre-wave ISAR for threat detection, Adv. Radio Sci., 5, 2007.

Cumming, I. G. and Wong, F. H.: Digital Processing of Synthetic Aperture Radar Data, Artech House, London, 2005.

Dallinger, A., Bertl, S., Schelkshorn, S., and Detlefsen, J.: SAR Techniques for the Imaging of Humans, in: EUSAR 2006, 6th European Conference on Synthetic Aperture Radar, Electronic Proceedings, VDE Verlag GMBH, Dresden, 2006.

Detlefsen, J., Dallinger, A., Schelkshorn, S., and Bertl, S.: UWB Millimeter-Wave FMCW Radar using Hilbert Transform Methods, pp. 46–48, Horizon House Publications, 2006.

Soumekh, M.: Fourier Array Imaging, Prentice Hall, Englewood Cliffs, N. J., 1994.

16

Spectral Signature Analysis – BIST for RF Front-Ends

D. Lupea, U. Pursche, and H.-J. Jentschel

Technische Universität Dresden, Institut für Verkehrsinformationssysteme, Mommsenstr. 13, D-01062 Dresden

Abstract. In this paper, the Spectral Signature Analysis is presented as a concept for an integrable self-test system (Built-In Self-Test – BIST) for RF front-ends is presented. It is based on modelling the whole RF front-end (transmitter and receiver) on system level, on generating of a Spectral Signature and of evaluating of the Signature Response. Because of using multi-carrier signal as the test signature, the concept is especially useful for tests of linearity and frequency response of front-ends. Due to the presented method of signature response evaluation, this concept can be used for Built-In Self-Correction (BISC) at critical building blocks.

1 Introduction

Presently, a rapidly growing integration density and more complex design structures of systems in mobile telecommunications are the characteristics of microelectronics. Taking into account the enormous cost pressure, the reduction of costs for test overhead is an important item. For that purpose Built-In-Self-Test (BIST) is applied as a method of integrating suitable test structures on the chip. A special challenge is the application of BIST in the RF- and mixed signal domain. In contrast to the test of digital systems, analog systems have only a few inputs and outputs and their internal states exhibit low time constants. From that follows, that a test with a high coverage is possible with small effort. Indeed, the more complicated task is the generation of the test stimulus. Therefore, this paper is concerned mainly with the problem of test stimulus generation. The object of investigations is the RF front-end. The text is structured as follows. Section 2 shortly outlines the selection of a test strategy. Section 3 describes a theoretical approach to evaluate the test signature response. In Sect. 4 our proposal for an universal front-end BIST is outlined. Section 5 presents a simulation environment for generating and analysing signature test stimuli and for modelling different devices under test with adjustable non-idealities. Conclusions are drawn in Sect. 6.

2 Test strategy

Possible strategies (Bushnell and Agrawal, 2001) to implement BIST for an RF front-end can be divided in two categories. Well known from the literature is the separate test of different single building blocks of the RF front-end (Huang et al., 2000). In this case every block corresponds to a device under test (DUT) and a special test signature will be formed according to the DUT's requirements. In addition, an appropriate processing of the DUT's signature response is necessary. The advantage of this principle is a high test coverage due to the usage of a special test stimulus that take into account all test conditions of the specific DUT. The test signature can be specially designed. The main disadvantage is the high test overhead. It results from the necessity to design a special test set-up for each block to be tested. Moreover, it will be assumed that all building blocks used to generate the test stimulus and to convert the response are ideal. Therefore any imperfections and failures of these building blocks will cause corresponding failures in the test

The second strategy (Hafed et al., 2000; Hafed and Roberts, 2000) consists in testing the whole transceiver front-end as a complete system (Fig. 1). In this case, the DUT's structure corresponds to a chain of building blocks of a transmitter and a receiver which are connected in a loop at the antennas. This principle is known from point-to-point radios as loopback technique (Nowakowski et al., 2001). The test signature will be injected in the transmitter's baseband interface and the signature response of the DUT will be evaluated on the receiver's baseband interface. Hence, all blocks of the transceiver's RF front-end are included in the DUT. The main advantage of this principle, compared to block testing, is the lower effort. There is a very small test overhead that is not depending on the architecture and the technology of the DUT. Therefore the results of the test will not be strongly in-

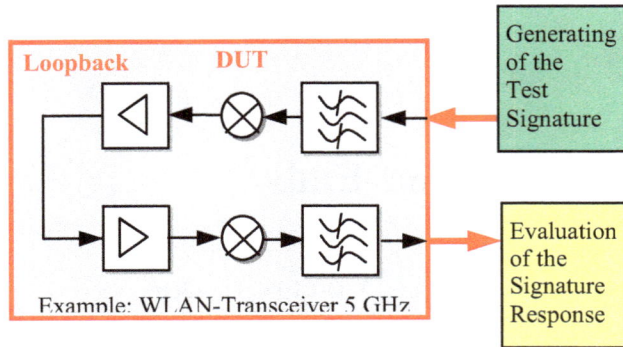

Fig. 1. System testing.

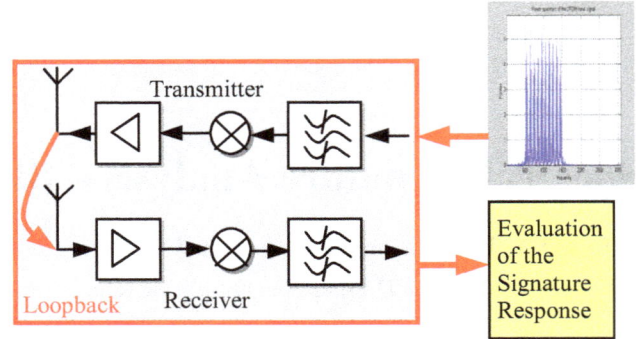

Fig. 2. OFDM test signature.

fluenced by failures in the test overhead. Also this principle has a higher flexibility. Adapting the test signature to different requirements can easily carried out by changing the algorithm used for signature generation in the DSP. This allows flexibility in respect of different architectures or technologies (Roberts and Lu, 1995). There exist also disadvantages of this principle. Mainly, the test coverage is lower due to the fact, that the complete transceiver is tested as a whole. Bad spectral properties of the transmitted signal could possible not detected due to a masking effect by the receiver's selectivity. Furthermore, because of the higher complexity of the test signature generation, an additional DSP is required (Mahoney, 1987). In addition both, transmitter and receiver, must be already implemented on silicon.

The argument of low costs for test overhead becomes increasingly important in the case of a quick production test in the high frequency range, such as a 5 GHz-system. Extensions towards to implement an optimisation of the signal path using BIST is also possible. In contrast to known concepts for block-orientated self-correction with low operating frequency, the system approach allows to optimise building blocks with high operation frequencies.

3   Test signature generation and analysis

Among other approaches test cost reduction is possible by scaling down the quantity of tests and employing low cost testers. In principle such requirements can be fulfilled by applying a complex and optimised test stimulus corresponding to the device under test (DUT). A sufficiently sophisticated post processing system to analyse and evaluate the DUT response is also required.

Testing methodologies of this type are called signature testing. Compared to conventional specification testing, signature testing has the following advantages:

– Multiple DUT specifications can be analysed using a single response acquisition

– Reduced overhead due to the single test stimulus and a single test configuration.

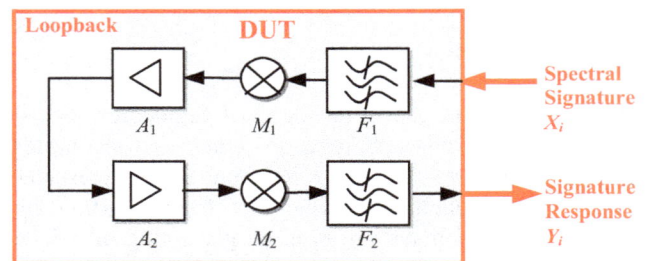

Fig. 3. Chain of stages in a typical architecture.

– Test instruments are less complicated and cheaper.

In the following a multitone OFDM signal (Nee and Prasad, 2000) will be used as spectral test signature. The power spectral density of this signal is depicted in Fig. 2. The motivation for the choice of the OFDM signal results from the fact, that a general characterisation of each block in a RF-Front-End is possible by its impulse response. This corresponds to the characterisation of the block by a transfer function in the frequency domain. On the other hand a impulse as a time domain signal would be approximated by a bandwidth-limited noise in frequency domain (Al-Qutayri, 2000). Therefore we propose as the general test signature a multitone OFDM signal.

3.1   Test signature synthesis

An approach to generate a time domain test signature that allows the detection of different conventional measured specification parameters is presented in Voorakaranam et al. (2002). Our investigations are focused in particular on WLAN-Front-Ends. Therefore the DUT is a transceiver, and modelling using transfer-functions can be applied. Usually the practical performance of a transceiver is characterised by means of some special parameters like gain, noise figure, IIP3 etc. These parameters are not depending on the specific realisation of the circuit. They are abstract, characterising a block on its behavioural level. In this sense these parameters are closely related to transfer-functions. This underlines our motivation to apply the spectral signature analysis as a

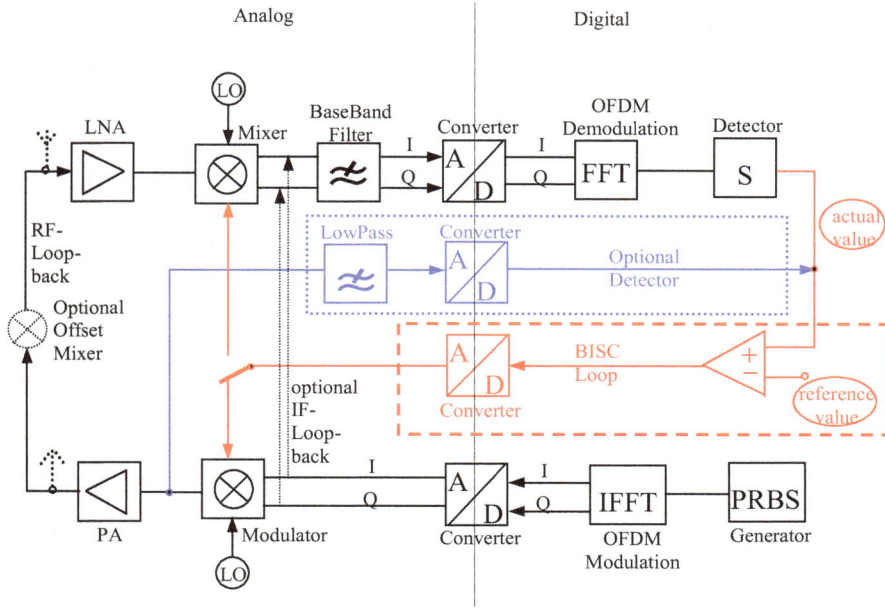

Fig. 4. Block diagram.

suitable BIST-concept and to use an OFDM signal as a test stimulus (Lee et al., 1999).

Let $X(f)$ be the test stimulus. It is of length n and consists of m discrete spectral lines at frequencies f_j, $j = 0, 1, ..., m - 1$ different from zero, equidistantly spaced on the frequency axis. We assume wl is the index of the first spectral component f_i in $X(f)$ that is unequal zero. Also we assume that $wu = wl + m - 1$. Then we get

$$X(f) = (0, ..., 0, X(f_{wl}), ..., X(f_{wu}), 0, ..., 0) \equiv$$
$$\equiv (0, ..., 0, X_{wl}, ..., X_{wu}, 0, ..., 0) \tag{1}$$

Using $X(f)$ as the input signal of the DUT we get the output signal $Y(f)$.

$$Y(f) = (Y(f_0), ..., Y(f_{n-1})) = (Y_0, ..., Y_{n-1}) \tag{2}$$

This spectral signature response Y is composed of n spectral lines on discrete frequencies f_i, $i = 0, 1, ..., n - 1$. Generally it holds $n \geq m$ because of the non linear behaviour of the DUT resulting in intermodulation products. In the case of the spectral signature analysis the property of interest concerns the relation between integrated power at the input and the output of the DUT. That means, the ratio between the input X_j at frequency f_j, and the resulting output Y_i at any other frequency f_i, is of interest. This ratio is known as the transfer factor $T_{i,j} = \frac{Y_i}{X_j}$. Using transfer factors of this kind gives the possibility of easy modelling the frequency spreading behaviour of a nonlinearity.

This is important for modules of the amplifier type. Regarding the definitions in (1) and (2) we can define a matrix $\mathbf{A_k}$ with transfer factors as matrix elements corresponding to

k^{th} non-linear stage

$$\mathbf{A_k} = \begin{bmatrix} T_{0,0} & \cdots & T_{0,n-1} \\ \vdots & \ddots & \vdots \\ T_{n-1,0} & \cdots & T_{n-1,n-1} \end{bmatrix} \tag{3}$$

In principle the matrix $\mathbf{A_k}$ describes how the energy is transmitted from the nth frequency input to the signature response Y. It follows that in the case of an ideal and distortion-free amplifier with constant gain over frequency there is no spreading of energy from one frequency to another. Therefore all elements beside of the main diagonal in $\mathbf{A_k}$ are zero.

When the module of interest in the lth stage is of a non-ideal and non-linear mixer type we get the matrix $\mathbf{M_l}$.

$$\mathbf{M_l} = \begin{bmatrix} T_{0,0} & \cdots & T_{0,n-1} \\ \vdots & \ddots & \vdots \\ T_{n-1,0} & \cdots & T_{n-1,n-1} \end{bmatrix} \tag{4}$$

An ideal mixer with constant conversion gain over frequency and an ideal sideband suppression shifts energy only to the frequency difference between the output frequency of the lth local oscillator and the discrete frequency of the input signal. Therefore all elements except parallels of the main diagonal are zero.

A different situation is given in the case of a module of filter type. In this case the selectivity of the filter is the most important parameter. It can be easy modelled assuming the filter is linear. In this situation the matrix $\mathbf{F_j}$ can be simplified. Only elements in the main diagonal are non-zero. They corresponds to the complex value of the transfer function of signal values at frequencies $f_0, ..., f_{n-1}$:

$$\mathbf{F_j} = \begin{bmatrix} T(f_0) & \cdots & 0 \\ \vdots & \ddots & \vdots \\ 0 & \cdots & T(f_{n-1}) \end{bmatrix} \tag{5}$$

In practice the behaviour of the filter is non-linear to a certain amount. This situation can be settled modelling the non-linear filter by a serial of a non-linear amplifier and a linear filter.

The concept of modelling by means of transfer functions could be applied to the whole transceiver (Fig. 3). Exciting the DUT with the spectral signature X the corresponding signature response Y will be formed by all modules included in the chain.

When we assume that the modules of the chain are connected trough non-reactive paths, there exists a simple approach to model the whole chain using the models for the components developed above. In this case it is possible to define the dependencies of the Y from the X by multiplying the matrices. Because of the non-linear properties of the amplifiers and the mixers, the sequence of matrices in the corresponding product is depending on the architecture of the transceiver chain. Therefore it holds:

$$Y = f(X) = F_2 M_2 A_2 A_1 M_1 F_1 X \tag{6}$$

The expression (6) could be simplified. Since all matrices are of the order (n, n), it is possible to merge all factors

$$\mathbf{T} = F_2 M_2 A_2 A_1 M_1 F_1 \tag{7}$$

The matrix $\mathbf{T}$ is valid for a defined level of X.

Obviously, the frequency domain approach introduced here is advantageous in comparison to a time domain approach described in Voorakaranam et al. (2002). In particular, here is no need to calculate very large matrices like in the time domain approach (Voorakaranam et al., 2002). There the order of the matrices is depending on the number of circuit elements. In our approach only to know the transmission matrices of all stages of the system for the finite number of discrete frequencies is needed. In respect of an optimisation of the signal path extension, only sensitivity matrices of stages corresponding to tuneable parameters are of interest.

3.2   Test signature response analysis

The intention of the analysis of the stimulus response consists in the calculation of the so called "distillation quality" of the receiver. That means, the capability of the receiver, to separate wanted signals from all other unwanted signals inside of a chosen frequency band had to be determined.

The wanted components of the spectral signature response inside of the chosen band are determined by

$$Y_w = \sum_{i=wl}^{wu} Y_i. \tag{8}$$

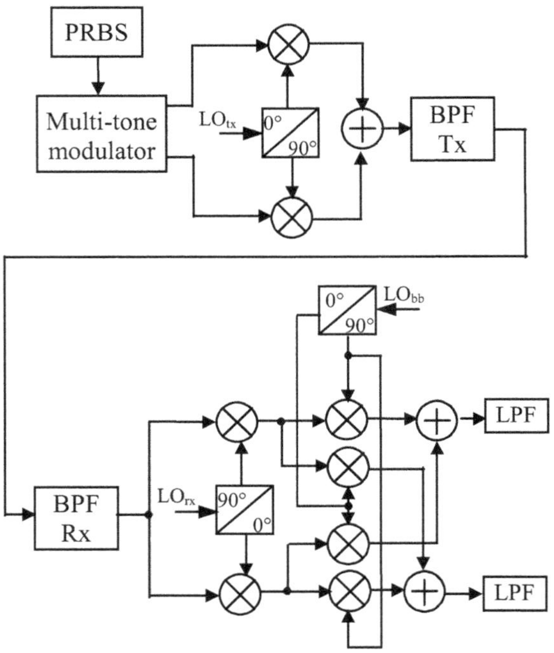

Fig. 5. The simulated transceiver chain.

For the unwanted components of the signature response holds

$$Y_{uw} = \sum_{i=0}^{wl-1} Y_i + \sum_{i=wu+1}^{n-1} Y_i \tag{9}$$

For further analysis, the so called global channel selectivity GCS in the frequency domain will be defined. This is the ratio between the power of the wanted signal inside the chosen frequency band $f_{wl}...f_{wu}$ and the sum of the power of the wanted and the unwanted signal:

$$GCS = 10 \lg \left(\frac{Y_w}{Y_w + Y_{uw}} \right). \tag{10}$$

Using GCS has two advantages. It takes into account filter behaviour out of tune characterised by slopes of the inband frequency response. In addition, it takes into account interferers and spurs outside the wanted band coming from intermodulations. By uses of GCS the so called disturbance figure S can be determined. It calculates in terms of an I/O relation the degradation of the ratio between wanted and unwanted spectral components. In other terms this means, that the ratio between the global channel selectivity at the input GCS_{in} and the output GCS_{out}, is calculated, if the test signature passes the DUT.

$$S = GCS_{in} - GCS_{out}. \tag{11}$$

If it is possible to generate a ideal test signature at the baseband input of the transmitter, then, $GCS_{in} = 0$ dB. Because of the continuous property of S, this figure can very effectively act as the actual value in a control loop necessary for self-correction instead of a more "digital" parameter like BER.

The separation between the wanted and the unwanted spectral components is possible because of the orthogonality of the OFDM carriers. By using the FFT in the OFDM demodulator each carrier will be separated from all others and its amplitude and phase will be determined. This corresponds to the principle of the Fourier Voltmeter (FVM) (Bushnell and Agrawal, 2001; Mahoney, 1987). For CDMA it is shown in Lee et al. (1999), how the measurement of the pilot channel strength as the wanted signal and the total signal strength for calculating SNR, noise figure and other parameters can be carried out. To adopt this principle for OFDM signals requires, that measurement of pilot and total channel strength will be replaced by the measurement of each carrier amplitude using FVM for BIST and BISC purposes.

4 Aspects of implementation

We propose the block diagram depicted in Fig. 4 in order to implement the loopback spectral signature analysis for a transceiver front-end. The RF loopback (Nowakowski et al., 2001) is marked by the dotted optional offset mixer. The usage of that mixer depends on the frequency planning of the target application. An attenuator can be also necessary. Beside this, the block diagram depicts a dotted IF loopback. The IF loopback is an additional option. Its application is limited to cases with identical IF in the transmit- ter and the receiver. Using more then one loopback can be reasonable for a self correction process. The optional detection path in the dotted box indicates, that for certain reasons additional effort may be necessary for the detection of the test signature responses. This is especially the case, when the demands in respect of the spectral purity at the transmit antenna are very high and low adjacent channel leakage is required for a transceiver system. In such a case the loopback test of a system of transceiver type is not sufficient in respect of spectral purity. Additional effort must be spent to test and adjust the compatibility with other transceivers not included in the test set-up. The path drawn in the dashed BISC loop box indicates the additional effort for a self-correction control loop. The loopback spectral signature test allows the on-wafer testing for SoC solutions. When the RF front-end is separated from the baseband chip, only on-board testing is possible. In both cases the input test signal can be generated by the baseband chip, if available, or by the tester that emulates the signal generation algorithm of the DSP.

5 Verification

To demonstrate the functionality of the proposed method a transceiver chain has been tested. The input signal is a multi carrier signal centred on 325 Hz with the bandwidth 450 Hz and a carrier spacing equal to 50 Hz. The input signal is converted up in frequency by an image reject mixer and filtered with a second order Butterworth band-pass filter. The mixer and the filter models the transmitter. The receiver is modelled by another second order band-pass filter, a quadrature

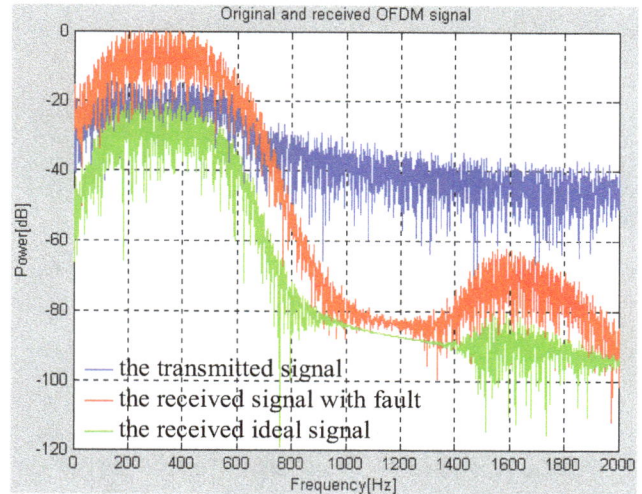

Fig. 6. The input signal, the output signal for linearity error of the band-pass filter and the expected ideal output.

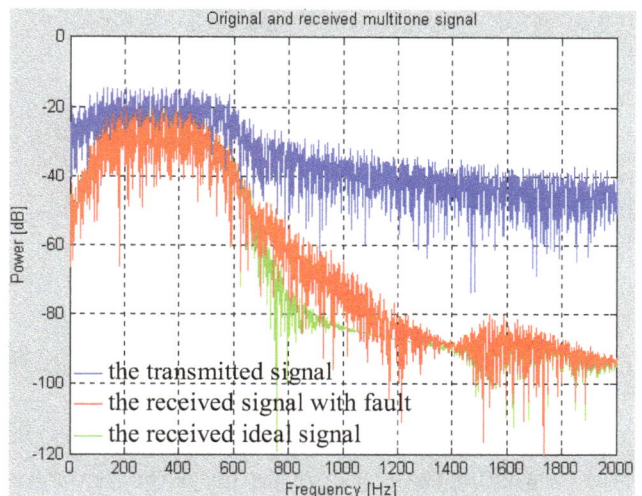

Fig. 7. The input signal, the output signal for phase error of the quadrature mixer and the expected ideal output.

mixer used to generate I and Q paths, a four multiplier image reject mixer and a sixth order Chebyshev low- pass filter for channel selection. The simulated transceiver chain is presented in Fig. 5. Several non-idealities can be introduced in the simulated chain. The most important are non-linearity, amplitude and phase mismatch in the quadrature mixers and frequency errors of the filters. As example the effect of the filter non- linearity on the GCS has been simulated. The power spectral densities of the signals at the input and output of the chain are presented in Fig. 6. The GCS has been measured against the input signal power. The measured values are given in Table 1.

Another example presents the effect of a non-rejected image signal on the GCS in the quadrature down-converter mixer. The power spectral densities of the signals at the input and output of the chain are presented in Fig. 7. The GCS

Table 1. GCS variation with the input signal level

Input power [dBm]	GCS[dB]
5	−23
2	−9
1	−1
0.5	6
0.1	19
0.05	23
0.01	17
5×10^{-3}	12
1×10^{-3}	−1
5×10^{-4}	−7
1×10^{-4}	−21
5×10^{-5}	−27
1×10^{-5}	−35

Table 2. GCS variation with the phase error

Phase error[°]	GCS [dB]
−0.2°	33
−0.1°	28
0°	26
0.1°	24
0.2°	23
0.3°	22
0.4°	21
0.5°	20.5
0.6°	20
0.7°	19
0.8°	18.8
0.9°	18.3

variation with the phase error of the image rejection mixer are given in Table 2.

6 Conclusions

In this paper a method called "Spectral Signature Analysis" has been presented. The method consists of two parts: Generation of a test signature and analysis of the signature response of the DUT. This approach allows an optimisation of the signature for self-correction. A block diagram and a simulation environment realised with MATLAB have been presented also. Simulation results show that the proposed method is not only suitable for BIST of RF Front-Ends, but also for BISC.

Acknowledgements. This work has been supported by the German Government (BMBF) under Grant No. 01M3040. In addition, this work has been supported by Nokia Research Centre in Bochum, Infineon AG in Munich and Melexis GmbH in Erfurt.

References

Bushnell, M. L. and Agrawal, V. D.: Essentials of Electronic Testing for Digital, Memory and Mixed-Signal VLSI Circuits, Boston a.o., Kluwer Acad. Publ. 2nd Ed. 2001.

Mahoney, M.: DSP-Based Testing of Analog and Mixed-Signal Circuits, Washington DC, IEEE Computer Soc. Pr. 1987.

Roberts, G. W. and Lu, A. K.: Analog Signal Generation for Built-In-Self-Test of Mixed-Signal Integrated Circuits, Boston a.o., Kluwer Acad. Publ. 1995.

Voorakaranam, R., Cherubal, S., and Chatterjee, A.: A Signature Test Framework for Rapid Production Testing of RF Circuits, Proc. of the Design, Automation and Test in Europe Conference (DATE 2002), Paris, 4–8 March 2002, pp. 186–191, 2002.

Lee, Ch.-Y., Panton, W., Granata, G., and Rajkotia, A.: Measurement of Noise Figure, G/T, and E_b/N_0 using RSSI, Proc. of the IEEE MTT-S Symposium on Technologies for Wireless Applications, Anaheim, CA, 17–18 June 1999, pp. 101–103, 1999.

Al-Qutayri, M. A.: System Level Testing of Analog Functions in a Mixed-Signal Circuit, 7th IEEE International Conference on Electronics, Circuits and Systems (ICECS 2000), Beirut, Libanon, 17–20 Dec 2000, Vol. 2, pp. 1026–1029, 2000.

Hafed, M., Abaskharoun, N., and Roberts, G. W.: A Stand-Alone Integrated Test Core for Time and Frequency Domain Measurements, Proc. of the International Test Conference (ITC 2000), Atlantic City, NJ. 3–5 Oct 2000, pp. 1031–1040, 2000.

Hafed, M. M. and Roberts, G. W.: A Stand-Alone Integrated Excitation/Extraction System for Analog BIST Applications, Proc. of the 2000 IEEE Custom Integrated Circuits Conference (CICC'2000), Orlando, FL, 21–24 May 2000, pp. 83–86, 2000.

Nowakowski, J-F., Bonhoure, B., and Carbonero, J. L.: A New Loopback GSM/DCS Bit Error Rate Test Method On Baseband I/Q Outputs, Proc. of the IEEE 57th Automatic RF Techniques Group (ARFTG 57), Phoenix, AZ, 25 May 2001, pp. 113–117, 2001.

Nee, R. v. and Prasad, R.: OFDM Wireless Multimedia Communications, Boston a.o., Artech House, 2000.

Huang, J.-L., Ong, C.-K., and Cheng, K.-T.: A BIST Scheme for On-Chip ADC and DAC Testing, Proc. of the Design, Automation and Test in Europe Conference (DATE 2000), Paris, 27–30 March 2000, pp. 216–220, 2000.

17

Performance analysis of general purpose and digital signal processor kernels for heterogeneous systems-on-chip

T. von Sydow, H. Blume, and T. G. Noll

Electrical Engineering and Computer Systems, RWTH Aachen University, Schinkelstr. 2, 52062 Aachen, Germany

Abstract. Various reasons like technology progress, flexibility demands, shortened product cycle time and shortened time to market have brought up the possibility and necessity to integrate different architecture blocks on one heterogeneous System-on-Chip (SoC). Architecture blocks like programmable processor cores (DSP- and GPP-kernels), embedded FPGAs as well as dedicated macros will be integral parts of such a SoC. Especially programmable architecture blocks and associated optimization techniques are discussed in this contribution. Design space exploration and thus the choice which architecture blocks should be integrated in a SoC is a challenging task. Crucial to this exploration is the evaluation of the application domain characteristics and the costs caused by individual architecture blocks integrated on a SoC. An ATE-cost function has been applied to examine the performance of the aforementioned programmable architecture blocks. Therefore, representative discrete devices have been analyzed. Furthermore, several architecture dependent optimization steps and their effects on the cost ratios are presented.

1 Introduction

In today's high-performance and computational intensive systems for e.g. video processing or wireless communication a high degree of flexibility and highest computational capabilities have to be provided. But the computational demands are and will be well beyond the performance of programmable processor kernels (Davis et al., 2001; Keutzer et al., 2000). Future generations of e.g. communication standards strengthen this computational gap. On the other hand in addition to these computational demands a high degree of flexibility is required. With a sufficient degree of flexibility it is possible to add new features and to adapt the changing demands of future systems without designing a new plat-form and thus increasing the product life time. Therefore, reconfigurable logic and programmable devices which are able to meet these requirements are important components in modern heterogeneous SoCs. Generally, dedicated hardware implementations offer orders of magnitude better computational performance at orders of magnitude lower power dissipation. But flexibility of those implementations is restricted to weak programmability.

Altogether, a well-balanced architecture of a sophisticated, high performance SoC has to include different types of these architecture blocks in order to provide the required performance (throughput rate) at reasonable costs (area, power dissipation) on one hand and ensuring sufficient flexibility on the other.

Partitioning a system in system blocks and mapping these system blocks on adequate architecture blocks is a challenging task discussed e.g. in the corresponding paper within this proceedings (Feldkämper et al., 2002). In this contribution an ATE-cost function which is the product of chip area, clock period and energy per calculated sample has been applied to determine quantitative results for architecture blocks whose features in terms of several parameters are depicted qualitatively in Fig. 1. In this paper we will focus on the metrics of programmable kernels. In the following especially features of programmable architecture blocks (see Sect. 2) will be examined. Therefore, an exemplary device from each category (GPP, DSP, ASIP) is considered.

2 Optimization

Several optimization steps have been applied individually to ensure that the considered programmable architecture blocks nearly perform optimally in terms of throughput rate and power dissipation. In the following, three exemplary devices and associated optimization techniques are presented.

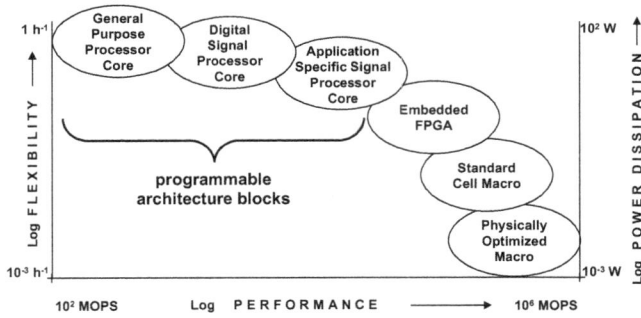

Fig. 1. Trade-off between flexibility, performance and power consumption.

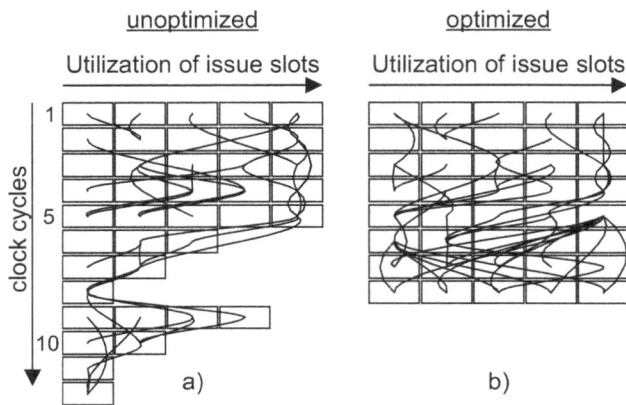

Fig. 2. Critical path diagram of the non-optimized (**a**) and optimized (**b**) add/reg operation.

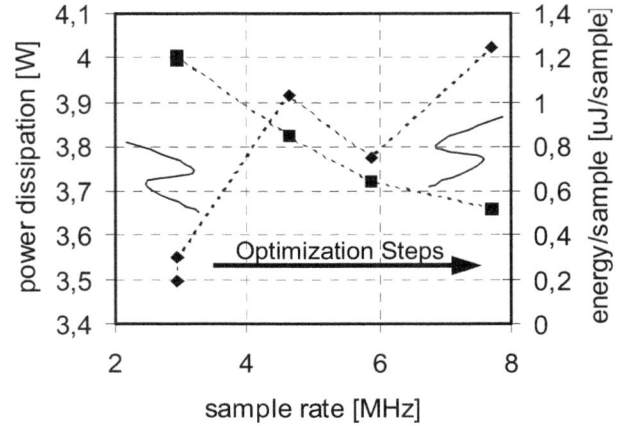

Fig. 3. add/reg operation for the TM1000 (absolute power dissipation).

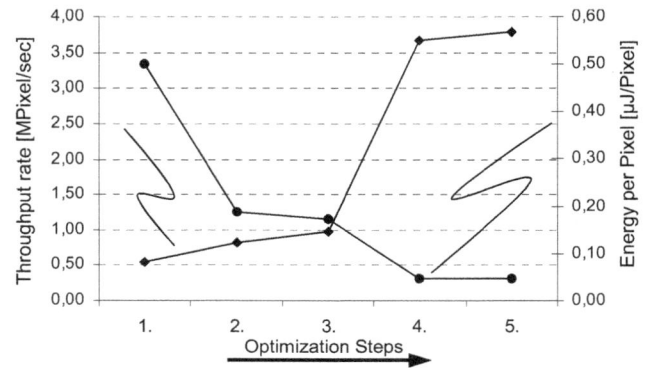

Fig. 4. Optimization steps applied to TI TMS320C6711 and associated effects.

2.1 ASIP

The Trimedia (Trimedia website) is a so-called Application Specific Instruction Set Processor (ASIP) developed for applications out of the multimedia domain. Therefore, video and audio interfaces already have been integrated. Additionally, the Trimedia ASIP contains a coprocessor most suitable for video and audio signal processing tasks. The processor core is based on a VLIW architecture including five issue slots. Hence, up to five different instructions could be executed per cycle.

Figure 2 shows the so-called critical path diagram of a loop. This loop of an exemplary add/reg-operation, which consists of basic arithmetical and register transfer operations applied in (Feldkämper et al., 2000) amounts for 88% of the total execution time. The utilization of the available issue slots with ASIP specific operations is depicted. Operations are visualized by boxes and the interdependencies between them are depicted by lines. The critical path diagram depicted in Fig. 2b could be achieved by performing ASIP suited software optimization steps such as software pipelining, common subexpression elimination and application of custom operations. Also power dissipation was examined for different code versions (Fig. 3). With each optimiza-

tion step the execution time decreases which was mainly accomplished by an improvement of the instruction level parallelism. This leads to an increase of the power dissipation. However, the consumed energy per sample decreases with each optimization step. This is due to the fact, that the number of achieved output samples per time increases more decisively than the power dissipation. The energy per sample is the metric which has to be chosen instead of the power consumption.

2.2 DSP

The TMS320C6711 (TMS320C6711 website) is a high sophisticated DSP. The underlying processor core is a VLIW-architecture providing eight issue slots. Thus, it is able to execute up to eight different instructions per cycle.

In the following the effects achieved by applying several optimization steps are illustrated. The examined basic operator of this example is a median filter based on an odd-even transposition network (Pitas, Venetsanopoulos; 1990). In Fig. 4 the throughput rate and the relative energy per sample are depicted over the applied optimization steps.

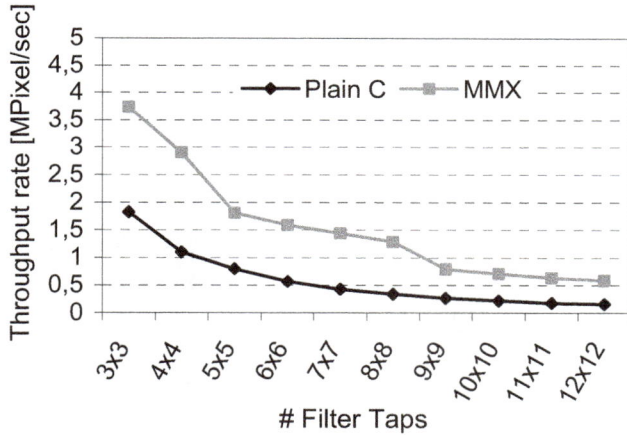

Fig. 5. Comparison Plain C vs. MMX implementation.

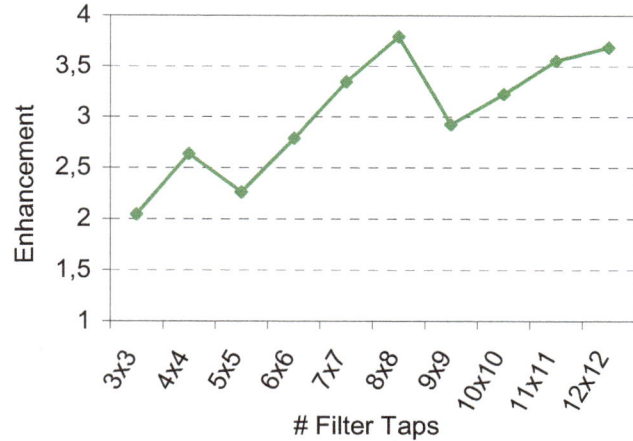

Fig. 6. Enhancement using MMX instructions.

The following optimization steps have been applied:

1. Plain C implementation without any optimization

2. Data type adaptation: Change-over memory data types to register data types

3. Adaptation of the utilized algorithm (Gupta, Evripidou; 1993)

4. Function inlining

5. Utilization of suitable keywords (e.g. compiler directives, restricted pointer etc.).

As can be seen from Fig. 4 the achieved throughput rate increases with every optimization step. Especially step 4 has been most suitable to gain performance. The energy per sample decreases with every optimization step. The adaptation of data types has been the outstanding optimization step in terms of reducing energy per sample: Memory data types have been changed in register data types. Thus, unnecessary memory accesses generally consuming more energy have been avoided.

2.3 GPP with MMX

Due to the high clock speeds and Multimedia Extensions (MMX) modern GPPs are also suitable to meet the high demands caused by complex signal processing algorithms. In this work the Pentium MMX in a low power version (Pentium website) has been analyzed. In contrast to the processors described before this processor core architecture is based on a superscalar concept.

In the following the effects which arise from utilizing the MMX instruction set extensions applied on a 2D-FIR filter with variable coefficients are shown (see Figs. 5 and 6). Due to the variable coefficients real multiplication have to be applied. The MMX implementation outperforms the plain C implementation independent of the number of filter taps. One interesting aspect are the discontinuities which can be

seen in Fig. 6. This is caused by the utilization of the SIMD instruction $PMADDWD$ (packed multiply and add word). The filter window is processed line by line. This instruction is able to handle four 16 bit values at once. Thus, due to the line by line processing mode and the ability to handle four values, e.g. a 5×5 filter mask is a disadvantageous case, because one line consists of five values. Due to that, two $PMADDWD$ instructions have to be utilized whereas one $PMADDWD$ instruction calculates just one input sample. Hence, optimal filter windows are multiples of four.

3 Results

In the following a cost evaluation of all aforementioned programmable architecture blocks concerning different basic operations is presented (see Fig. 7). The costs of other architecture blocks like FPGAs, physically optimized macros and semi-custom solutions are also depicted in this context. Thus, it is possible to compare different architecture blocks providing different architecture specific characteristics. Considering the energy per sample as well as throughput rate and the required area, the cost function motivated in detail in (Feldkämper et al., 2002) was applied here to exemplary basic operations. In this figure the cost ratios between different kinds of architecture blocks can be seen. The chosen ASIP is the aforementioned Trimedia.

For each operation the costs are normalized to that of the physically optimized implementation. In terms of programmable architecture blocks it has been distinguished between differential energy per sample (free computational resources) and absolute energy per sample (first instantiation) (see Feldkämper et al., 2002). For the differential results the cost ratio between a physically optimized and an ASIP based implementation spans from at least 4 to 7 orders of magnitude. FPGA based implementations show costs which mostly lie between the hard macro implementations and the software implementations on an ASIP. One interesting aspect is the implementation of operations like filtering on dedicated

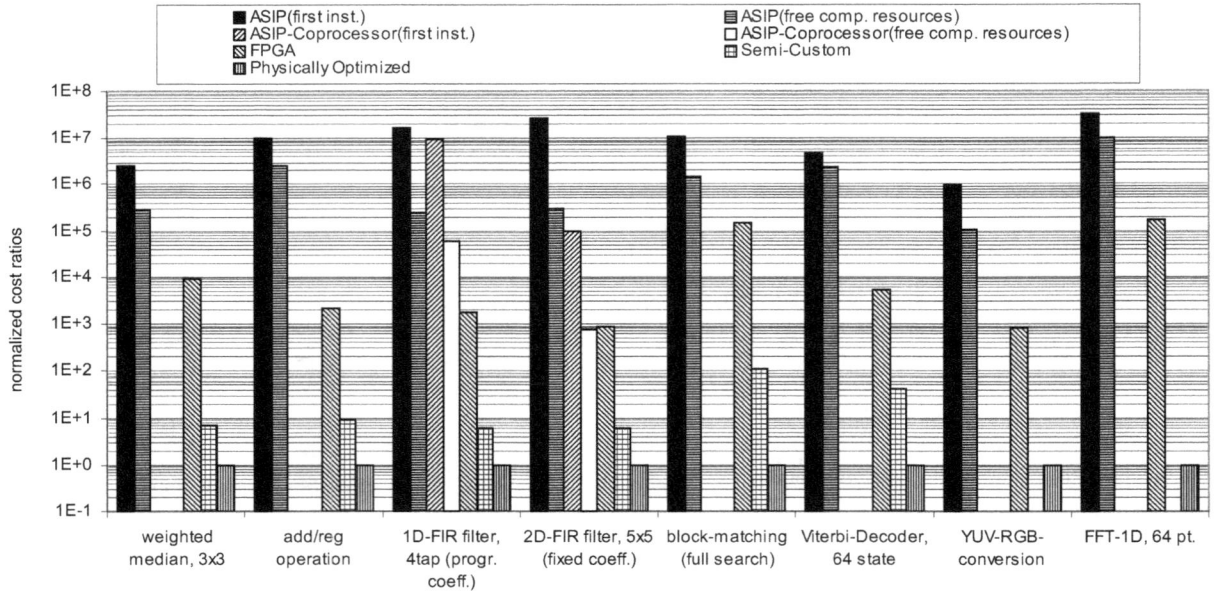

Fig. 7. Exemplary comparison of normalized ATE-cost ratios.

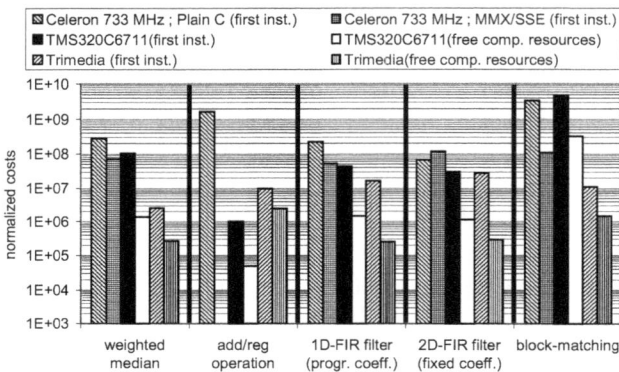

Fig. 8. Comparison of cost ratios for programmable processors.

coprocessor structures which are available on the applied Trimedia ASIP. Results for the FIR filter executed on the coprocessor device lead to minor costs compared to the execution on the ASIP CPU. This example shows what dramatic performance gains are achievable if programmable processors are integrated together with dedicated and optimized coprocessor devices. Coprocessors can relieve the ASIP CPU in order to attain this valuable resource for other algorithms which require less computational performance but demand for a high degree of flexibility. Considering absolute power consumption the total cost ratio between ASIP and physically optimized implementation even increases by up to two additional orders of magnitude.

In Fig. 8 the normalized costs are depicted for implementations on the discussed programmable processors including the aforementioned optimization techniques.

Implementations on the DSP architecture TMS320C6711

and the ASIP Trimedia are compared to implementations on the described General Purpose (GP) processor. For the GP processor, it is further differentiated between plain C implementations and optimized software implementations applying the MMX coprocessor unit. The implementation costs for the first instantiation of operations differ by up to three orders of magnitude. Several aspects can be emphasized:

– As to be expected, the implementation costs on the GP processor are higher than on the DSP or ASIP architecture.

– The MMX unit reduces the implementation costs on a GP processor by about one order of magnitude for those operations they are suited for (e.g. median, FIR filtering).

– For both Trimedia and TMS320C6711 the costs between first and second instantiation differ by about 1.5 orders of magnitude.

– The DSP and ASIP implementations differ according to their suitability of the instruction set architecture (ISA) concerning the operation to be performed. For example, the blockmatching operation mainly consists of calculating the sum of pixel differences. This is optimally supported by a Trimedia command (ume8uu). As the TMS320C6711 architecture does not support this type of operation the resulting cost is worse.

4 Conclusions

Programmable architecture blocks are easily adaptable to changing constraints by customizing program code. Thus, they provide a high degree of flexibility. Additionally today's programmable devices yield high throughput rates. Nevertheless, in many cases they are not able to meet the required constraints in terms of e.g. power consumption and throughput rate. As mentioned in the introduction Systems-on-Chip (SoC) will be well-balanced architectures consisting of different architecture blocks. In this context due to their inherent flexibility programmable architecture blocks will be one integral part of modern SoCs.

It has been shown here, that the ATE cost ratios between different programmable architecture blocks vary up to five orders of magnitude. It is most decisive for the costs how the underlying instruction set and thus the underlying processor architecture matches the requirements of the particular operator. Differences in terms of ATE-costs between dedicated physically optimized macros and programmable devices can amount up to seven orders of magnitude.

References

Blume, H., Feldkämper, H., Hübert, H., and Noll, T. G.: Analyzing heterogeneous system architectures by means of cost functions: A comparative study for basic operations, Proc. ESSCIRC, pp. 424–427, 2001.

Celeron/Pentium MMX, website, http://www.intel.com

Davis, R., Zhang, N., Camera, K., Chen, F., Markovic, D., Chan, N., Nikolic, B., and Brodersen, R.: A Design Environment for High-Throughput, Low Power Dedicated Signal Processing Systems, Proc. CICC, pp. 545–548, 2001.

De Hon, A.: The Density Advantage of Configurable Computing, IEEE Computer, pp. 41–49, April 2000.

Feldkämper, H., Blume, H., and Noll, T. G.: Study of heterogeneous and reconfigurable architectures in the communication domain, Kleinheubacher Tagung, Oct. 2002.

Feldkämper, H. T., Schwann, R., Gierenz, V., and Noll, T. G.: Low Power Delay Calculation for Handheld Ultrasound Beamformers, Proc. IEEE Ultrasonics Symposium, 22.–25. October 2000, pp. 1763–1766, 2000.

Gupta, R. and Evripidou, P.: Design and implementation of an efficient general-purpose median filter network, Digital Signal Processing, 3, pp. 64–72, 1993.

Hausner, J. : Integrated Circuits for Next Generation Wireless Systems, Proc. ESSCIRC, pp. 26–29, 2001.

Keutzer, K., Malik, S., Newton, A., Rabaey, J., and Sangiovanni-Vincentelli, A.: System-Level-Design: Orthogonalization of Concerns and Platform-Based Design, IEEE Transaction on Computer-Aided Design of Integrated Circuits and Systems, Vol. 19, No. 12, pp. 1523–1543, Dec. 2000.

Pitas, I. and Venetsanopoulos, A.: Nonlinear Digital Filters, Kluwer, 1990.

TMS320C6711 datasheets, website, http://www.ti.com

Trimedia SDE Documentation Set; 1998, http://www.semiconductors.philips.com/platforms/nexperia/media_processing/

18

A novel bottom-left packing genetic algorithm for analog module placement

L. Zhang and U. Kleine

Otto-von-Guericke University of Magdeburg, IESK, PO Box 4120, D-39016 Magdeburg, Germany

Abstract. This paper presents a novel genetic algorithm for analog module placement. It is based on a generalization of the two-dimensional bin packing problem. The genetic encoding and operators assures that all constraints of the problem are always satisfied. Thus the potential problems of adding penalty terms to the cost function are eliminated, so that the search configuration space decreases drastically. The dedicated cost function covers the special requirements of analog integrated circuits. A fractional factorial experiment was conducted using an orthogonal array to study the algorithm parameters. A meta-GA was applied to determine the optimal parameter values. The algorithm has been tested with several local benchmark circuits. The experimental results show this promising algorithm makes the better performance than simulated annealing approach with the satisfactory results comparable to manual placement.

1 Introduction

The significant tendency of system-on-chip (SoC) intensifies the booming market share of mixed-signal integrated circuits (ICs). Although most of the functions in such an integrated system are performed with digital circuitry, analog circuits are always needed as an interface to the external, continuous-valued world. The design of digital portion can be tackled with modern cell-based tools (Wang et al., 2000) for synthesis, mapping and physical design. The analog counterpart, however, is still routinely designed by hand. The layout of analog circuits is intrinsically more difficult than the digital one. To address this problem, an automated layout tool called ALADIN (Automatic Layout Design Aid for Analog Integrated Circuits) (Zhang et al., 2000) is currently being developed for analog experts who can and must bring their specific knowledge into the synthesis process in order to create high quality layouts. The focus of this paper is on the placement phase. Due to analog constraints such as matching requirements, it is usually preferred to build more or less complex clusters of devices, hereafter called modules or macro-cells, which are parameterized for the processed sub-circuits. The objective of placement is to position the modules appropriately so that the chip area and the total wire length of the interconnections are minimized under certain constraints.

Many heuristic strategies for module placement based on iterative improvement have been published so far, such as force-directed, min-cut, passive resistive optimization, simulated annealing (SA) (Su et al., 2001; Plas et al., 2001) and genetic algorithm (GA) (Shahookar and Mazumder, 1990; Esbensen and Mazumder, 1994). Among them, SA and GA are the latest and most promising techniques. SA is widely used in the domain of both digital and analog (Cohn et al., 1994; Plas et al., 2001) circuits. Although it yields good placement solutions, it is a very time-consuming process. In contrast, GA has been mainly applied for digital circuits. This paper describes a GA application in analog circuit placement. It is organized as follows. Section 2 introduces the design flow with the use of DesignAssistant in which this placement approach is included. Section 3 describes the implementation of this adaptive strategy. Section 4 gives the parameter optimization with fractional factorial experiment and meta- GA. Section 5 shows experimental results and the conclusion is drawn finally.

2 DesignAssistant

A design environment called DesignAssistant (Wolf et al., 1998) is provided in ALADIN which eases analog designers for their silicon compilation. The DesignAssistant is integrated in a commercial design framework. Its Graphical User Interface (GUI) executes external programs, such as the module generator, the placer and the router, to create layouts automatically. In Fig. 1 the design flow in the DesignAssis-

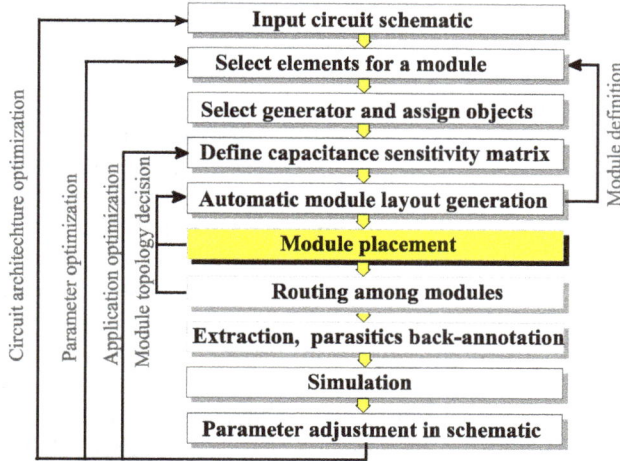

Fig. 1. Design Flow with DesignAssistant.

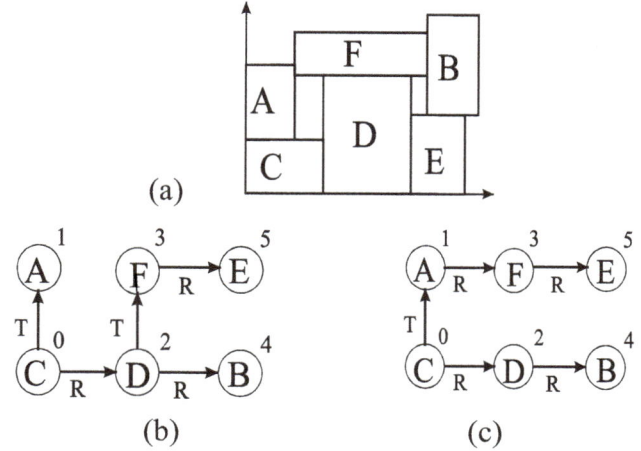

Fig. 2. Representation of BLGA with one phenotype (a), the corresponding two genotypes (b) (c).

tant is illustrated, where the shadowed block is the subject of this paper.

3 Algorithm implementation

The GA is a search strategy based on the mechanics of natural selection and natural reproduction in a biological system. It differs from the other stochastic search techniques by being able to encode and exploit past information efficiently during a search.

3.1 Genetic encoding

The conventional chromosomal representation of the GA is based on bit-string (Shahookar and Mazumder, 1990). A GA for the two-dimensional bin packing problem has been developed by Kroeger et al. (1991). Esbensen and Mazumder (1994) used this representation for the digital circuit placement. In this paper a bottom-left GA (BLGA) for the analog module placement is developed based on comprehensive extensions and modifications of the genetic encoding and operators found in the above work (Esbensen and Mazumder, 1994; Kroeger et al., 1991). In GA a distinction is made between genotype and phenotype of an individual. Here the genetic encoding is inspired by the two-dimensional bin packing problem, which is the problem of compactly packing a number of rectangular blocks into a bin with a fixed width and infinite height in such a way that the distance from the top edge of the highest placed block to the bottom edge of the bin is minimized. The standard algorithm for this problem places each block at a time at the downmost and then at the leftmost position. The placement algorithm is based on a generalization of this scheme. The solution space considered by the algorithm is restricted to the set of all possible BL-placements.

Assume that the given problem has n cells $c_1, ..., c_n$. An example genotype with $n = 6$ cells is shown in Fig. 2 to-

gether with the corresponding phenotypes. A binary tree (V, E), $V = \{c_1, ..., c_n\}$, in which the i'th node corresponds to the cell i, representing the absolute positions of all cells. Two kinds of edges exist: top-edges E_t and right-edges E_r, so that

$$E = E_t \cup E_r, \quad E_t \cap E_r = \emptyset \qquad (1)$$

Each node has at most one outgoing top-edge and at most one outgoing right-edge. All edges are oriented away from the root of the tree. Let $e_{ij} \in E$ denote an edge from c_i to c_j and let (c_i^{xl}, c_i^{yl}) and (c_i^{xr}, c_i^{yr}) denote the coordinates of the lower left and upper right corners of c_i, respectively. Then $e_{ij} \in E_t(E_r)$ means that cell c_j is placed above (or to the right of) c_i in the phenotype. That is,

$$\forall e_{ij} \in E : c_j^{yl} \geq c_i^{yr} \text{ if } e_{ij} \in E_t, \quad c_j^{xl} \geq c_i^{xr} \text{ if } e_{ij} \in E_r \qquad (2)$$

The tree is decoded as follows. The cells are placed one by one in a rectangular area with horizontal length W and infinite vertical length. Each cell is moved as far down and then as far left as possible. The cells are placed in ascending order according to their priorities defined by one-to-one function $P : V \rightarrow \{0...n - 1\}$. Any node has higher priority than its predecessor in the tree. In Fig. 2 the priorities are indicated at the top right hand of each node. The orientation (i.e., transformation and reflection) of each cell is defined by the function $O : V \rightarrow \{0, 1, 2...7\}$, which is also part of the genotype.

3.2 Genetic operators

Given two individuals α and β, the crossover operator generates a new feasible descendant individual γ. Two proposals are given. The operations are illustrated in Fig. 3. Throughout this section, a superscript specifies which individual the marked property belongs to. The experimental results are given in Sect. 5. In the first proposal, E^r (i.e., edge set of the

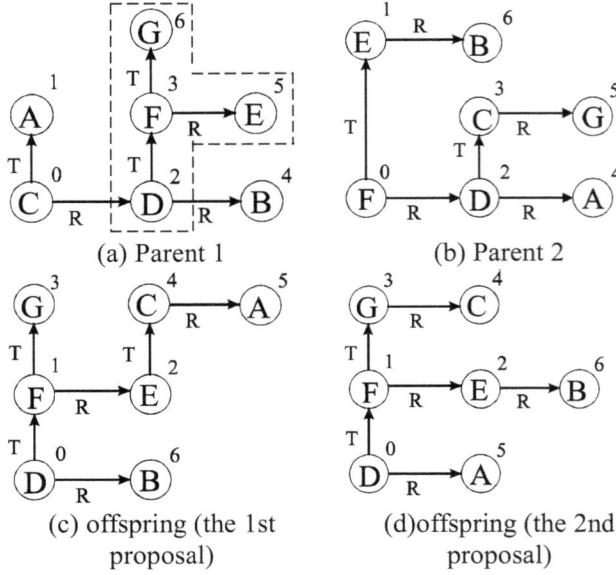

Fig. 3. Crossover operators.

descendant γ) is constructed as follows. From the cell tree of α, a connected subset

$$T_s = (V_s, E_s), V_s \subset V, E_s \subset E^\alpha \tag{3}$$

is chosen. T_s is chosen at random but subject to the constraint that decoding T_s in the order defined by P^α (i.e., priority of individual α), using

$$c \in V_s | \forall c' \in V_s/\{c\} : P^\alpha(c) < P^\alpha(c') \tag{4}$$

as root, causes no constraint violations. In Fig. 3, the chosen T_s is indicated by the dashed line. Initially E^γ is defined to be E_s. Hence, γ has inherited all cells in V_s from α. The remaining cells $V - V_s$ are then inherited from β by extension of E^γ. The cell tree of β is traversed in ascending order according to $^\beta$. At any node it is checked if the corresponding cell c belongs to V_s, that is, whether it has been placed in γ already. If so, the cell is skipped. Otherwise, c is added to the cell tree of γ by extending E^γ randomly. The orientation of any cell is inherited unaltered together with the cell itself. That is,

$$t^\gamma(c) = \begin{cases} t^\alpha(c) \ if \ c \in V_s \\ t^\beta(c) \ if \ c \in V - V_s \end{cases} \tag{5}$$

P^γ should correspond to the order in which the cells were placed when creating E^γ. Since P is a bijection, the following constraints in Eq. (6) uniquely determines P^γ:

$$\forall c_i \in V_s, \quad \forall c_j \in V - V_s : P^\gamma(c_i) < P^\gamma(c_j)$$
$$\forall c_i, \quad c_j \in V_s : P^\alpha(c_i) < P^\alpha(c_j) \Rightarrow P^\gamma(c_i) < P^\gamma(c_j)$$
$$\forall c_i, \quad c_j \in V - V_s : P^\beta(c_i) < P^\beta(c_j) \Rightarrow$$
$$P^\gamma(c_i) < P^\gamma(c_j) \tag{6}$$

The second proposal differs from the first one by the construction of the remaining cells $V - V_s$, which inherits from β by the ordered extension. In detail, the concatenation tries to follow the structure of β first of all. If impossible, randomly add to any free position of γ. In Fig. 3d the node B is added to the left of the node E, instead of the left of the node D in Fig. 3c. As well the node A is added to the left of the node D, instead of the left of the node C in Fig. 3c.

Five different mutation operators are developed. Each performs some random change in the given genotype. When performing each of these mutations, a part of the genotype has to be decoded to check if the mutated individual satisfies all constraints. A mutation is only performed if it does not cause any constraint violations. The purpose of the inversion operator is to weaken the linkage among genes. Given a genotype, the inversion operator computes a new genotype by rearranging the components in such a way that their mutual distances changes, while at the same time assuring that the corresponding phenotype is still the same. The inversion operator selects a subtree at random and moves it to another free position in such a way that no constraints are violated and so that the corresponding phenotype is still the same. An example of this is shown is Fig. 2b and c. This genotype tree is generated by moving the subtree rooted at the node F.

3.3 Cost function

The cost function is the goodness criterion of searched configurations. It consists of four parts, which are given in Eq. (7).

$$C = (\alpha_{all\_area} C_{all\_area} + \alpha_{N\_area} C_{N\_area} + \alpha_{P\_area} C_{P\_area}$$
$$+ \alpha_{A\_area} C_{A\_area} + \alpha_{D\_area} C_{D\_area}) + \alpha_{nets} C_{nets}$$
$$+ \alpha_{asp\_rat} C_{asp\_rat} \tag{7}$$

α^* is a weight factor for the corresponding cost C^*, which balances the importance of all the possible considerations according to different design requirements. The first is the area cost which is made up of the whole area, NWELL and PWELL region areas, analog and digital region areas. They will help to decrease the whole area. Moreover they could make NWELL and PWELL regions relatively concentrated in order to ease the fabrication. And analog and digital regions are separated from each other so that the constraints for mixed signal circuits can be imposed. The second is the net-length cost C_{nets}, in which priority coefficient can be specified for each net. For analog circuits the significance of different nets is distinct. Some sensitive nets, for instance differential input signal nets, should be as short as possible in order to decrease the parasitic capacitance and crosstalk. The more sensitive one net is, the higher its priority coefficient is defined. Linear or exponential operations can be chosen on the net priority. Five approaches, including half-perimeter, center-of-mass, complete graph, minimum spanning tree and minimum steiner tree, are developed for the net-length estimation so that analog designers can choose for the trade-off between accuracy and efficiency (Sait and Youssef, 1995).

```
Algorithm BLGA()
(M: the population size.)
Begin
1    input macro-cell geometry and net-list;
2    initialize the first population randomly;
3    evaluate the fitness;
4    while not (stopCriterion())
5        foreach (M * crossoverRate)
6            choose the first parent based on rank selection;
7            choose the second parent randomly;
8            do crossover to generate one offspring;
9        endfor
10       choose M individuals with the largest fitness among the
         combined set of parents and offspring;
11       foreach (M)
12           do mutation based on mutationRate;
13           do inversion based on inversionRate;
14       endfor
15   endwhile
16   output the best member;
End
```

Fig. 4. Outline of BLGA.

```
Algorithm meta-GA()
(M: the population size.)
Begin
1    set M as 20 and the generation sum as 100;
2    initialize the first population randomly;
3    evaluate the fitness;
4    while not (stopCriterion())
5        foreach (M)
6            make two random trials and select two parents from
             the population with the probability proportional to
             fitness;
7            perform crossover by selecting each parameter
             randomly from either parent with equal probability;
8            mutate offspring with 0.8 probability by selecting a
             parameter at random and adding to it a random
             number within the range of [0, 10];
9        endfor
10       choose M individuals with the largest fitness among the
         combined set of parents and offspring;
11   endwhile
12   select the fittest set of parameters from the final population;
End
```

Fig. 5. Outline of the meta-GA.

The third is the aspect-ratio cost C_{asp_rat}, which is used to control the shape of the final layout. Designers can define the ideal width-length ratio or exact width value. The more the real layout shape differs from the ideal definition, the more penalty is brought about. The graphic input window of the cost function is provided in the DesignAssistant.

3.4 Algorithm outline

The algorithm outline is depicted in Fig. 4. $stopCriterion()$ makes the evolution process terminated if no improvement has been observed for a predefined number of consecutive generations or a fixed number of generations is over. A fitness is the reciprocal of the corresponding cost.

4 Parameter optimization

Since the operators are of paramount importance to the overall performance of the algorithm, their parameters, including the population size, crossover, mutation and inversion rates, have to be investigated with care. The study of the correlation and sensitivity helps to shrink the regarded ranges and set up the exact optimal parameters in the problem of analog module placement.

4.1 Parameter analysis

The fractional factorial experiment is an important technique in Robust Design (Park, 1996). A Taguchi orthogonal array $L_{27}(3^{13})$ is employed to construct the fractional factorial experiment. The population size, crossover, mutation and inversion rates are taken as the factors. The reason why to choose such an array with 13 columns is to give a deliberate study about inter-correlation between two factors. The experiment design is depicted in Table 1, cr means crossover rate, mr means mutation rate, ir means inversion rate, M means population size and * means the combination of two factors. Three columns (4, 7 and 11) are left for error estimation.

For the crossover and inversion rates, three levels are chosen as (0.2, 0.55, 0.85). (0.05, 0.1, 0.2) are chosen for the levels of the mutation rate. Two groups of experiments were performed in order to cover a wide range for the population size, one with (10, 35, 60) and the other with (40, 70, 100). The cost and execution time are taken as the search target, where the cost is the primary consideration and execution time is secondary. As a result, the correlation between parameters is found quite weak. The population size becomes insignificant to the cost if it is more than 60. The crossover and mutation rates are sensitive for the search. Even though the inversion rate is insensitive to the cost, it is 0.25 optimally with the consideration of its effect on the execution time. So the optimal value or ranges for the four parameters are set as follows. The population size was set as 60, the mutation range was from 0 to 0.1, the crossover rate from 0.55 to 1 and the inversion rate from 0.20 to 0.55.

4.2 Parameter determination

Because the levels in the orthogonal array are limited, it is rough to depend on it determining the parameter exact values. Furthermore although the result in Sect. 4.1 gives the weak correlation between two parameters, it is preferable to take the correlation into account when determining parameter exact values. So a meta-GA (Shahookar and Mazumder, 1990) depicted in Fig. 5 was employed to determine the exact rate values. The individuals in the population of the meta-GA consist of three integers in the range of [0, 10], represent-

Table 1. Design of the fractional factorial experiment

Exp. No.	1	2	3	4	5	6	7	8	9	10	11	12	13
	cr	mr	cr*mr		ir	cr*ir		mr*ir	M	cr*M		mr*M	ir*M
1	1	1	1	1	1	1	1	1	1	1	1	1	1
2	1	1	1	1	2	2	2	2	2	2	2	2	2
...	...	...	...	...	...	...	...	...	...	...	...	...	...
27	3	3	2	1	3	2	1	2	1	3	1	3	2

Table 2. Comparison among distinct algorithms

		SA	BLGA1	BLGA2
	C_{mean}	11799	11823	11698
	C_σ	176	46	39
Circuit 1				
	$T_{mean}(s)$	846	213	222
	$T_\sigma(s)$	210	40	69
	C_{mean}	86653	83487	82122
	C_σ	3087	2925	2165
Circuit 2				
	$T_{mean}(s)$	5068	3897	2848
	$T_\sigma(s)$	783	163	1031
	C_{mean}	215720	220740	216350
	C_σ	13554	4224	3932
Circuit 3				
	$T_{mean}(s)$	1143	2586	3222
	$T_\sigma(s)$	4	280	461

Table 3. Comparison among distinct algorithms

	circuit 1	circuit 2	circuit 3	Average
cr	0.775	0.82	0.865	0.82
mr	0.01	0.02	0.01	0.013
ir	0.235	0.41	0.41	0.352
initCost	12453	82276	220330	
endCost	11731	77992	203704	

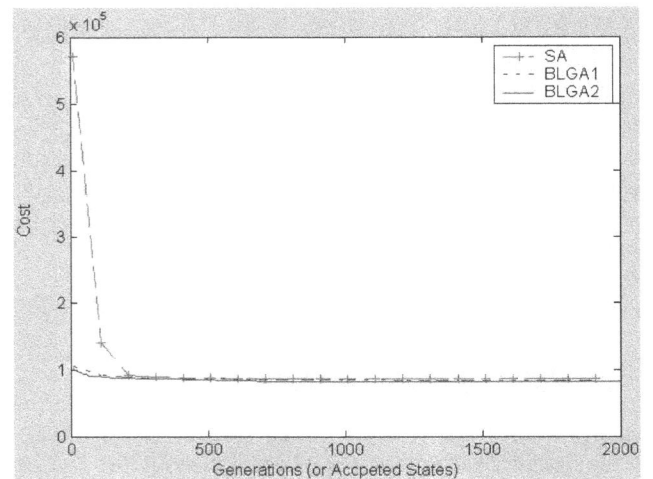

Fig. 6. Convergence schedules of different algorithms.

ing the crossover rate, inversion rate, and mutation rate of BLGA. The fitness of an individual (a BLGA with a certain parameter combination) is taken to be the fitness of the best placement that the meta-GA can find in the entire run, using these parameters. In meta-GA, the population size was 20, and the algorithm was run for 100 generations. The crossover probability was 1.

5 Experimental results

Because so far the analog benchmark circuits are unavailable for the synthesis purpose, three local circuits are used to evaluate the above algorithms. Each algorithm is executed for ten times so that the mean and standard deviation are used for evaluation. The cost value is superior to the execution time. As the focus here is on the comparison of algorithms, the simple half-perimeter estimation is applied for all the trials. In order to demonstrate the efficiency of GA, one optimization with SA was also performed. The results are depicted in Table 2 including SA, BLGA1 means with the first crossover proposal and BLGA2 means with the second crossover proposal.

The result shows SA is generally poorer than BLGA while need more execution time. The BLGA2 works marginally better than BLGA1. So finally the second proposal, i.e., the close inheritance crossover scheme is applied. In order to

keep the diversity during the evolution, the selection based on fitness rank is applied instead of the fitness-based such as roulette wheel selection (Kroeger et al., 1991). The convergence schedule is depicted in Fig. 6. Since the costs of BLGA1 and BLGA2 are the best cost in each generation, the variance amplitude is smoother than SA in the whole view.

The representation of BLGA improves the searching efficiency so that the search need not cover a wide scope as SA but with more accuracy. The results with the Meta-GA for the three circuits are given in Table 3. Finally the best parameter set is the crossover rate of 0.82, the mutation rate of 0.023 and the inversion rate of 0.235. The program is written in $C++$ running under Solaris-UNIX in a Sun-Ultra60 workstation. The weight factors in the cost function are set as follows: $\alpha_{all\_area} = 2$, $\alpha_{N_a rea} = 0.2$, $\alpha_{P_a rea} = 0.2$, $\alpha_{A_a rea} = 0.2$, $\alpha_{D_a rea} = 0.2$, $\alpha_{nets} = 10$, and $\alpha_{asp\_rat} = 5$.

(a)

(b)

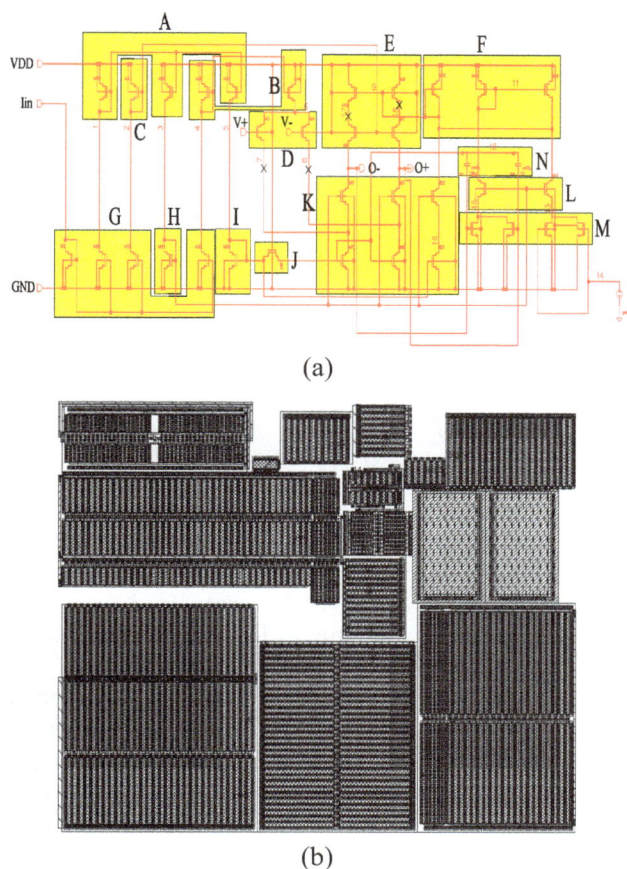

Fig. 7. Schematic (**a**) and placement layout (**b**) with BLGA2 of a common mode feedback optional amplifier.

Figure 7a gives the schematic of the third circuit, a common mode feedback optional amplifier, in which the partitioning is indicated by rectangles. The corresponding placement result with the BLGA2 is shown in Fig. 7b, which is comparable to the manual placement. In DesignAssistant the relative position of modules are recorded and passed to the router as the input. The router then compacts all the modules with the detailed routing.

6 Summary

In this paper a new technique with genetic algorithm to solve the analog module placement has been introduced. By using the notion of bin-packing, a genetic encoding has been developed in which most constraints of the problem are implicitly represented. As a consequence, each individual always satisfies constraints. The advantage of the proposed strategy is that it allows a more accurate estimation of the layout quality with the configuration space shrunk dramatically, since the use of penalty terms have been avoided. Special constraints for analog integrated circuits are included in the cost function. The fractional factorial experiment with an orthogonal array is employed in order to study the algorithm parameters.

A meta-GA is applied to determine the exact parameter values. The experimental results show that this approach with the optimized parameters contributes high design efficiency with the satisfactory result, which is comparable to the manual counterpart.

References

Wang, M., Yang, X., and Sarrafzadeh, M.: Dragon2000: Fast Standard-cell Placement for Large Circuits, Proc. of the IEEE International Conference on Computer-Aided Design (ICCAD), pp 260–263, November 2000.

Zhang, L., Kleine, U., Roewer, F., Rudolph, T., and Wolf, M.: A Novel Design Tool for Analog Integrated Circuits, Proc. of the First Joint Symposium on Opto- & Microelectronic Device and Circuits, pp.146–149, April 2000.

Su, L., Buntine, W., Newton, A. R., and Peters, B. S.: Learning as Applied to Stochastic Optimization for Standard Cell Placement, IEEE Trans. on Computer-aided Design of Integrated Circuits and Systems, 20, 4, pp. 1499–1513, April 2001.

Varanelli, J. M. and Cohoon, J. P.: A two-stage simulated annealing methodology, Proc. the Fifth Great Lakes Symposium on VLSI, pp. 50–53, 1995.

Cohn, J. M., Garrod, D. J., Rutenbar, R. A., and Carley, L. R.: Analog Device-leve Layout Automation, Boston, Kluwer Academic Publishers, 1994.

van der Plas, G., Debyser, G., Leyn, F., Lampaert, K., Vandenbussche, J., Gielen, G., Sansen, W., Veselinovic, P., and Leenaerts, D.: AMGIE – A Synthesis Environement for CMOS Analog Integrated Circuits, IEEE Trans. on Computer-aided Design of Integrated Circuits and Systems, 20, 9, pp. 1037–1058, Sept. 2001.

Shahookar, K. and Mazumder, P.: A Genetic Approach to Standard Cell Placement Using Meta-Genetic Parameter Optimization", IEEE Trans. Computer-Aided Design, 9, 5, pp. 500–511, May 1990.

Esbensen, H. and Mazumder, P.: SAGA: A Unification of the Genetic Algorithm with Simulated Annealing and Its Application to Macro-Cell Placement, Proc. of The 7th International Conference on VLSI Design, pp. 211–214, 1994.

Kroeger, B., Schwenderling, P., and Vornberger, O.: Genetic Packing of Rectangles on Transputers , Transputing 91, 2, IOS Press, 1991.

Wolf, M., Kleine, U., and Schafer, F.: A Novel Design Assistant for Analog Circuits, Proc. Asia and South Pacific Design Automation Conference, pp. 495–500, Feb. 1998.

Sait, S. M. and Youssef, H.: VLSI Physical Design Automation (Theory and Practice), McGraw-Hill, 1995.

Park, S. H.: Robust Design and Analysis for Quality Engineering, Chapman & Hall, London, 1996.

19

Adiabatic circuits: converter for static CMOS signals

J. Fischer, E. Amirante, A. Bargagli-Stoffi, and D. Schmitt-Landsiedel

Institute for Technical Electronics, Technical University Munich, Theresienstrasse 90, D-80290 Munich, Germany

Abstract. Ultra low power applications can take great advantages from adiabatic circuitry. In this technique a multiphase system is used which consists ideally of trapezoidal voltage signals. The input signals to be processed will often come from a function block realized in static CMOS. The static rectangular signals must be converted for the oscillating multiphase system of the adiabatic circuitry. This work shows how to convert the input signals to the proposed pulse form which is synchronized to the appropriate supply voltage.

By means of adder structures designed for a $0.13\,\mu m$ technology in a 4-phase system there will be demonstrated, which additional circuits are necessary for the conversion. It must be taken into account whether the data arrive in parallel or serial form. Parallel data are all in one phase and therefore it is advantageous to use an adder structure with a proper input stage, e.g. a Carry Lookahead Adder (CLA). With a serial input stage it is possible to read and to process four signals during one cycle due to the adiabatic 4-phase system. Therefore input signals with a frequency four times higher than the adiabatic clock frequency can be used. This reduces the disadvantage of the slow clock period typical for adiabatic circuits. By means of an 8 bit Ripple Carry Adder (8 bit RCA) the serial reading will be introduced. If the word width is larger than 4 bits the word can be divided in 4 bit words which are processed in parallel. This is the most efficient way to minimize the number of input lines and pads. At the same time a high throughput is achieved.

1 Introduction

Adiabatic circuitry can save energy in the frequency domain up to about 100 MHz which is interesting for digital signal processing. The family considered in this work is the Positive Feedback Adiabatic Logic (PFAL) (Vetuli et al., 1996; Blotti et al., 2000) because it has very low energy dissipation

and shows the best performance at high frequencies. The needed signal waveforms are equivalent to the signals for Efficient Charge Recovery Logic (ECRL) and 2N-2N2P (Moon et al., 1996; Kramer et al., 1995). These families use symmetrical trapezoidal supply voltages and dual rail encoded input/output signals. A comparison of ECRL, 2N-2N2P and PFAL was performed in Amirante et al. (2001) regarding to the frequency dependent energy dissipation and the robustness against parameter variations. The energy dissipation can amount 13% or less than that of static CMOS which has the fundamental limit of $\frac{1}{2}CV_{DD}^2$.

The goal of this paper is to present the integration of adiabatic function blocks in a standard CMOS environment. The important circuits for this matter are the converters. They consist not only of a stage which converts the signal into the right pulse shape and phase but also of a synchronizing circuit. The design of the entire conversion stage is comparable to the one for static CMOS buffer stages. For synchronizing the incoming signal from the static CMOS block to the supply clock of the adiabatic circuits the 4-phase system will provide the timing information. Thus the benefits of this system will be explained first before the synchronized converter is presented. Adder structures vary in the way the data can be applied to their input. The Carry Lookahead Adder (CLA) which is the best choice for internal data (Amirante et al., 2002) needs parallel input data. That means the input data are driven by the same phase. The Ripple Carry Adder (RCA) with its cascaded full adders can also handle serial data. This is useful for off-chip data sources. The number of pads is often limited so that a serial input is necessary. By the right arrangement of the converters, the RCA can be used for the serial input without any additional buffer stages.

2 4-phase system

We have chosen ECRL, 2N-2N2P and PFAL because of their favorable properties with respect to design and implementation. They are robust against parameter variations and their

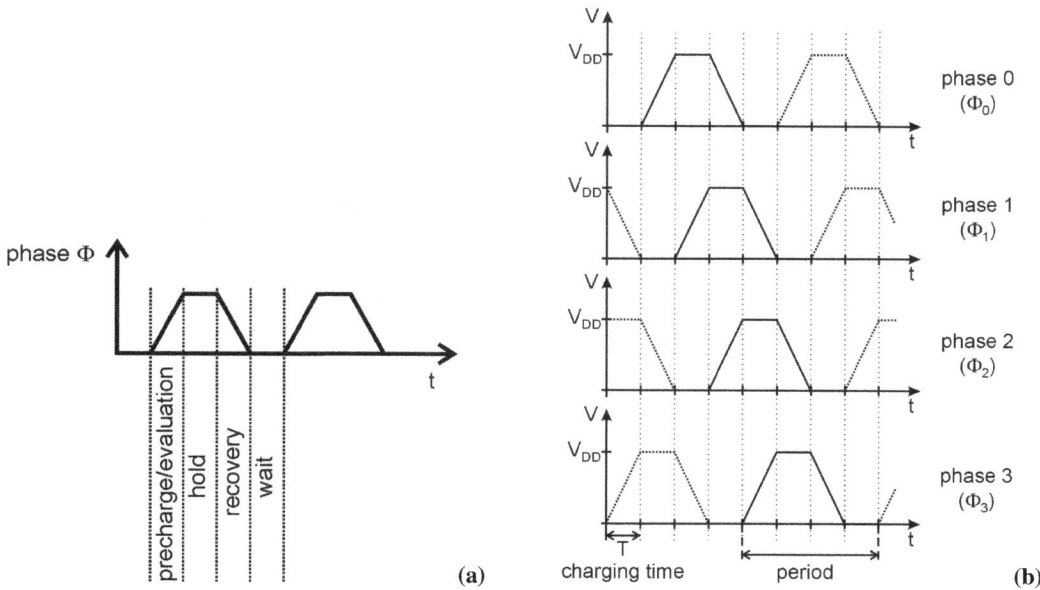

Fig. 1. 4-phases system, (**a**) one phase with the four states: evaluation, hold, recovery and wait, (**b**) 4 phases and their 90 degree shift.

energy dissipation is low even at high frequencies. Because of the dual rail encoding the oscillator sees a constant load and the design of high efficiency oscillators has been simplified. A presentation of an oscillator design is given in Bargagli-Stoffi et al. (2002). The ideal waveform for the supply voltage is shown in Fig. 1a. It can be divided in four states of equal length. The precharge/evaluation state is characterized by the increasing voltage ramp, which allows to load the capacitance efficiently regarding to energy dissipation. During the hold state the output value is valid and can be evaluated by the next stage. By decreasing the voltage a recovery of energy is achieved. During the wait state the supply voltage is zero and no leakage current flows. As mentioned before the evaluation of one stage and the holding of the former stage have to take place at the same time so that two subsequent phases are shifted by 90 degrees (Fig. 1b). Altogether there are four phases, and during one period four functions can be executed. Thus the effective throughput rate can be up to four times higher than the adiabatic clock frequency. In order to reach the minimum energy dissipation the value of 90 degree shift should be achieved exactly by proper oscillator design . Because of this accurate relation between the phases the synchronization signal can be extracted from them.

3 Converter

3.1 Basic input converter without synchronization

The prerequisite of the presented converter is the presence of the supply clocks with their accurate trapezoidal waveform. These signals will also act as supply for the intrinsic converter (Fig. 2a). Thus the pulse shaping will be provided by the oscillator and the static signals only need to control

switches. If the output node follows the supply clock it will be at a HIGH level in adiabatic circuitry. Otherwise if the output stays at ground the signal is at the LOW level. In order to generate an adiabatic high level a transmission gate is the best choice as pull-up tree. The provided trapezoidal supply clock can pass the transmission gate without disturbances, such as voltage steps due to the threshold voltage. The dimensioning of the transmission gate should follow that for static CMOS. The equivalent resistance must fit the requirements. In adiabatic circuitry the delay of the stages is not important but the energy dissipation which is strongly dependent on the resistance. The dimensioning of the transmission gate or the minimization of the equivalent resistance is a tradeoff between area consumption and energy dissipation. In order to avoid a deformation of the adiabatic signal, the static CMOS input signal of the converter must not be switched during the whole precharge, hold and recovery states. Thus the static signal switches only during the wait state, when the supply clock is at ground. To generate an adiabatic low level, the output has just to be clamped at ground. Therefore a n-channel transistor is sufficient as pulldown tree. Because the resistance of the n-channel transistor does not influence the energy dissipation its dimension can be chosen minimal.

3.2 Input converter with synchronization

In order to produce a dual rail encoded trapezoidal signal with the phase Φ_1 two basic converters are needed (Fig. 2b). They are driven by a static CMOS D-register generating the input and the inverted input signal. To avoid metastability cascaded registers are used. A comparator which compares the preceding supply voltage phase Φ_0 with half of V_{DD} generates the clock for the positive edge-triggered registers.

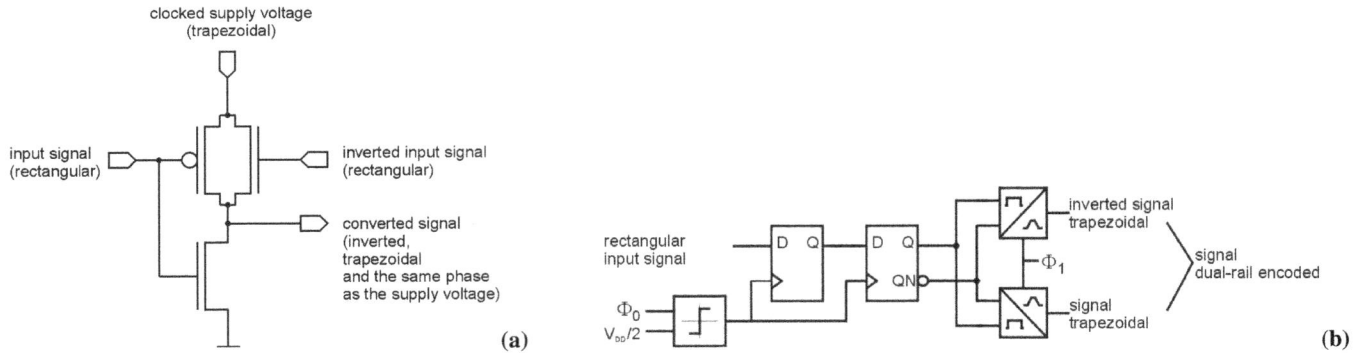

Fig. 2. (a) Basic converter; a rectangular signal is converted into an inverted trapezoidal one by a trapezoidal adiabatic supply voltage, **(b)** converter with synchronization circuit to generate dual rail encoded trapezoidal signals.

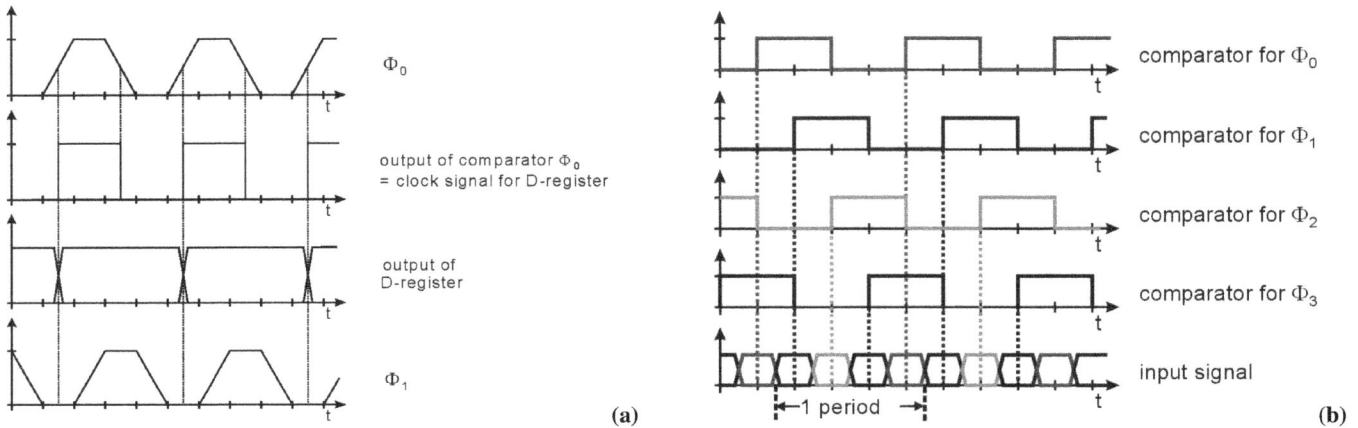

Fig. 3. (a) Timing diagram of a converter stage, the supply voltage Φ_0 is compared to $V_{DD}/2$ generating the clock for the registers, the register switches only in the waiting state of the supply voltage Φ_1, **(b)** taking advantage from the 4-phase system, the serial reading can be done with four times the frequency of the trapezoidal clock.

As mentioned above the static input signal must switch only during the wait state of the supply voltage. The synchronization is explained in Fig. 3a for an adiabatic signal having the phase Φ_1. To have a robust synchronization the switching point should be in the middle of the wait state of Φ_1 that is simultaneous to the precharge state of the preceding phase (Φ_0) . The clock signal can easily be generated by comparing Φ_0 with $V_{DD}/2$. The simplest comparator for this use is a static CMOS buffer. Of course the delay of the buffer and the register must not exceed half of the wait state which is one eighth of the adiabatic clock period. In most cases, this is fulfilled even with minimal dimensioned gates because of the rather long adiabatic clock period. Because of the exact relation between two phases mentioned in Sect. 2, the whole margin amounts one eighth of the period. To improve the delay an inverter can be used instead of a buffer. But in this case $V_{DD}/2$ must be compared with the next stage Φ_2 which is in the recovery stage or a negative edge-triggered register must be used.

The comparators generate clock signals with the same 90 degree shift as the oscillator (see Fig. 3b). An input bitstream applied to all these converters will be sampled by each stage at another specified time. Thus a 4bit word can be transmitted per adiabatic period and the input rate is increased by a factor of four. Of course every output bit of such a 4bit converter has its own phase. If all output data should have the same phase additional buffer stages must be inserted.

3.3 Output converter

The conversion of an adiabatic signal to a static one can be achieved in a similar way. A D-register is used to sample the adiabatic output signal in the middle of the hold state. The clock for this sampling operation is generated by a comparator and the proper phase. The output of the D-register is a valid CMOS signal. A four to one multiplexer can follow to obtain a serial output corresponding to the input signal.

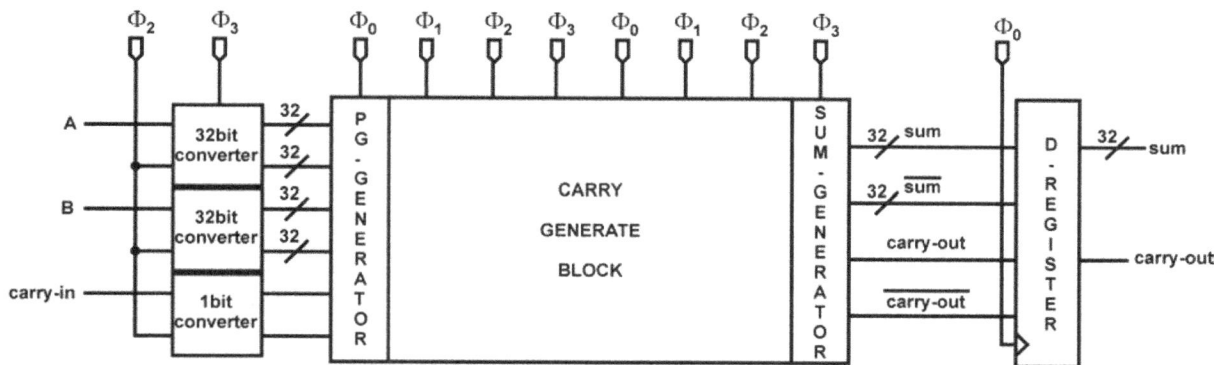

Fig. 4. Schematic of an 32bit-carry lookahead adder (32-bit CLA) with input and output converters. The terms of the sum A and B are converted by the first stage into 64 dual rail encoded adiabatic signals. The output conversion is performed by D-registers.

4 Adder structures

Many adder structures for static CMOS are demonstrated in Weste, Eshraghian (1992). Most of them can be adapted for adiabatic circuitry. Adders realized in adiabatic circuitry are pipelined structures. Therefore it is possible to apply new data to the input every cycle although it can take longer than one period to get the result. Adiabatic adder structures differ mainly from static CMOS in how long it takes to perform the whole addition.

4.1 Carry Lookahead Adder (CLA)

The carry lookahead adder consists of three parts (Fig. 4). The first stage generates the 'propagate' and the 'generate' signals. These signals are used in the second part to calculate the carry information. In the last stage the sum is built. The important part is the carry generation because by a proper implementation the number of phases can be reduced. The algorithm shows its whole potential if the word width is larger than 8 bit especially for adiabatic circuits. In order to support the adiabatic system design the output signals should all have the same phase which can be evaluated by the subsequent input stage. Then you do not have to care about the phases and automatic placement will work like in static CMOS. Hence one function block should consist of a multiple of four stages. If the word width is smaller than 32 bit additional buffer stages must be integrated to have the proposed output phase. One addition of 32bit will take two adiabatic periods but in every period new data can be fed to the input stage because of the intrinsic pipelined structure of adiabatic circuitry. For internal addition of wide data words the CLA is the preferred choice.

In Fig. 4 also the integration of a CLA into a static CMOS environment is shown. The synchronized converters must provide the parallel input data with the phase Φ_3 so that the first stage with the phase Φ_0 can evaluate it. Therefore 32 converters per term of the sum are synchronized by the phase Φ_2. The output signals have the phase Φ_3. If they should be converted to static signals positive edge triggered D-registers

are used whose clock signal is generated by a comparison of the subsequent phase Φ_0 and $V_{DD}/2$.

4.2 Ripple Carry Adder (RCA)

A ripple carry adder is built by cascading several full adders. Therefore the time needed for the addition is proportional to the word width, e.g. an 8 bit RCA consists of eight full adders and the addition will take two adiabatic periods. Like in the CLA in every period new data can be fed into the pipeline. So the response time between applying data to the input and receiving the sum at the output is a major drawback of adiabatic RCAs if the word width gets large.

The RCA is especially suitable with serial inputs. The 4bit word converters of Sect. 3.2 transform the data in 4 signals which can directly drive the input stage of the RCA without any additional buffer stages. The minimal number of pads is achieved if the data consists of one bitstream for each term of the sum (Fig. 5a). This bitstream is divided by an internal demultiplexer into 4 bit words which are converted by the 4 bit word converters. However, as long as the most significant bits (MSBs) are applied to the adder input no valid data are present at the input of the least significant bits (LSBs) and vice versa. Therefore, time which could have been used for processing is wasted and the throughput decreases with larger word width. In order to maximize the throughput the data must be partitioned into parallel bitstreams with forefield multiplexing. This results in two pads in the case of an 8bit RCA (Fig. 5b). First the 4 LSBs are applied to the first pad. During the next period the 4 MSBs appear at the second pad. Because of the pipelined structure of adiabatic circuitry the LSBs of the next term of the sum can drive the first pad. Of course this principle can be expanded for larger word width.

The method using an internal demultiplexer was implemented on a testchip with a $0.13\,\mu m$ technology. The requirements were minimal number of pads and no additional buffer stages. The purpose of the testchip is to measure the energy consumption of the 8 bit adders. Therefore the use of additional energy consuming adiabatic circuits should be

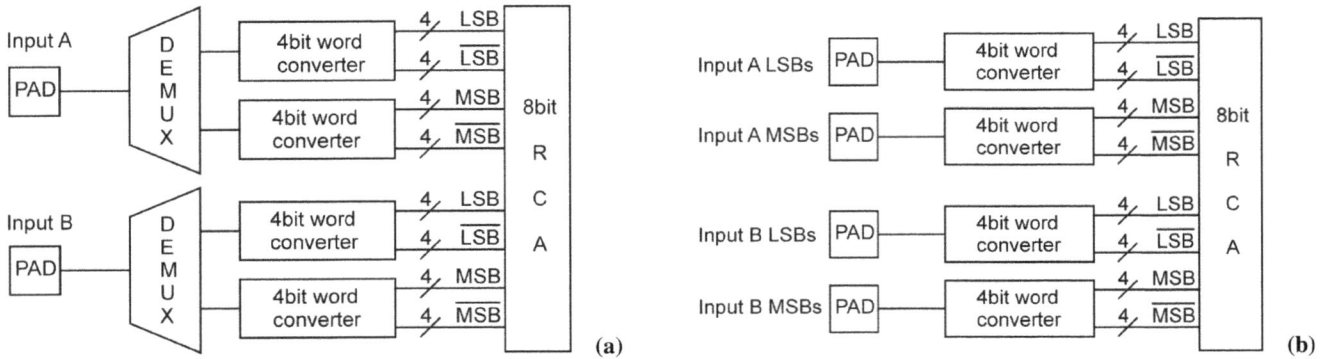

Fig. 5. Addition of 4 bit words with a 8 bit RCA, (**a**) minimizing number of pads by using internal demultiplexer and 4 bit word converter, (**b**) maximal throughput without internal demultiplexing.

minimized or avoided. The testchip has not been measured yet but simulations performed with extracted parasitics from the layout show that the presented converter principles will work at frequencies beyond 100 MHz. The interface between the static CMOS and the adiabatic circuits can be designed fast enough, and therefore robust enough, to meet the requirements in the frequency domain in which adiabatic circuitry saves a large amount of energy.

5 Conclusion

In this work principles are presented which allow to integrate adiabatic function blocks into static CMOS circuits. A synchronized converter is used to generate the dual rail encoded trapezoidal signals which are necessary for the considered adiabatic families PFAL, ECRL and 2N-2N2P. Inside the converter a comparator is used to extract the clock signal for synchronizing the input signal. This clock signal can be used for every timing issue between the two circuitries. The robustness of this signal is an issue of oscillator and comparator design. The oscillator must provide the circuit with four phases which have an exact difference of 90 degrees, otherwise the energy dissipation increases. The comparator's delay is well known and can be adjusted regarding to the timing conditions. By combining the serial input data to 4bit words, the input rate can be increased by a factor of four. This principle can be used to minimize pads at the input. It is also worth noticing that the throughput rate of adiabatic circuits is up to four times higher than the adiabatic frequency because of the 4-phase system.

The embedding of adiabatic circuits into a static CMOS environment was a major concern for the use in real application. This issue has now been solved. Adiabatic circuits allow to save energy in a frequency range typical for digital signal processing applications. Therefore adiabatic circuitry can be a suitable ultra low power strategy for digital signal processing in the frequency domain up to hundreds of MHz.

Acknowledgements. The authors would like to thank Infineon Technologies AG, in particular Dr. C. Heer, Dr. J. Berthold und Dr. R. Nadal for their help with the design and for useful discussions and advices.

This work is funded by the German Research Foundation (DFG) under the grant SCHM 1478/1-2.

References

Amirante, E., Bargagli-Stoffi, A., Fischer, J., Iannaccone, G., and Schmitt-Landsiedel, D.: Variations of the Power Dissipation in Adiabatic Logic Gates, Proceedings of the 11th International Workshop on Power And Timing Modeling, Optimization and Simulation, PATMOS'01, Yverdon-les-Bains, Switzerland, pp. 9.1.1–9.1.10, September 2001.

Amirante, E., Bargagli-Stoffi, A., Fischer, J., Iannaccone, G., and Schmitt-Landsiedel, D.: Adiabatic 4-bit Adders: Comparison of Perfomance and Robustness against Technology Parameter Variations, 45th Midwest Symposium on Circuits and Systems, MWSCAS'02, Tulsa Oklahoma, USA, August 2002.

Bargagli-Stoffi, A., Amirante, E., Fischer, J., Iannaccone, G., and Schmitt-Landsiedel, D.: Resonant 90 Degree Shifter Generator for 4-Phase trapezoidal adiabatic logic, U.R.S.I. Kleinheubacher Tagung 2002, Miltenberg, Germany, 30 Sep–02 Oct 2002.

Blotti, A., Di Pascoli, S., and Saletti, R.: Simple model for positive-feedback adiabatic logic power consumption estimation, Electronics Letters, 36, 2, 116–118, 2000.

Kramer, A., Denker, J. S., Flower, B., and Moroney, J.: 2nd order adiabatic computation with 2N-2P and 2N-2N2P logic circuits. Proceedings of the International Symposium on Low Power Design, 191–196, 1995.

Moon, Y. and Jeong, D.: An Efficient Charge Recovery Logic Circuit. IEEE Journal of Solid-State Circuits, 31, 4, 514–522, 1996.

Vetuli, A., Pascoli, S. D., and Reyneri, L. M.: Positive feedback in adiabatic logic, Electronics Letters, 32, 20, 1867–1869, 1996.

Weste, N. H. E. and Eshraghian, K.: Principles of CMOS VLSI design: a systems perspective, 2nd edition, Addison-Wesley Publishing Company, 1992.

20

Exploration of dual supply voltage logic synthesis in state-of-the-art ASIC design flows

T. Mahnke[1]**, W. Stechele**[1]**, M. Embacher**[2]**, and W. Hoeld**[2]

[1]Institute for Integrated Circuits, Technical University of Munich, Germany
[2]National Semiconductor GmbH, Fuerstenfeldbruck, Germany

Abstract. Dual supply voltage scaling (DSVS) for logic-level power optimization at the has increasingly attracted attention over the last few years. However, mainly due to the fact that the most widely used design tools do not support this new technique, it has still not become an integral part of real-world design flows. In this paper, a novel logic synthesis methodology that enables DSVS while relying entirely on standard tools is presented. The key to this methodology is a suitably modeled dual supply voltage (DSV) standard cell library. A basic evaluation of the methodology has been carried out on a number of MCNC benchmark circuits. In all these experiments, the results of state-of-the-art power-driven single supply voltage (SSV) logic synthesis have been used as references in order to determine the true additional benefit of DSVS. Compared with the results of SSV power optimization, additional power reductions of 10% on average have been achieved. The results prove the feasibility of the new approach and reveal its greater efficiency in comparison with a well-known dedicated DSVS algorithm. Finally, the methodology has been applied to an embedded microcontroller core in order to further explore the potentials and limitations of DSVS in an existing industrial design environment.

1 Introduction

The total power dissipation of digital CMOS circuits is composed of static and dynamic components. While static power contributes significantly to the total power in certain applications that are inactive for long periods of time, it is still dominanted by dynamic power in the majority of applications.

The dynamic power P_{dyn} is composed of the capacitive power P_{cap} and the short-circuit power P_{sc}. The capacitive power P_{cap} is due to currents charging or discharging the node capacitances and can be written as

$$P_{cap} = \alpha_{01} \cdot f_{clk} \cdot C_{\text{node}} \cdot V_{DD}^2 \ , \qquad (1)$$

where α_{01} is the switching activity, f_{clk} is the clock frequency, C_{node} is the node capacitance, and V_{DD} is the sup-

ply voltage. Although P_{cap} usually accounts for the largest portion of the total dynamic power, P_{sc} must not be neglected. The short-circuit power is caused by currents flowing through simultaneously conducting n- and p-channel transistors. A first-order approximation of P_{sc} is

$$P_{sc} = \alpha_{01} \cdot f_{clk} \cdot (\beta/12) \cdot t_T \cdot (V_{DD} - 2V_t)^3 \ , \qquad (2)$$

where β is an effective transconductance, t_T is the input signal transition time and V_t is the threshold voltage.

A very efficient means of reducing P_{dyn} is supply voltage scaling. However, since gate delay increases with decreasing V_{DD}, globally lowering V_{DD} degrades the performance. At the logic level, dual supply voltage scaling (DSVS) can be used for lowering V_{DD} only in non-timing-critical paths, thus keeping the overall performance constant (Chen at al., 2001; Usami and Horowitz, 1995; Usami et al., 1998a,b; Yeh et al., 1999b).

In our work, we modified an existing power-driven logic synthesis methodology such that DSVS is supported in addition to state-of-the-art optimization techniques. This approach enabled us to carry out DSVS in a conventional logic synthesis environment, while all previously published work required proprietary tools that comprise dedicated DSVS algorithms. In our discussion of the experimental results, we use the results of state-of-the-art power-driven single supply voltage (SSV) logic synthesis as reference values in order to reveal the true additional benefit of DSVS.

The remainder of the paper is structured as follows. In Sect. 2, a short overview of state-of-the-art logic-level power optimization is given. In Sect. 3, we introduce the DSVS technique. Our novel power-driven logic synthesis methodology is described in Sect. 4. Results of the optimization of a number of benchmark circuits and an embedded microcontroller are presented in Sects. 5 and 6. Finally, we provide concluding remarks.

2 State-of-the-art in power-driven logic synthesis

In conventional logic synthesis methodologies, P_{dyn} can typically be minimized by means of gate sizing, equivalent pin swapping and buffer insertion (Synopsys Inc., 1998).

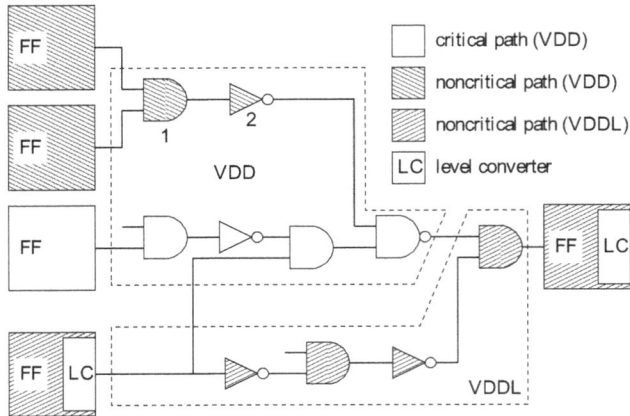

Fig. 1. A typical dual supply voltage (DSV) circuit structure.

Down-sizing primarily aims at reducing C_{node} and, thus, P_{cap} by using smaller slower cells in non-timing-critical paths, but also reduces short-circuit currents and, hence, P_{sc} at the sized gates. On the other hand, increasing the size of a gate shortens the signal transition time t_T at its output, which in turn reduces P_{sc} at the gates driven by the sized cell. Alternatively, extra buffers can be inserted at heavily loaded nodes in order to shorten t_T. Equivalent pin swapping takes advantage of the fact that functionally equivalent input pins of logic gates often exhibit different power characteristics. With pin swapping, high activity nets are connected to power-efficient input pins with priority.

In our experiments, we made extensive use of the above mentioned techniques when we created the SSV reference designs.

3   Dual supply voltage scaling (DSVS)

The purpose of DSVS is to reduce the supply voltage for gates in noncritical paths from the nominal value V_{DD} to a lower value V_{DDL} (Chen at al., 2001; Usami and Horowitz, 1995; Usami et al., 1998a,b; Yeh et al., 1999b). Figure 1 illustrates a typical (DSV) circuit structure. In DSV circuits, low voltage cells must not directly drive high voltage cells. Otherwise, quiescent currents occur at the driven gates. This is the reason why gates 1 and 2 in Fig. 1 are operated at V_{DD} although they are part of a noncritical path. Level-converting cells can be inserted where transitions from V_{DDL} to V_{DD} are required (Usami et al., 1998a). However, these cells introduce additional delay and cause power and area overhead. In order to minimize this overhead, we enable level conversion only at the input and output nodes of combinational blocks as depicted in Fig. 1.

Other difficulties are the distribution of two supply voltages across the chip and the layout synthesis. One possible solution to these problems is placing low and high voltage cells in separate rows. This can be realized on the basis of conventional cell layouts but requires proprietary tools (Usami et al., 1998a). Another possibility is the use of two sep-

arate power rails for V_{DD} and V_{DDL} in each row. This requires modification of the layouts of all cells. However, low and high voltage cells can then be mixed within rows and, hence, placement and routing can be carried out using standard tools (Yeh et al., 1999a).

4   Dual supply voltage logic synthesis methodology

4.1   Tools for dual supply voltage scaling

All known approaches to DSVS are based on dedicated algorithms (Chen at al., 2001; Usami and Horowitz, 1995; Usami et al., 1998a,b; Yeh et al., 1999b), and not one of these algorithms has been integrated into standard tools yet. However, DSVS can be carried out without the need for any dedicated algorithm which is evident from the following simple arguments. At the logic level, standard cells are distinguished only by functionality, delay, power, input capacitance and area. Typical cell-library-based gate sizing algorithms, such as the one presented by Coudert (1997), revert only to these properties when picking cells that implement certain functionalities while minimizing the power consumption subject to delay constraints. Knowing that a reduction of the supply voltage for a cell changes only its delay and power, we conclude that cell-library-based gate sizing algorithms should be able to handle functionally equivalent low and high voltage cells in the same way as cells of different size. Cell-library-based gate sizing algorithms are readily available with standard tools such as Synopsys' Power Compiler (SPC). Thus, instead of developing yet another dedicated DSVS algorithm, we forced SPC to perform DSVS along with gate sizing. Fortunately, the tool allows input and output pins of cells to be classified such that only pins of the same class will be interconnected. While this feature was originally introduced for coping with high I/O and low core voltages, it also allows us to solve the level conversion issue discussed in Sect 3. Power analysis was carried out at the logic level using Synopsys' Design Power (SDP).

4.2   Design flow and optimization strategy

Provided that a suitably modeled DSV library exists, delay-constrained power optimization can be performed following the three-step strategy illustrated in Fig. 2. After reading the original design, delay-constrained logic synthesis is carried out (STEP 1). At this stage, low voltage (VDDL) and level-converting (LC) cells are disabled. After capturing switching activities during gate-level simulation, state-of-the-art delay-constrained power optimization comprising the techniques mentioned in Sect. 2 is carried out (STEP 2), which results in a timing- and power-optimized SSV implementation. Finally, power optimization is repeated with low voltage and level-converting cells enabled (STEP 3). This leads to a timing- and power-optimized DSV implementation.

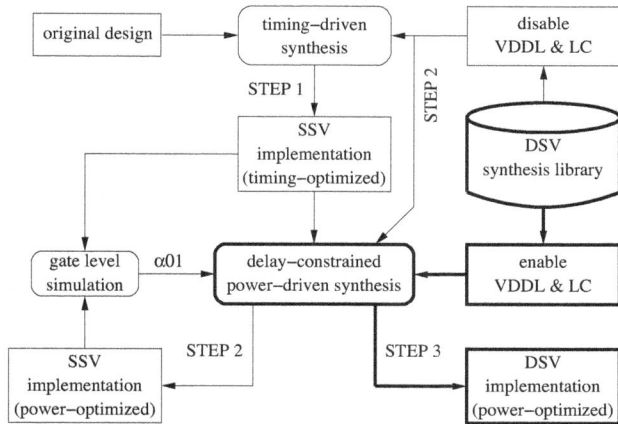

Fig. 2. Design flow comprising DSVS in addition to state-of-the-art power optimization.

4.3 Dual supply voltage synthesis library

The key to DSVS exploiting gate sizing algorithms is a suitably modeled standard cell library. We developed a DSV synthesis library from a commercial library realized in 0.25 μm CMOS and characterized at supply voltages of 1.8 V and 2.5 V. It has been shown elsewhere that for a given V_{DD} an optimal V_{DDL} exists (Chen at al., 2001; Usami and Horowitz, 1995; Usami et al., 1998a,b). On the other hand, the optimal choice of V_{DDL} depends largely on the circuit to be optimized (Chen at al., 2001). Note that in our experiments we always used the voltage levels given above, which were defined by the library vendor, and forwent the costly procedure of determining an optimal voltage pair for each circuit, which was used by Chen at al. (2001) and Usami et al. (1998a).

The DSV library contains inverters, buffers, (N)ANDs, (N)ORs, X(N)ORs and D-flip-flops in up to five different sizes each. For each cell, high and low voltage synthesis models are provided and a level-converting flip-flop (DF-FLC) similar to the one used by Usami et al. (1998b) was included in order to enable level conversion at the inputs and outputs of combinational blocks as described in Sect. 3. Furthermore, we classified the input and output pins of all cells such that ouput pins of low voltage cells are not allowed to drive input pins of high voltage cells.

For SDP to properly calculate the power consumption in the presence of two supplies, we modeled P_{dyn} for each cell individually. While cell-internal look-up tables are normally used for modeling only cell-internal dynamic power (Ackalloor and Gaitonde, 1998), we used them for modeling all the dynamic power. For a more detailed discussion of tool-specific DSV library modeling issues see Mahnke et al. (2002a).

The 0.25 μm CMOS library was used for implementing MCNC benchmark circuits (see Sect. 5). For the implementation of the embedded microcontroller (see Sect. 6), we developed another DSV library based on National Semiconduc-

Table 1. Results of the optimization of combinational benchmarks. Col. 2/3: circuit complexity. Col. 4: power reduction due to SSV power optimization. Col. 5/6: additional power reduction due to the use of a second supply voltage

	Number of		Power red.	Power red.	
	cells	I/O	SSV	DSVS	CVS
apex6	742	135/99	-26%	-10%	-8%
c432	184	36/7	-26%	-3%	±0%
c880	416	60/26	-22%	-12%	-4%
c1908	275	33/25	-30%	-7%	-6%
c3540	1077	50/22	-25%	-5%	-1%
c5315	1413	178/123	-23%	-12%	-9%
c6288	3040	32/32	-20%	-6%	-2%
c7552	1462	207/108	-17%	-9%	-5%
i10	1879	257/224	-28%	-14%	-11%
i5	373	133/66	-28%	-5%	-5%
my_adder	169	33/17	-24%	-13%	-6%
pair	1509	173/137	-26%	-9%	-8%
rot	708	135/107	-28%	-13%	-10%
x3	718	135/99	-28%	-20%	-11%
x4	368	94/71	-35%	-12%	-9%
avg.	–		-26%	-10%	-6%

tor's 0.18 μm CMOS technology. This library contains high voltage (1.8 V) and low voltage (1.3 V) synthesis models of all logic cells that exist in the original SSV library. Furthermore, high voltage, low voltage and level-converting scan-flip-flops are provided in three different sizes each.

5 Evaluation of the methodology

We applied our methodology to MCNC benchmark circuits (CBL, 2002) subject to reasonably strict delay constraints. In the following discussion, we use the results of state-of-the-art power-driven SSV logic synthesis (see Sect. 2) as reference values in order to reveal the true additional benefit of DSVS. In this paper, we restrict the discussion to a selection of combinational benchmark circuits. For the results of the optimization of sequential benchmark circuits and for a more detailed discussion of delay constraints see Mahnke et al. (2002b).

We optimized the power consumption of 15 combinational MCNC benchmark circuits, firstly, using the state-of-the-art methodology for power-driven logic synthesis (SSV optimization, STEP 1 and STEP 2 in Fig. 2) and, secondly, using our DSVS methodology (STEP 3). The results are summarized in Table 1. Column five shows the advantage of our methodology over SSV power optimization. On average, the final power consumption was 10% lower if DSVS was used. In the best case, the improvement was 20%.

In order to judge the quality of our methodology in comparison with previously published DSVS algorithms, we also implemented the clustered voltage scaling (CVS) algorithm developed by Usami and Horowitz (1995). We performed power optimization using the established SSV methodology first, followed by CVS. Column six of Table 1 shows that the additional power reduction due to CVS was only 6% on average and only 11% in the best case. This is significantly less than the additional power reduction that we achieved using our DSVS methodology.

The fact that we observed less power reduction than other researchers reported can be explained on the basis of a slack distribution analysis. We performed static timing analysis on a number of MCNC benchmark circuits after timing-driven synthesis, after SSV power optimization and after DSV power optimization, thereby assigning to every gate in the netlists the slack of the longest path that contains the respective gate. The results are given in Fig. 3a. In this bar graph, the slack normalized to the delay of the critical path and devided into seven intervals is shown on the horizontal axis. The normalized slack values contained in the figure denote the upper limits of the intervals. There are three bars associated with each slack interval. In each group of three bars, the left bar corresponds to the situation after timing-driven synthesis, the middle bar represents the results of SSV power optimization and the right bar describes the situation after DSV power optimization. The height of each bar is proportional to the number of cells that have a slack in the respective interval.

A similar analysis was carried out by Chen at al. (2001) on their selection of MCNC benchmark circuits after timing-driven synthesis. From the results, which are reproduced in Fig. 3b, Chen et al. concluded that there was a large potential for power reduction using the DSVS technique because of the large number of noncritical cells.

However, from a comparison of the two bar graphs, it is evident that, after timing-driven synthesis (see left bars in Fig. 3a), the benchmark circuits were more timing critical in our work. Consequently, there was less potential for power-delay-tradeoff. This discrepancy must be accredited to the capabilities of the tools used for timing-driven synthesis. Moreover, the extensive use of SSV power optimization techniques, particularly the use of gate sizing, significantly increased the number of critical cells (see middle bars in Fig. 3a) and, hence, reduced the optimization potential even further. As a result, the increase of the number of critical cells during DSV power optimization (see right bars in Fig. 3a) and the additional power reduction was comparatively small.

6 Application to an embedded microcontroller

We ported our methodology to National Semiconductor's standard ASIC design environment and applied it to the 16-bit CompactRISC™ (CR16) microprocessor core module. The CR16 is usually implemented as part of embedded microcontroller systems, which typically include a numerous peripheral modules such as bus controllers, timers, interrupt controllers, memory controllers, memory (e.g. cache, RAM, ROM) and a variety of interfaces (e.g. USB, I2C, Mi-

crowire). Recently developed applications comprising such microcontrollers are a keyboard and power management controller for notebooks and information appliances, DECT and Bluetooth baseband controllers, and a digital color image processor.

In our work, we synthesized the CR16 core module to National Semiconductor's $0.18\,\mu$m CMOS technology for operation at a nominal supply voltage of $1.8\,V$. The timing-driven synthesis was performed subject to the strictest timing constraints. For the DSV power optimization, the second supply voltage was set to $1.3\,V$. Some important characteristics of our experimental CR16 implementation are a clock frequency of 100 MHz and a complexity of approximately 14000 cells. Following National Semiconductor's common standards for CR16 implementations, the module was prepared for the scan test method and gated clocks were used for dynamic power reduction. In order to make scan testing of the DSV implementation possible, we developed level-converting flip-flops that support the scan test method.

In a first set of experiments, we performed timing-driven synthesis followed by SSV and DSV power optimization assuming a high voltage clock signal. The results show that this module had only limited optimization potential. The SSV power optimization, for instance, reduced P_{dyn} by only 11% and DSVS yielded only 4% additional power reduction.

In a second set of experiments, we extended the voltage scaling approach to the clock network in order to achieve additional power reduction and, thus, improve the optimization results. For this purpose, we disabled the use of high voltage flip-flop cells, so that the SSV implementations contained only level-converting flip-flops and the DSV implementations contained only low voltage and level-converting flip-flops. Under these circumstances, the signal level in the clock network could be safely reduced from V_{DD} to V_{DDL}.

The substitution of conventional high voltage flip-flops with their level-converting counterparts generally creates delay and power overheads, since the level-converting cells are slower and consume more cell-internal dynamic power. In the case of the CR16 core module, the performance penalty was only 2% while the power overhead was 5%. On the other hand, the large number of level converters improved the efficiency of DSVS in the logic, i.e. the dynamic power was reduced by 7% instead of 4%. This partially compensated for the power overhead.

Overall, the dynamic power of the DSV implementation with a low voltage clock was 5% lower than the dynamic power consumption of the power-optimized SSV implementation with a high voltage clock. Clearly, the use of clock voltage scaling did not lead to a significant improvement over DSVS without clock voltage scaling. The reason for this is that, with gated clocks, the clock network accounted for only 7% of the total dynamic power and, hence, even a significant reduction of the power in the clock network (e.g. 50%) results in very little reduction of the total power (e.g. 3%).

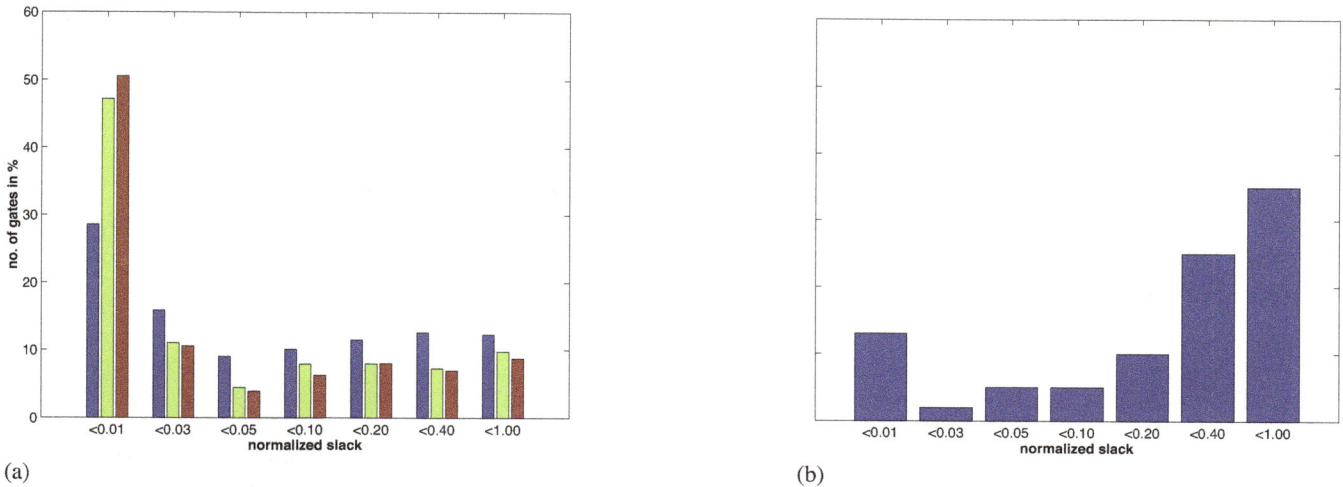

(a)   (b)

Fig. 3. Slack statistics (**a**) for a subset of ten combinational benchmark circuits implemented and optimized using the methodology presented in this paper, and slack statistics (**b**) for 16 combinational benchmark circuits published by Chen at al. (2001).

7 Conclusions

We have shown that DSVS can be carried out exploiting existing gate sizing tools, provided that a suitably modeled DSV standard cell library exists. The required DSV synthesis library file can easily be created from two conventional SSV libraries. The only costly task remaining is the design of the level-converting flip-flop cells, which, of course, is required by any DSV design methodology. Consequently, our methodology can be adopted with a modicum of effort.

A slack distribution analysis has shown that timing-driven synthesis using standard synthesis tools usually generates netlists that contain a large number of timing-critical cells. Moreover, the number of critical cells further increases if state-of-the-art SSV power optimization is used. Therefore, the true additional benefit of DSVS is smaller than claimed by other researchers. On average, we observed an additional power reduction of 10%.

For a comparison with related work, the selection of circuits, the quality of the timing-driven synthesis, the technology and the library, the supply voltages and the use of state-of-the-art power optimization techniques had to be taken into account. For this reason, we implemented a previously published DSVS algorithm and applied it to benchmark circuits within our synthesis environment. The results revealed a greater efficiency of our approach.

Our methodology supports clock voltage scaling for power reduction in the clock network. The performance penalty due to level-converting flip-flops in critical paths was small in the case of the CR16 microprocessor core. However, clock voltage scaling turned out to be inefficient for a design that makes extensive use of gated clocks.

References

Ackalloor, B. and Gaitonde, D.: An overview of library characterization in semi-custom design, IEEE Custom Integrated Circuits Conf., 305–312, 1998.

Chen, C., Srivastava, A., and Sarrafzadeh, M.: On gate level power optimization using dual-supply voltages, IEEE Trans. VLSI Systems, 9, 616–629, 2001.

Collaborative Benchmarking Laboratory (CBL), LGSynth93 benchmark set used in conjunction with 1993 MCNC International Workshop on Logic Synthesis, http://www.cbl.ncsu.edu, June 2002.

Coudert, O.: Gate sizing for constrained delay/power/area optimization, IEEE Trans. VLSI Systems, 5, 465–472, 1997.

Mahnke, T., Embacher, M., Stechele, W., et al.: Power optimization through dual supply voltage scaling using Power Compiler, Proc. European Synopsys Users Group Meeting, 2002a.

Mahnke, T., Stechele, W., and Hoeld, W.: Dual supply voltage scaling in a conventional power-driven logic synthesis environment, Proc. Int. Workshop on Power and Timing Modeling, Optimization and Simulation (PATMOS), 146–155, 2002b.

Synopsys Inc.: Design power and power compiler technology backgrounder, 1998, http://www.synopsys.com/products/power/power_bkg.html, 2001.

Usami, K. and Horowitz, M.: Clustered voltage scaling technique for low-power design, Proc. Int. Symp. Low-Power Design, 3–8, 1995.

Usami, K., Igarashi, M., Minami, F., et al.: Automated low-power technique exploiting multiple supply voltages applied to a media processor, IEEE Journal of Solid-State Circuits, 33, 463–472, 1998.

Usami, K., Igarashi, M., Ishikawa, T., et al.: Design methodology of ultra low-power MPEG4 codec core exploiting voltage scaling techniques, Proc. 35th DAC, 483–488, 1998.

Yeh, C., Chang, M.-C., Chang, S.-C., et al.: Gate-level design exploiting dual supply voltages for power-driven applications: Proc. Design Automation Conf., 68–71, 1999a.

Yeh, C., Kang, Y.-S., shieh, S.-J., et al.: Layout techniques supporting the use of dual supply voltages for cell-based designs, Proc. Design Automation Conf., 62–67, 1999b.

21

An Adiabatic Architecture for Linear Signal Processing

M. Vollmer and J. Götze

University of Dortmund, Information Processing Lab., Germany

Abstract. Using adiabatic CMOS logic instead of the more traditional static CMOS logic can lower the power consumption of a hardware design. However, the characteristic differences between adiabatic and static logic, such as a four-phase clock, have a far reaching influence on the design itself. These influences are investigated in this paper by adapting a systolic array of CORDIC devices to be implemented adiabatically.

We present a means to describe adiabatic logic in VHDL and use it to define the systolic array with precise timing and bit-true calculations. The large pipeline bubbles that occur in a naive version of this array are identified and removed to a large degree. As an example, we demonstrate a parameterization of the CORDIC array that carries out adaptive RLS filtering.

1 Functional Simulation of Adiabatic Logic

Adiabatic logic families such as Positive Adiabatic Logic (PFAL, Vetuli et al., 1996) use voltage ramps in order to charge/discharge the capacitances in an energy effcient way. In contrast, traditional static CMOS loads/unloads the capacitances with steep voltage slopes. In addition to two phases where the clock is *high* or *low*, respectively, adiabatic logic uses two more phases of the same duration where the clock transitions in a ramp (Fig. 1, signals ϕ_1 and ϕ_2). See Fischer et al. (2003) for a more detailed description.

The optimum operating frequency of adiabatic logic tends to be lower than that of static CMOS, leading to the desire for highly parallel designs. We will present such a design for linear signal processing in the sequel.

Due to the way basic functional blocks such as *not*, *and* or larger entities like a half-adder are implemented, they are inherently synchronized with the power clock: they sample their inputs in the rising phase of their clock and their outputs are valid during the immediately following high phase.

When basic blocks are connected, they automatically form a pipeline. For example, when two blocks are connected serially, they form a pipeline that can consume one input every clock-cycle and produces one output every clock-cycle with a delay of two phases. The second block needs to sample its inputs while the outputs from the first block are valid. Thus, its clock needs to be in the rising phase while the clock of the first block is in its high phase. In general, an adiabatic circuit therefore needs to provide four synchronized global power clocks such that in every phase all of the four phase kinds (low, rising, high, falling) are available.

Figure 1 shows this situation for two inverters. It also shows the dual rail encoding that adiabatic logic families use: for every input signal, one also needs to provide the logically inverted signal. For a logic *one*, a signal follows its associated power clock, for a logic *zero*, it stays at ground.

To verify the lower energy dissipation of adiabatic logic as compared to static logic, one needs to perform transient simulations on the transistor and wire level with, e.g., SPICE. When designing larger circuits, like the array of CORDICs in the sequel, it is advantageous to first concentrate only on the functional aspects. This allows a simulation to complete much faster and thus mistakes can be made and corrected more quickly.

We have therefore created a simple set of conventions for VHDL that allow the description of logic blocks that are implicitly synchronized with a four phase clock. Figure 2 shows the simplifications relative to Fig. 1 and Fig. 3 shows simulated wave forms. The dual rail encoding is not modeled, and there is only one global clock net. Signals are valid during two phases since VHDL events happen at phase transitions and the sampled signal must be stable at this point. Nevertheless, the phase relations of two blocks can be observed in simulated waveforms and misalignments can be detected.

Figure 4 shows the VHDL code for a adiabatic inverter and Fig. 5 shows how to instantiate the two inverters of Fig. 2. The `phase` generic of a component aligns it with one of the four phases of the global clock.

Building on this conventions, a parameterizable array of CORDIC cells has been developed that can be programmed to carry out a number of signal processing tasks.

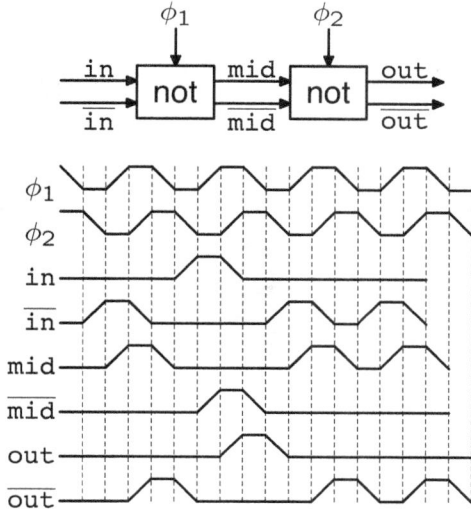

Fig. 1. Two adiabatic inverters.

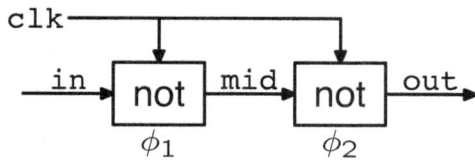

Fig. 2. Two adiabatic inverters in VHDL.

Fig. 3. Simulated wave forms for Fig. 2.

2 Systolic Architecture

The presented architecture is an array of locally connected CORDIC devices that resembles the familiar triangular array for computing a QR decomposition (Haykin, 1996). It is depicted in Fig. 6 together with its inputs and outputs.

Such a device is able to find a transformation Θ and apply it to matrices V_1, V_2, W_1, W_2 of real numbers such that

$$\Theta \begin{bmatrix} V_1 & V_2 \\ W_1 & W_2 \end{bmatrix} = \begin{bmatrix} V_1' & V_2' \\ 0 & W_2' \end{bmatrix}. \tag{1}$$

In this equation, V_1, which must be upper triangular, and V_2 represent the values in the internal registers of the architecture. The matrices W_1 and W_2 represent the input values. The transformation Θ is chosen such that the W_2 matrix is annihilated. After applying Θ, the matrices V_1' (again upper triangular) and V_2' represent the new values of the internal registers, and the matrix W_2' represents the outputs.

```
architecture default of adi_inv is
begin

    process
    begin
      o <= 'X';
      loop
        wait on clk;
        if clk = eval_phase (phase) then
          o <= not i;
        elsif clk = reco_phase (phase) then
          o <= 'X';
        end if;
      end loop;
    end process;

end default;
```

Fig. 4. VHDL code for an adiabatic inverter.

```
inv1: entity work.adi_inv
    generic map (
      phase => ph2)
    port map (
      clk => clk,
      i => s_in,
      o => s_mid);

inv2: entity work.adi_inv
    generic map (
      phase => ph2+1)
    port map (
      clk => clk,
      i => s_mid,
      o => s_out);
```

Fig. 5. VHDL code for instantiating two inverters.

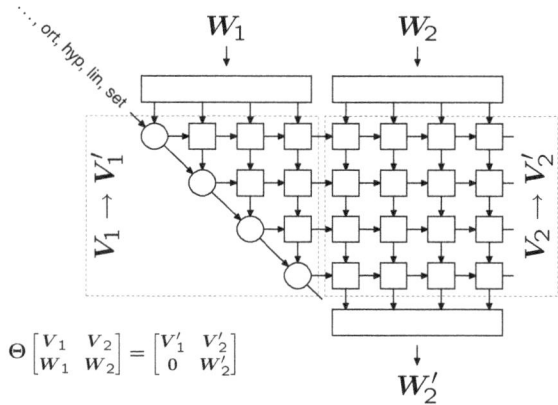

$$\Theta \begin{bmatrix} V_1 & V_2 \\ W_1 & W_2 \end{bmatrix} = \begin{bmatrix} V_1' & V_2' \\ 0 & W_2' \end{bmatrix}$$

Fig. 6. The adiabatic architecture.

The CORDICs can be programmed at run-time to construct transformations Θ with different additional properties, see the next section. A circular cell in Fig. 6 represents a *vector* CORDIC: it finds a elementary 2×2 transformation that annihilates a single element of W_1. A square cell represents a *rotation* CORDIC that applies the elementary transformation found by the circular cells in its row.

The CORDIC cells have an internal feedback loop as depicted in Fig. 7. In a way, these loops create the internal

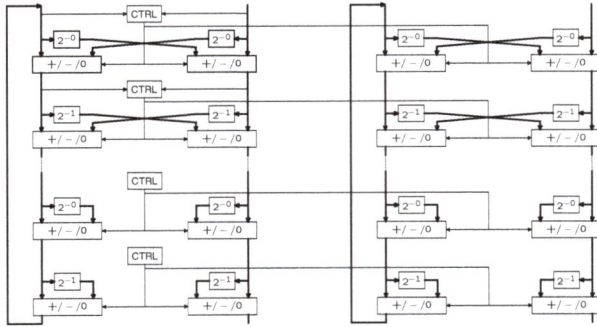

Fig. 7. The CORDIC cells. Left: vector, right: rotation.

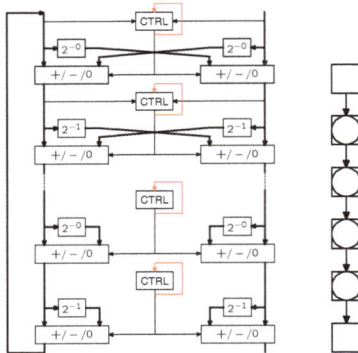

Fig. 8. Area-optimized device.

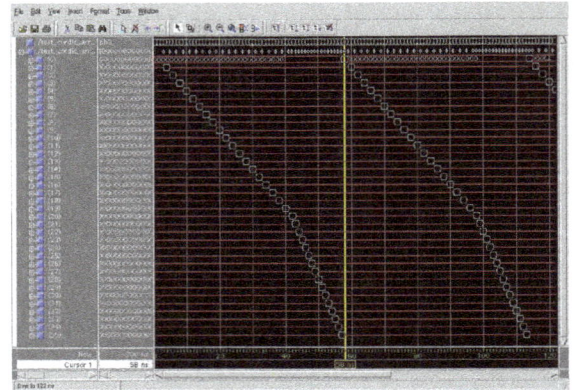

Fig. 9. Pipeline bubbles.

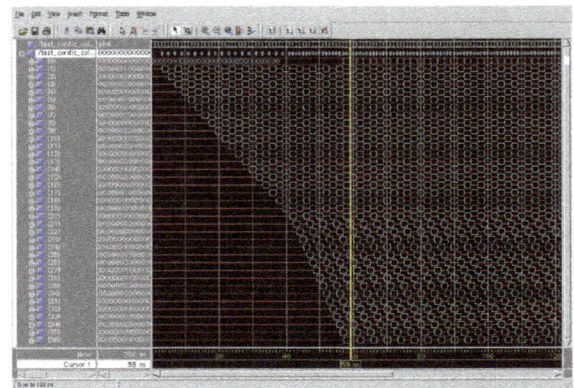

Fig. 10. No pipeline bubbles.

registers that store $\mathbf{V}_1$ and $\mathbf{V}_2$. In a static design, there will be a real register to form the loop, but in an adiabatic design, the adders themselves are pipelined and behave inherently as registers.

The more micro-rotation stages there are in the CORDICs, the longer the pipeline inside each cell becomes. Waiting for the result to be fed back to the input creates large pipeline bubbles. Figure 9 depicts this effect. It shows simulated waveforms of the signals between the stages of one CORDIC. It can be seen that one must wait for a value to slowly ripple through all stages until the next meaningful computation can be started.

To get around this unfortunate situation, one can observe that a given stage in a given CORDIC cell can do the work of its right neighbor during the time it would otherwise sit idle. So instead of distributing the control information to the right, a circular cell can feed this information back to itself and assume the role of its neighbors in subsequent cycles. Those neighbors can be removed from the design. Figure 8 shows the result: a single column of CORDIC devices that can be in both the *vector* and *rotation* modes. Figure 10 confirms that the bubbles have disappeared. The length of the original bubble determines how many columns can be collapsed into one.

3   Linear Transformations for Signal Processing

As mentioned in the previous section, properties of the transformation matrix $\boldsymbol{\Theta}$ in Eq. (1) can be controlled at run-time such that a number of *modes* are created that the array can be in. The modes and possible applications of them are listed below.

– In the *orthogonal* mode, $\boldsymbol{\Theta}$ is chosen to be orthogonal, such that the following relationships can be established:

$$\mathbf{V}_1'^{H}\mathbf{V}_1' = \mathbf{V}_1^{H}\mathbf{V}_1 + \mathbf{W}_1^{H}\mathbf{W}_1$$
and
$$\mathbf{V}_1'^{H}\mathbf{V}_2' = \mathbf{V}_1^{H}\mathbf{V}_2 + \mathbf{W}_1^{H}\mathbf{W}_2. \tag{2}$$

Thus, this mode can be used to carry out a QR decomposition $\mathbf{X} = \mathbf{Q}\mathbf{R}$ of an arbitrary matrix by letting $\mathbf{V}_1 = \mathbf{0}$ and $\mathbf{W}_1 = \mathbf{X}$. Then we will find $\mathbf{V}_1' = \mathbf{R}$. Additionally, we will find $\mathbf{V}_2' = \mathbf{Q}^{H}\mathbf{W}_2$ which allows us to solve the least squares problem $\min_{\boldsymbol{w}} \|\mathbf{X}\boldsymbol{w} - \boldsymbol{y}\|$ by setting $\mathbf{W}_2 = \boldsymbol{y}$ and using a subsequent transformation in the *linear* mode (see below).

This mode can also be used to perform a QR updating step, which is most concisely expressed as starting from an upper triangular $\mathbf{R}_1$ such that $\mathbf{R}_1^H\mathbf{R}_1=\mathbf{X}_1^H\mathbf{X}_1$ and efficiently finding an upper triangular $\mathbf{R}_2$ such that $\mathbf{R}_2^H\mathbf{R}_2=\mathbf{X}_1^H\mathbf{X}_1+\mathbf{X}_2^H\mathbf{X}_2$. This can be achieved by letting $\mathbf{V}_1=\mathbf{R}_1$ and $\mathbf{W}_1=\mathbf{X}_2$, leading to $\mathbf{V}_1'=\mathbf{R}_2$. $\mathbf{V}_2$ and $\mathbf{W}_2$ can be used in the same manner as above to update the right hand side such that the solution to a least-squares problem can be updated.

Note that the QR updating step can be used repeatedly to compute an upper triangular $\mathbf{R}$ such that $\mathbf{R}^H\mathbf{R}=\sum_i \mathbf{X}_i^H\mathbf{X}_i$ without having to access the internal registers, apart from the initialization $\mathbf{V}_1=\mathbf{0}$. In fact, the device effectively computes the QR decomposition of $\mathbf{X}$ by successively updating the solution one row at a time until all rows of $\mathbf{X}$ have been accounted for.

Channel estimation, channel equalization, data detectors and adaptive filters can use the methods mentioned above, for example.

— In the *linear* mode, $\boldsymbol{\Theta}$ has the form

$$\boldsymbol{\Theta} = \begin{bmatrix} \mathbf{I} \\ \boldsymbol{\Delta} \ \mathbf{I} \end{bmatrix} \tag{3}$$

leading to the computation of the *Schur Complement*

$$\mathbf{W}_2' = \mathbf{W}_2 - \mathbf{W}_1\mathbf{V}_1^{-1}\mathbf{V}_2.$$

This can be used to compute matrix-matrix multiplications, matrix inverses, and solutions to systems of linear equations, and combinations thereof. For example, the first step of solving the least squares problem $\|\mathbf{X}\boldsymbol{w}-\boldsymbol{y}\|$ has left the device in the state $\mathbf{V}_1=\mathbf{R}$ and $\mathbf{V}_2=\mathbf{Q}^H\boldsymbol{y}$ (see above). The solution can be completed with a transformation in the linear mode by letting $\mathbf{W}_1=-\mathbf{I}$ and $\mathbf{W}_2=\mathbf{0}$. Then $\mathbf{W}_2'=\mathbf{R}^{-1}\mathbf{Q}^H\boldsymbol{y}=(\mathbf{X}^H\mathbf{X})^{-1}\mathbf{X}^H\boldsymbol{y}$. By letting $\mathbf{X}$ be square, this method can clearly be used to solve arbitrary systems of linear equations with an arbitrary number of right hand sides.

To compute the arbitrary matrix-matrix product $\boldsymbol{AB}$, one can start with $\mathbf{V}_1=\mathbf{0}$, $\mathbf{V}_2=\mathbf{0}$ and input $\mathbf{W}_1=\mathbf{I}$, $\mathbf{W}_2=\boldsymbol{B}$ in a orthogonal run, leading to $\mathbf{V}_1=\mathbf{I}$, $\mathbf{V}_2=\boldsymbol{B}$. A subsequent linear run with $\mathbf{W}_1=\boldsymbol{A}$, $\mathbf{W}_2=\mathbf{0}$ will yield $\mathbf{W}_2'=\boldsymbol{AB}$.

This mode is useful for filters and other signal transformations such as an FFT, for example.

— In the *hyperbolic* mode, $\boldsymbol{\Theta}$ fulfills

$$\boldsymbol{\Theta}^H\mathbf{J}\boldsymbol{\Theta}=\mathbf{J} \quad \text{with} \quad \mathbf{J}=\begin{bmatrix} \mathbf{I} \\ & -\mathbf{I} \end{bmatrix}. \tag{4}$$

This mode can be used to compute a QR downdating step, which can reverse a updating step. Similar to the updating step described above, an application of the device in hyperbolic mode will give us the upper triangular $\mathbf{R}_2$ such that $\mathbf{R}_2^H\mathbf{R}_2=\mathbf{X}_1^H\mathbf{X}_1-\mathbf{X}_2^H\mathbf{X}_2$ when starting from $\mathbf{R}_1^H\mathbf{R}_1=\mathbf{X}_1^H\mathbf{X}_1$.

The hyperbolic mode can also be used to carry out one step of the *Schur Algorithm* and can thus be used to efficiently compute the QR decomposition of a matrix with a Toeplitz-derived structure (Kailath and Chun, 1994). These matrices appear in time-invariant single- and multi-user systems (Vollmer et al., 1999, 2001).

— The *set* mode is provided to initialize the array. It performs the assignments

$$\mathbf{V}_1' = diag(\mathbf{W}_1) \quad \text{and} \quad \mathbf{V}_2' = \mathbf{0}$$

where $\mathbf{W}_1 \in \mathbb{R}^{1\times n}$ is a row-vector whose elements are put on the diagonal of $\mathbf{V}_1'$.

Initializing the diagonal of $\mathbf{V}_1$ is useful to compute the *best linear estimator* in white noise, for example, which is similar to a least-squares solution and is given by the formula

$$\boldsymbol{w} = (\mathbf{X}^H\mathbf{X} + \sigma^2\mathbf{I})^{-1}\mathbf{X}^H\boldsymbol{y}.$$

Setting $\mathbf{V}_1=\sigma\mathbf{I}$, $\mathbf{V}_2=\mathbf{0}$ via a run in the set mode, and then inputting $\mathbf{W}_1=\mathbf{X}$, $\mathbf{W}_2=\boldsymbol{y}$ for a run in the orthogonal mode will lead to $\mathbf{V}_1=\mathbf{R}$ such that $\mathbf{R}^H\mathbf{R}=\sigma^2\mathbf{I}+\mathbf{X}^H\mathbf{X}$ and $\mathbf{V}_2=\mathbf{R}^{-H}\mathbf{X}^H\boldsymbol{y}$. This can be transformed into the desired solution with a final run in the linear mode: as previously, $\mathbf{W}_2=-\mathbf{I}$ and $\mathbf{W}_2=\mathbf{0}$ will lead to $\mathbf{W}_2'=\mathbf{V}_1^{-1}\mathbf{V}_2=\mathbf{R}^{-1}\mathbf{R}^{-H}\mathbf{X}^H\boldsymbol{y}=\boldsymbol{w}$.

— The *copy* mode finally can be used to retrieve $\mathbf{V}_2$ in case it is needed, such as with the Schur algorithm or when the $\mathbf{Q}$ factor of a QR decomposition is needed explicitly. It sets

$$\mathbf{V}_1' = \mathbf{V}_1, \quad \mathbf{V}_2' = \mathbf{V}_2, \quad \mathbf{W}_2' = \mathbf{V}_2.$$

4 Example: Adaptive RLS Filter

A adaptive RLS filter alternates between estimating and equalizing the transmission channel. The filter that equalizes the channel is modeled as a FIR filter and the estimation is performed while a known training sequence $\boldsymbol{y}_1$ is transmitted. The coefficients $\boldsymbol{w}$ of the equalization filter are chosen such that $\|\mathbf{X}_1\boldsymbol{w}-\boldsymbol{y}_1\|$ is minimized where $\mathbf{X}_1$ is the convolution matrix of the signal received during the training period. The convolution matrix $\mathbf{X}$ of a sequence $\{\ldots, x_{i-1}, x_i, x_{i+1}, \ldots\}$ has a Toeplitz structure:

$$\mathbf{X} = \begin{bmatrix} x_i & x_{i-1} & x_{i-2} & \cdots \\ x_{i+1} & x_i & x_{i-1} & \cdots \\ x_{i+2} & x_{i+1} & x_i & \cdots \\ \vdots & \vdots & \vdots & \end{bmatrix}$$

The equalization is then performed by computing $\boldsymbol{y}_2=\mathbf{X}_2\boldsymbol{w}$ where $\mathbf{X}_2$ is the convolution matrix of the received signal during the payload period.

The CORDIC device presented above is well suited to carry out this task. The estimation phase first sets

$\mathbf{V}_1=\mathbf{0}$, $\mathbf{V}_2=\mathbf{0}$ and then carries out the QR decomposition of $\mathbf{W}_1=\mathbf{X}_1$ and $\mathbf{W}_2=\mathbf{y}_1$ as explained above, giving $\mathbf{V}_1=\mathbf{R}$ and $\mathbf{V}_2=\mathbf{Q}^H\mathbf{y}$. The equalization phase runs a subsequent linear mode transformation with $\mathbf{W}_1=-\mathbf{X}_2$, $\mathbf{W}_2=\mathbf{0}$, computing $\mathbf{W}_2'=\mathbf{X}_2\mathbf{R}^{-1}\mathbf{Q}^H\mathbf{y}_1=\mathbf{X}_2\boldsymbol{w}=\boldsymbol{y}_2$.

The convolution matrices $\mathbf{X}_1$ and $\mathbf{X}_2$ are constructed implicitly by connecting the outputs of a delay-line to the inputs of the device. The device can be simply switched from the training mode to the filter mode by inputting zeros instead of the training sequence and switching the mode from *orthogonal* to *linear*.

In order to allow the filter to gradually forget the past, it is customary to change the scaling factor in of each CORDIC such that each elementary orthogonal rotation reduces the length of the involved vector by a factor of 0.97, say.

5 Conclusions

The well-known systolic QR array can be generalized to also be able to compute a wide variety of linear signal processing tasks. Implementing this generalized array with adiabatic logic offers opportunities for significant low-level optimizations that find uses for hardware resource that would otherwise sit idle. The array has been simulated with a bit-true and phase-true VHDL model by making use of a general VHDL package that allows the description of adiabatic logic on a functional level.

The result is a highly parallel, highly efficient data flow processor that can compute things like matrix/matrix products, matrix inverses, solutions to systems of linear equations, QR decompositions, least-squares solutions to overdetermined systems of equations, QR up- and down-dating steps, the core tasks of the Block-Schur algorithm, and Best Linear (Unbiased) Estimates.

References

Fischer, J., Amirante, E., Bargali-Stoffi, A., and Schmitt-Landsiedel, D.: Adiabatic circuits: converter for static CMOS signals, in: Kleinheubacher Berichte 2002, Advances in Radio Sciences, 247–251, 2003.

Haykin, S.: Adaptive Filter Theory, Prentice Hall, third edn., 1996.

Kailath, T. and Chun, J.: Generalized Displacement Structure for Block-Toeplitz, Toeplitz-Block, and Toeplitz-Derived Matrices, SIAM J. Matrix Anal. Appl, 15, 114–128, 1994.

Vetuli, A., Pascoli, S. D., and Reyneri, L. M.: Positive feedback in adiabatic logic, in: Electronics Letters, vol. 32, 1867–1869, 1996.

Vollmer, M., Haardt, M., and Götze, J.: Schur algorithms for Joint Detection in TD-CDMA based mobile radio systems, Annals of Telecommunications (special issue on multi user detection), 54, 365–378, 1999.

Vollmer, M., Haardt, M., and Götze, J.: Comparative Study of Joint-Detection Techniques for TD-CDMA Based Mobile Radio Systems, IEEE J. Select. Areas Commun., 19, 1461–1475, 2001.

22

Improving the positive feedback adiabatic logic family

J. Fischer, E. Amirante, A. Bargagli-Stoffi, and D. Schmitt-Landsiedel

Institute for Technical Electronics, Technical University Munich, Theresienstrasse 90, D-80290 Munich, Germany

Abstract. Positive Feedback Adiabatic Logic (PFAL) shows the lowest energy dissipation among adiabatic logic families based on cross-coupled transistors, due to the reduction of both adiabatic and non-adiabatic losses. The dissipation primarily depends on the resistance of the charging path, which consists of a single p-channel MOSFET during the recovery phase. In this paper, a new logic family called Improved PFAL (IPFAL) is proposed, where all n- and p-channel devices are swapped so that the charge can be recovered through an n-channel MOSFET. This allows to decrease the resistance of the charging path up to a factor of 2, and it enables a significant reduction of the energy dissipation. Simulations based on a $0.13\mu m$ CMOS process confirm the improvements in terms of power consumption over a large frequency range. However, the same simple design rule, which enables in PFAL an additional reduction of the dissipation by optimal transistor sizing, does not apply to IPFAL. Therefore, the influence of several sources of dissipation for a generic IPFAL gate is illustrated and discussed, in order to lower the power consumption and achieve better performance.

1 Introduction

The technology progress together with the system on a chip (SoC) approach leads to an increasing number of transistors on a chip. As a result, even in modern technologies with reduced supply voltage the dynamic power dissipation of integrated circuits continues to increase. As long as the system is mains-operated the cooling of the chip and the generation of a constant DC voltage represent the main challenges. In battery-operated or mobile systems, the dynamic power increase leads to reduced operating time because the capacity of batteries does not increase in the same way. In future, also the leakage losses will become a major concern especially for mobile systems.

Low power digital signal processors work at frequencies up to 200 MHz. In this frequency range, adiabatic circuits allow a significant reduction of the energy dissipation breaking the fundamental limit of static CMOS ($\frac{1}{2}CV_{DD}^2$). The adiabatic logic families using cross-coupled transistors show very low energy consumption and robust operation. One of the first logic families proposed was the Efficient Charge Recovery Logic (ECRL) (Moon et al., 1996). Using not only cross-coupled p-channel transistors but a complete latch consisting of cross-coupled inverters leads to the logic family 2N-2N2P (Kramer et al., 1995) and the Positive Feedback Adiabatic Logic (PFAL) (Vetuli et al., 1996; Blotti et al., 2000). Among these three implementations, PFAL exhibits the lowest energy dissipation. A comparison of these logic families can be found in Amirante et al. (2001).

In PFAL, the main dissipation of energy occurs in the p-channel transistors used in the latch. To enhance the conductivity of the p-channel devices their width can be increased. However, there is an upper bound for the device width because the cross-coupled transistors also act as a load capacitance (Fischer et al., 2003). In this paper, we propose an approach where the charging resistance is lowered not by optimal device sizing but replacing p-channel transistors with n-channel transistors. As a consequence, all transistors are swapped and the maximum value of the supply voltage V_{DD} is chosen as reference potential instead of ground. This paper describes the properties of this new logic family which we call Improved PFAL (IPFAL). The energy dissipation is reduced over the whole frequency range compared to PFAL. It is shown that circuit optimization in dependence of the fanout is different from the original PFAL. As the charging path is less dissipative other mechanisms exhibit significant contributions to the total energy loss.

2 Adiabatic logic families

Static CMOS circuits use pull-up and pull-down networks connected to the DC voltages V_{DD} and ground. As a consequence, the load capacitance C_L is charged or discharged

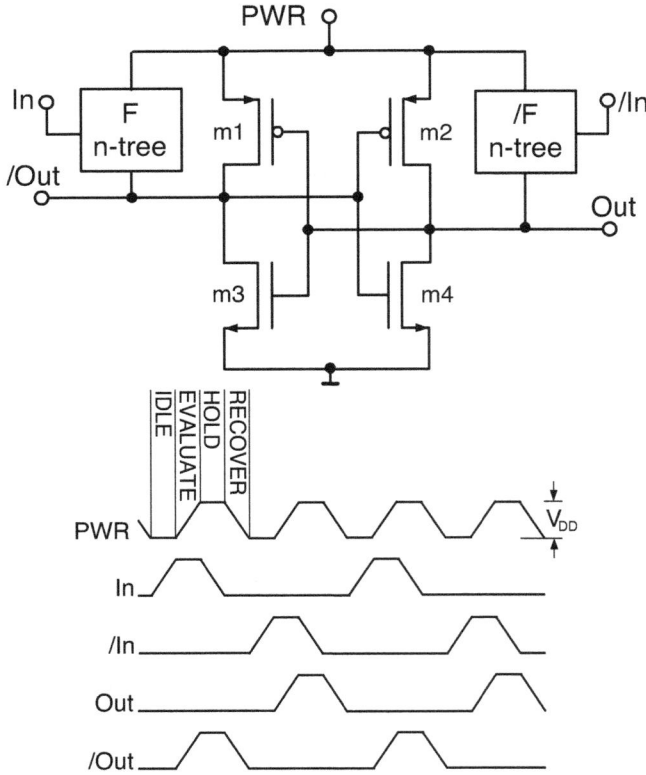

Fig. 1. Top: Timing diagram of a PFAL inverter with the oscillating power supply PWR, the dual rail encoded input signal (In and /In) as well as the dual rail encoded output signal (Out and /Out). Bottom: General schematic of the Positive Feedback Adiabatic Logic (PFAL).

with a voltage step, leading to an energy consumption of $\frac{1}{2}C_L V_{DD}^2$. In adiabatic circuits instead, the switching operation is optimized with respect to energy dissipation. An oscillating power supply is used, as shown in Fig. 1. The loading and unloading of the load capacitances is performed with voltage ramps which result in a minimized voltage drop over the switching transistors. The voltage ramp is provided by the oscillating power supply PWR. The evaluate (charge) and the recover (discharge) phase should take the same time T_{charge} as the hold phase in which the valid output signals are evaluated by the succeeding stages. An idle phase of the same duration time is inserted, in order to achieve a symmetric trapezoidal waveform, which is advantageous for signal generation. A system is powered by four power supply signals with a phase shift of 90°. The energy dissipation per switching operation of a gate supplied by a trapezoidal waveform with $T_{\text{charge}} \gg R_{\text{charge}} C_L$ can be determined according to the equation

$$E_{adiab} = \frac{R_{\text{charge}} C_L}{T_{\text{charge}}} C_L V_{DD}^2 \qquad (1)$$

where R_{charge} is the resistance of the charging path.

Moreover, energy can only be recovered as long as the transistor in the charging path is conducting. When the sup-

ply voltage goes below the threshold voltage V_{th}, the transistor is turned off and a residual charge remains on the output node. This charge is discharged non-adiabatically at the beginning of the next cycle when the new logic value is evaluated. This term does not depend on the operating frequency and represents the main part of the non-adiabatic dynamic losses (Fig. 2, top). The energy dissipated in such a non-adiabatic discharge process is equal to

$$E_{Vth} = \frac{1}{2} C_L V_{th}^2 \qquad (2)$$

Other parts of the non-adiabatic dynamic losses derives from coupling effects whose contributions to the energy dissipation depend on the topology.

Among the MOSFET-only logic families, PFAL shows the best properties in the high frequency range (Amirante et al., 2001, 2002). Figure 1 shows the general schematic of a PFAL gate. The input n-channel transistors evaluating the logic function F are connected between the oscillating power supply and the output nodes. The cross-coupled inverters (latch) drive the dual-rail encoded output signals Out and /Out. The timing of the signals is explained with an inverter as example (timing diagram of Fig. 1). The power supply PWR has a phase shift of 90° compared to the dual-rail encoded input signals In and /In. When the input signal In is Low (/In is High) the output signal Out follows the oscillating power supply PWR whereas /Out stays at ground and vice versa.

The figure of merit for the evaluation of adiabatic circuits is the energy dissipation per cycle. In Fig. 2 (bottom), the energy dissipation per cycle of a 8bit PFAL Ripple-Carry Adder (RCA) is compared to the corresponding implementation in static CMOS. At high operating frequency, the energy consumption of static CMOS is frequency independent whereas in adiabatic circuits the dissipation is inversely proportional to the charging time according to Eq. (1) and therefore directly proportional to the operating frequency. In the low frequency range, both circuits are affected by leakage currents. In PFAL, the off-resistance of the n-channel transistors connected to V_{SS} (m3 and m4) determines the leakage losses. This part of the energy dissipation characteristic is inversely proportional to the frequency. Up to a operating frequency of 100 MHz, the 8bit PFAL RCA exhibits a large energy saving factor compared to the static CMOS implementation. At $f = 20\,\text{MHz}$, a energy saving factor up to 7 can be achieved, as was confirmed by measurements in a $0.13\,\mu\text{m}$ CMOS technology (Amirante et al., 2003).

3 Improving PFAL

The energy dissipation of PFAL is simulated in a $0.13\,\mu\text{m}$ CMOS technology with BSIM 3V3.2 parameters by means of the inverter chain shown in Fig. 3. Only the energy dissipation of the third stage was taken into account. The first two stages provide a realistic input signal, whereas the last two represent the load capacitance.

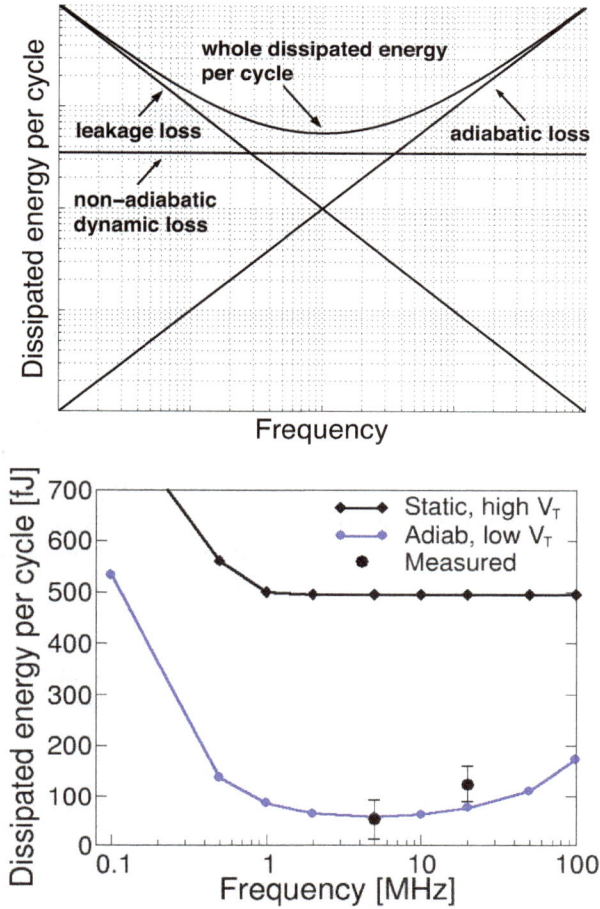

Fig. 2. Top: Different sources of the energy dissipation in adiabatic logic gates. Bottom: Comparison of a static CMOS implementation of an 8-bit Ripple Carry Adder (RCA) with high V_T transistors to an adiabatic implementation in PFAL with low V_T transistors, both simulated, and measurement result for the adiabatic implementation. The 8-bit PFAL RCA dissipates less energy up to an operating frequency of about 100 MHz (Amirante et al., 2003).

The resistance of the charging path R_{charge} is mainly determined by the p-channel transistor. Since the voltage drop over the transistor is very low the MOSFET works in the linear region. For this operating point the charging path resistance can be approximated by the long channel approximation of the p-channel transistor on-resistance $R_{p,on}$:

$$R_{\text{charge}} \approx R_{p,on} \approx \frac{1}{\mu_p C_{OX} \frac{W_p}{L} V_{GST,avg}} \qquad (3)$$

where μ_p is the hole mobility, C_{OX} the oxide capacitance per unit area, W_P the transistor width, L the channel length and $V_{GST,avg}$ the gate overdrive voltage averaged over the time.

The on-resistance of the p-channel transistors $R_{p,on}$ can be improved choosing a larger width during the design. Using this approach, the charging path resistance can be decreased, however at the cost of increased internal capacitances. Therefore, to increase the p-channel transistor width is a useful ap-

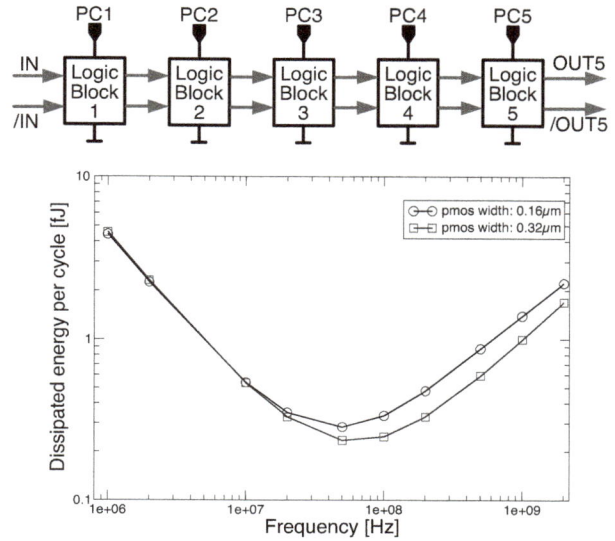

Fig. 3. Top: Energy dissipation is investigated by means of an inverter chain driven by the oscillating power supplies PC1 to PC5. The dissipation of the third stage is determined. Bottom: The energy dissipation of a PFAL inverter strongly depends on the sizing of the p-channel transistors. The n-channel transistor width is kept minimum sized. The fanout value amounts to 10.

proach only if the internal capacitances remain lower than the external load capacitance. A detailed description of the device sizing rules for PFAL gates can be found in (Fischer et al., 2003). Figure 3 shows the energy loss of an inverter with a fanout of 10 in dependence of the p-channel transistor width. For such a high external load capacitance, the energy dissipation is lowered by a factor of 1.5 with double transistor width. At $f = 500 \, \text{MHz}$, the dissipation amounts to $0.88 \, fJ$ with a width of $0.16 \, \mu\text{m}$ compared to $0.60 \, fJ$ using a width of $0.32 \, \mu\text{m}$.

In digital circuits with a low fanout value, larger transistor widths give rise to an increase of the energy dissipation and therefore no improvement is possible by width scaling. As an alternative optimization approach, we propose to replace the p-channel transistors with n-channel devices and vice versa. As reference potential, now the maximum value of the supply voltage V_{DD} is chosen instead of ground (see Fig. 4). This new logic family is called Improved PFAL (IPFAL). The waveform of the oscillating power supply does not change, but now the highest potential corresponds to the idle phase because it is equal to the reference potential. In IPFAL, ramping down the supply voltage means charging the capacitances (evaluate phase) whereas the discharging occurs while the supply voltage is ramping up (recover phase). The hold phase takes place when the voltage reaches its lowest value, that is when the difference to V_{DD} is maximum. Compared to PFAL, this means a swap of the single phases with respect to the supply voltage value. In terms of logic values and in analogy to PFAL, a charged capacitance represents a logical 1 whereas an unloaded capacitance stands for a logical 0. Therefore, the lowest potential represents a logical 1 whereas the highest stands for a logical 0.

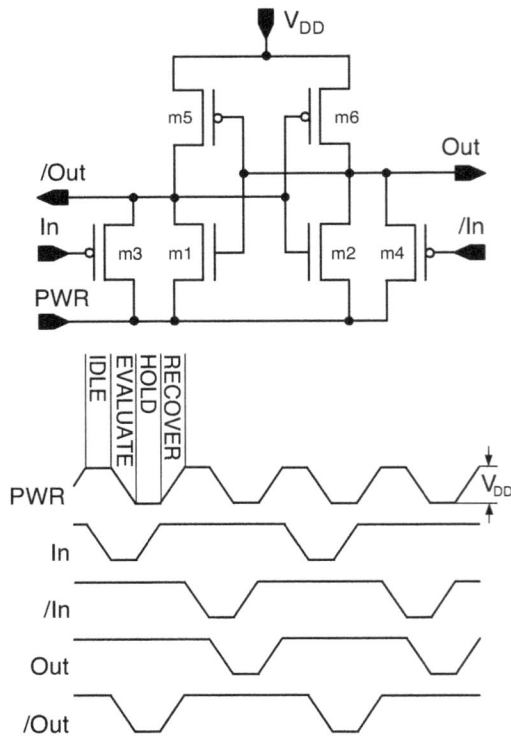

Fig. 4. Schematic (top) and timing diagram (bottom) of an Improved PFAL (IPFAL) inverter.

4 Comparison of the energy dissipation

Figure 5 shows the energy dissipation of PFAL and IPFAL with a fanout of 1 using minimum sized transistors. As expected, the higher conductivity of the n-channel transistors leads to a decrease of the adiabatic energy loss at high frequencies due to the reduced charging path resistance. At $f = 500\,\text{MHz}$, the energy dissipation of PFAL amounts to $0.22\,fJ$ compared to $0.15\,fJ$ in IPFAL. In spite of the expected gain factor of 2 due to the the mobility difference between holes and electrons, only a gain factor of 1.5 is observed showing that additional loss mechanisms play a role in IPFAL. The lower energy dissipation in IPFAL at low frequencies is due to the dependence of the leakage current on the transistors connected to the reference potential (PFAL Fig. 1: m3 and m4; IPFAL Fig. 4: m5 and m6). For equally sized devices, the leakage current is lower in p-channel than in n-channel MOSFETs. Therefore, IPFAL shows a lower energy dissipation than PFAL also in the frequency range dominated by the leakage current. At $f = 1\,\text{MHz}$, the energy dissipation of PFAL amounts to $0.46\,fJ$ compared to $0.25\,fJ$ in IPFAL.

For other fanout values, PFAL offers a simple sizing rule, as only the width of the p-channel transistor in the charging path has to be adapted to the increased external load capacitances. The n-channel transistors m3 and m4 are kept minimum sized, because they provide a good clamp of the output nodes to ground even for this transistor width. In IPFAL, when width of the n-channel transistor in the charg-

ing path is increased, the p-channel transistors m5 and m6 must also provide a good clamp of the output nodes to the reference voltage V_{DD}, which is not achieved keeping the p-channel transistor width to the minimum size. Therefore, when the width of the n-channel devices is enlarged, also the p-channel transistor width have to be increased. In Fig. 5 (bottom), three sizing examples for IPFAL with a fanout value of 10 are compared: minimum sized transistors ($W_p = W_n = W_{\text{minimum}}$), $W_p = 2 \cdot W_n = 2 \cdot W_{\text{minimum}}$ and $W_p = 2 \cdot W_n = 4 \cdot W_{\text{minimum}}$, where W_p is the width of the transistors m5 and m6. Over the whole frequency range IPFAL exhibits lower energy dissipation than PFAL. In IPFAL, the lowest energy dissipation in the high frequency range is achieved with minimum sized transistors. Increasing the n-channel transistor width leads to a decreased charging path resistance. However, due to the larger p-channel transistors the internal capacitances are increased, thus compensating the energy decrease. Hence, it results an increase of the total energy dissipation for larger transistor widths. At low operating frequencies, the circuit with the largest transistor widths exhibit the lowest energy dissipation. For these values of the transistor width, the threshold voltages take their highest values (narrow width effect) and the leakage current is minimized.

As a result, the simple sizing rule used to dimension PFAL gates can not be adopted for IPFAL. Contrary to optimizing just the n-channel transistors in the charging path, a load-dependent gate optimization of the n- and p-channel transistors must be performed due to the several concurrent loss mechanisms and to the $V_{th}(L)$ characteristics of a technology.

5 Conclusions

In modern technologies, adiabatic logic families using only MOSFETs show the lowest energy dissipation. Among these logic families, the Positive Feedback Adiabatic Logic (PFAL) gives the largest energy saving factor with respect to static CMOS circuits. Performance improvements of adiabatic logic gates can be achieved decreasing the charging path resistance. In PFAL, the charging path resistance is determined by p-channel transistors. Widening these transistors leads to a resistance decrease but also to a load capacitance increase due to the cross-coupling of the transistors in the latch. Therefore, an upper bound for the p-channel device width is determined by the ratio of internal and external capacitances.

An Improved PFAL (IPFAL) is proposed, where the p-channel transistors are replaced by n-channel devices and vice versa. The maximum value of the supply voltage V_{DD} is the reference potential. Therefore, the highest potential represents a logical 0 whereas the logical 1 corresponds to the lowest potential. The energy dissipation of IPFAL is reduced by a factor of 1.5 with respect to PFAL over the whole frequency range. In the high frequency range, this additional energy savings was achieved reducing the charging

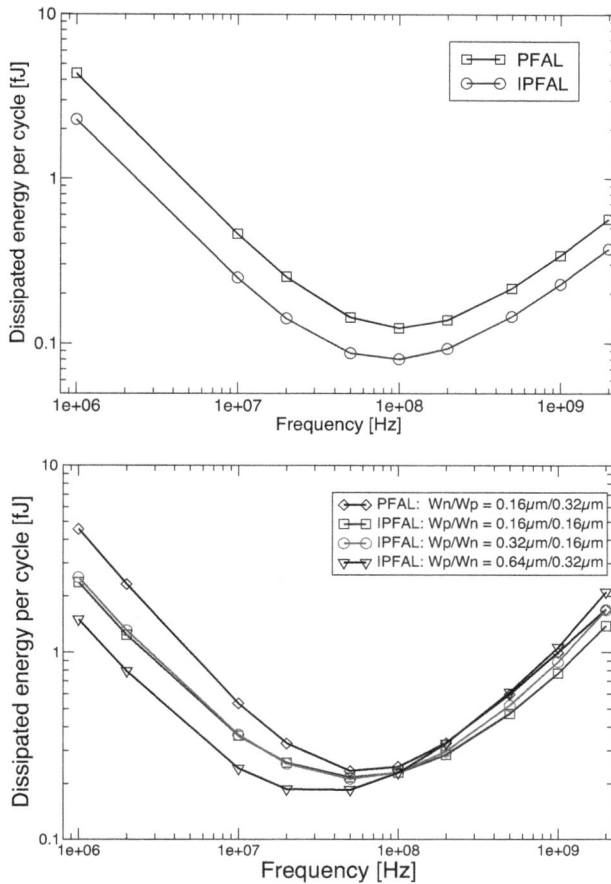

Fig. 5. Top: Comparison of the energy dissipated in PFAL and IPFAL inverter using minimum sized transistors ($W_n / W_p = 1$) and a fanout of 1. Energy dissipation is reduced by a factor of 1.5 in IPFAL because the charging device is a n-channel transistor whose conductivity is higher than that of a p-channel device with the same dimensions. Bottom: Energy dependence on the transistor sizing for PFAL and IPFAL inverters with a fanout value of 10. In PFAL, only the width W_p of the p-channel transistors in the charging path must be adapted to the increased external load. In IPFAL, widening only the n-channel transistors in the charging path is not useful due to concurrent contributions to the energy loss. Therefore, also the p-channel transistor width is increased.

Acknowledgement. This work is supported by the German Research Foundation (DFG) under the grant SCHM 1478/1-3.

References

Amirante, E., Bargagli-Stoffi, A., Fischer, J., Iannaccone, G., and Schmitt-Landsiedel, D.: Variations of the Power Dissipation in Adiabatic Logic Gates, Proceedings of the 11th International Workshop on Power And Timing Modeling, Optimization and Simulation, PATMOS'01, Yverdon-les-Bains, Switzerland, 9.1.1–9.1.10, 2001.

Amirante, E., Bargagli-Stoffi, A., Fischer, J., Iannaccone, G., and Schmitt-Landsiedel, D.: Adiabatic 4-bit Adders: Comparison of Perfomance and Robustness against Technology Parameter Variations, Proceedings of the 45th Midwest Symposium on Circuits and Systems, MWSCAS'02, Tulsa Oklahoma, USA, Vol. III, 644–647, 2002.

Amirante, E., Fischer, J., Lang, M., Bargagli-Stoffi, A., Berthold, J., Heer, Chr., and Schmitt-Landsiedel, D.: An Ultra Low-Power Adiabatic Adder Embedded in a Standard 0.13 μm CMOS enviroment, Proceedings of the 29th European Solid-State Circuits Conference, ESSCIRC'03, Estoril, Portugal, 599–602, September 2003.

Blotti, A., Di Pascoli, S., and Saletti, R.: Simple model for positive-feedback adiabatic logic power consumption estimation, Electronics Letters, 36, 2, 116–118, 2000.

Fischer, J., Amirante, E., Randazzo, F., Iannaccone, G., and Schmitt-Landsiedel, D.: Reduction of the Energy Consumption in Adiabatic Gates by Optimal Transistor Sizing, Proceedings of the 13th International Workshop on Power And Timing Modeling, Optimization and Simulation, PATMOS'03, Turin, Italy, 309–318, September 2003.

Kramer, A., Denker, J. S., Flower, B., and Moroney, J.: 2nd order adiabatic computation with 2N-2P and 2N-2N2P logic circuits, Proceedings of the International Symposium on Low Power Design, 191–196, 1995.

Moon, Y., and Jeong, D.: An Efficient Charge Recovery Logic Circuit, IEEE Journal of Solid-State Circuits, 31, 4, 514–522, 1996.

Vetuli, A., Di Pascoli, S., and Reyneri, L. M.: Positive feedback in adiabatic logic, Electronics Letters, 32, 20, 1867–1869, 1996.

path resistance by the use of n-channel transistors. At low operating frequencies, the energy dissipation is dominated by the leakage current through the p-channel transistors m5 and m6 (Fig. 4). Since p-channel transistors exhibit lower leakage currents than n-channel devices, IPFAL enables a decrease of the dissipation also at low operating frequencies.

For higher fanout values, PFAL gates can be adapted increasing the p-channel transistor width. Unfortunately, this simple sizing rule can not be exploited for IPFAL. Here a load-dependent gate optimization considering both n- and p-channel devices has to be performed. At low fanout values typical for digital systems, Improved PFAL exhibits the lowest energy dissipation among all adiabatic logic families, especially in the high frequency range.

23

A Comparison of current SDRAM types: SDR, DDR, and RDRAM

B. Klehn and M. Brox

Infineon Technologies AG, Munich, Germany

Abstract. The ever increasing demand for bandwidth of computer-systems lead to several standards of SDRAMs. This article compares SDR, DDRI, DDRII, and RDRAM systems. Besides the overall basic innovations, differences will be discussed. Topics like architecture, interfaces, and modules are described.

1 Introduction

Over the last years the clock rate of a microprocessor in a typical desktop PC has exceeded the 2 GHz number. To feed the processor and the other active components in such a system with data, the memory sub-system has to be able to provide the data sufficiently fast. To achieve this goal, multiple innovations had been introduced and integrated into ever more powerful devices. On the other hand, important basic concepts have remained in use over various generations. In this article, we intend to describe differences and commonalities of the various main-memory types and discuss reasons for the changes. In this respect, we will start with a discussion of the basic concept of a DRAM cell.

2 Basic DRAM functionality

The memory cell of a DRAM consists of a capacitor and a select transistor (Fig. 1). The binary information is stored as a charge on the capacitor. Cells are oriented along a matrix of parallel bitlines and perpendicular wordlines. The read-out of a memory cell is shown in Fig. 2. To prepare for the read-out, bitlines are pre-charged to an intermediate level, which generally is set at half of the array operation voltage. The read-out starts with decoding the externally supplied row-address into a physical wordline. Asserting this wordline opens the select transistor. As the voltage on the storage capacitor is either lower (logical '0') or higher (logical '1') the

voltage on the BL changes during this process. As an example, Fig. 2 shows the read-out of a '1' causing the bitline voltage to increase (trace BLt). Typical values for a storage capacitor are in between 20 fF and 40 fF, which is much lower than the value for the parasitic bitline capacitance (between 100 fF and 200 fF). Therefore, the voltage change on the bitline is relatively small – often in the order of some 10 mV, only. To be able to securely read the cell, the voltage change is amplified in the differential sense-amplifier which compares voltage on the bitline BLt with the voltage on a neighbouring, floating bitline BLc. At the end of a sense-amplification, all cells of one wordline are read out and read-out information has been amplified to digital values. Selection of the sense-amplifier to be read is performed through the column-decoder. Dependent on the column-address, one sense-amplifier gets connected to the data-lines over the array, re-amplified in a second amplifier stage and driven towards the output buffers (Fig. 3).

A common method to achieve a continuous data-stream from a DRAM is the so-called burst-mode. Here, multiple data are read-out from one line of sense-amplifiers one after another. The memory controller has to supply only the address of the first read access; the DRAM internally generates all the subsequent column-addresses. In its simplest implementation, however, this continuous stream would have to stop the moment a new wordline would have to be activated. Activation of a different wordline requires bringing down the old wordline, precharging the bitlines and asserting the new wordline.

In order to minimize this severe access penalty, all modern DRAM types are internally arranged as multiple quasi-independent memories. Fig. 3 shows a typical block-level layout for a 256 Mb-SDRAM. Memory cells are organized into four independent memory banks of 64 Mbit, each. Organization into multiple banks allows e.g. a wordline activation in bank 1 to appear concurrently to a read-operation in bank 0. Thus, as long as the row-address sequence supplied from the memory controller is well behaved, a continuous read data-stream can be provided. Yet, in typical applica-

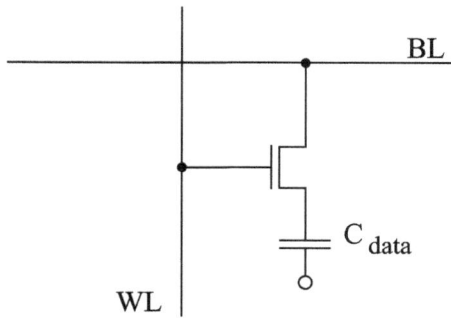

Fig. 1. Schematics of a memory cell build from a storage capacitor and a select transistor. The select transistor connects the capacitor with the bitline BL. Activation of the select transistor is performed through wordline WL.

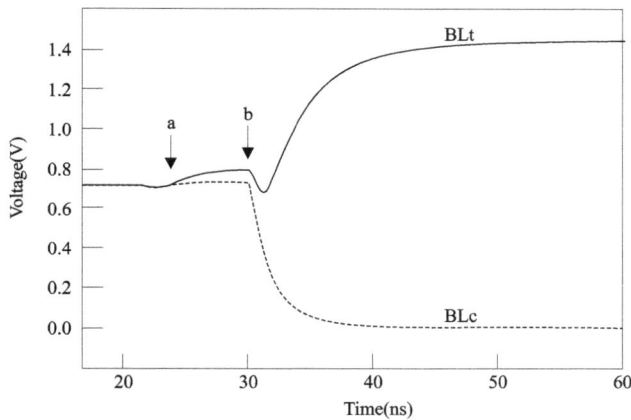

Fig. 2. Bitline voltage as a function of time during read-out of a memory cell. Example shows read-out of a '1' from a cell attached to BLt. At a, wordline gets asserted and charge from the memory cell flows onto the bitline. The neighbouring bitline BLc gets capacitively coupled high. At b, charge transfer from the cell is complete and the sense-amplifier can get activated. After sense-amplification at around 40 ns, digital levels are present on BLt and BLc.

tions, peak bandwidth cannot be sustained. Often word-lines need to be activated in the same bank. In this case the delay encountered by the array pre-charge cannot be hidden. Activation of wordlines, sensing and pre-charging are dominated by parasitic delays as heavily loaded, yet narrow lines in the array need to be driven. Shrinking device dimensions yielded only a performance improvement of a factor around 3 (see Fig. 4). This is rooted in the fact that devices as implemented in DRAMs need to follow other device design guide lines such as e.g. devices in logic processes. As the number of DRAMs in a memory system is often very large, the standby current has to be as low as possible, which forces one to use transistors with some 100 mV of threshold voltage. To close the gap between the processor and the memory, circuit techniques have to be employed, which will be discussed in the following section.

3 Clock synchronization

Older memory devices up to 66 MHz used an external asynchronous interface. The memory itself did not have an external clock input. Any activity was purely edge-triggered and commands could be issued at any arbitrary time as long as the internal time constants of the memory were met. For higher frequency operation, to help ensure reliable input of data, addresses, and commands, an external clock has been used. Typically, the memory samples the input lines at the rising edge of the clock. Commands can, thus, be issued to the memory at only multiples of the clock cycle time. Legal command combinations are described in detail in the datasheet of the manufacturer. Use of an external clock marked the introduction of the so-called SDRAM.

3.1 SDR

Standardized SDRAMs input commands and addresses at the rising edge of the clock. In a first development step, input and output of data was also possible at the rising clock-edge, only (Fig. 5). This concept permits a single data transmission per clock cycle and consequently led to the name of Single-Data-Rate-SDRAM (SDR-SDRAM). The clocksignal itself is single-ended. Output of the data itself was not specifically synchronized to this clock. The only requirement is that following the clock edge for a minimum time of t_{OH} the previous data is held valid, while after t_{AC} new data is available. To give typical values for a 133 MHz-SDR-SDRAM, $t_{OH} = 3$ ns, and $t_{AC} = 5.4$ ns. In between, no valid data is present on the bus.

3.2 DDR

The natural evolution of this concept towards higher bandwidth, is to allow input and output of data at twice the rate. This scheme as shown in Fig. 5 is called Double-Data-Rate (DDR). DDRI and DDRII-SDRAMs derive their name from this data transmission approach.

As illustrated in Fig. 5 for the same clock-frequency, the so-called data-eye (the amount of time data is valid) is only half as wide in DDR as in SDR. Increased precision can be gained through an improved clock system architecture as will be discussed in more detail below. An important device level change, however, is the introduction of an active alignment of the data-output to the clock. For this purpose, Delay-Locked-Loop (DLL) circuits were added to the DDR-devices. In a DLL, the received external clock is purposely delayed such that t_{AC} can be made as close to 0ns as possible. Comparing to SDR-SDRAMs, a typical value for a 133 MHz-DDR-SDRAM is -0.75 ns $< t_{AC} < 0.75$ ns. In a further change, the clock itself is improved. The single-ended clock of the SDR is replaced by a fully-differential clock. As a timing reference, the clock cross-point is defined which is more precisely to determine than the clock edge especially in the presence of significant system noise.

(a)

(b)

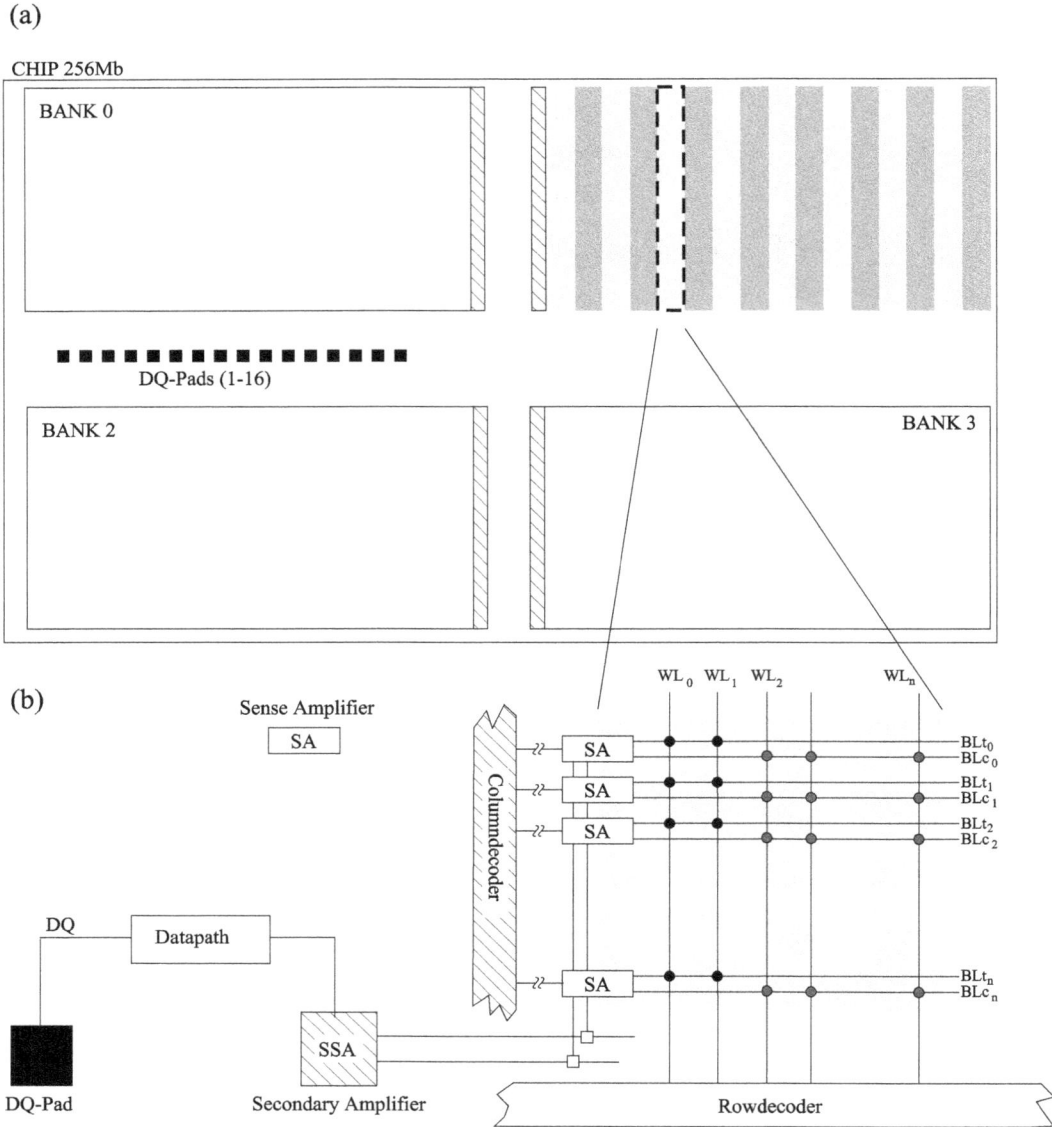

Fig. 3. **(a)** Sketch of the architecture of a typical 256 Mb-DRAM. Memory cells are laid out in four memory banks of 64 Mb. Each bank can be treated as a partially independent memory. **(b)** Sub-division of one 64 Mb-bank into 4 Mb-units. The row-decoder selects one WL. Through column-decoding, one sense-amplifier gets selected to drive data towards the data-path.

4 Prefetch

A memory using the DDR output-scheme has to deliver up to 16bit for each half clock cycle. This could be implemented by reading out the array twice per clock-cycle. At a clock rate of 200 MHz this would be equivalent to a read-out rate of 400 MHz which could be facilitated if the array is partitioned into very small sub-units with short bitlines and short wordlines. Short bitlines, however, require a high number of sense-amplifiers and short wordlines a high number of row-decoders on the chip. A more typical solution is to prefetch 2×16 bit in one array-access for one DDR clock-cycle (more precise: $2 \times$-prefetch). Under this condition, core frequency can be held constant at 200 MHz, while data-

rate achieves 400 Mb/s per pin. Figure 6 illustrates the principle, which also leads to a simple extension path towards even higher bandwidths while still keeping core frequency constant. RDRAM in contrast, uses an $8 \times$-prefetch to reach a data-rate of 1066 Mb/s per pin while running the core at a relaxed 133 MHz-frequency. Time-multiplexing circuitry in the chip has to guarantee only that prefetched data is driven out in the correct order.

5 Differences in the concepts

Table 1 gives an overview over clock-rates, bit-rates and core frequency for the various architecture as presented above. While prefetching is effective in limiting the core frequency

Table 1. Performance summary of the relevant architectures. Typical values for frequency is given

Typ	Clock(MHz)	Core(MHz)	Scheme	Datarate per pin (Mb/s)	Prefetch	Data-width
SDR	133	133	single	133	none	64
DDRI	133	133	double	266	2	64
DDRII	266	133	double	533	4	64
RDRAM	533	133	double	1066	8	16

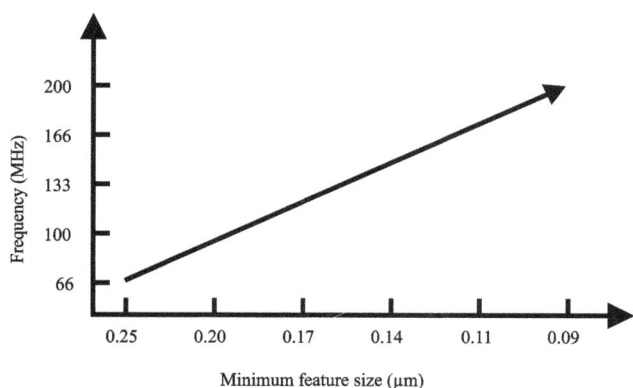

Fig. 4. Operation frequency improvement through smaller feature size. Improvement is around a factor of 3 going from $0.25\,\mu$m to $0.09\,\mu$m.

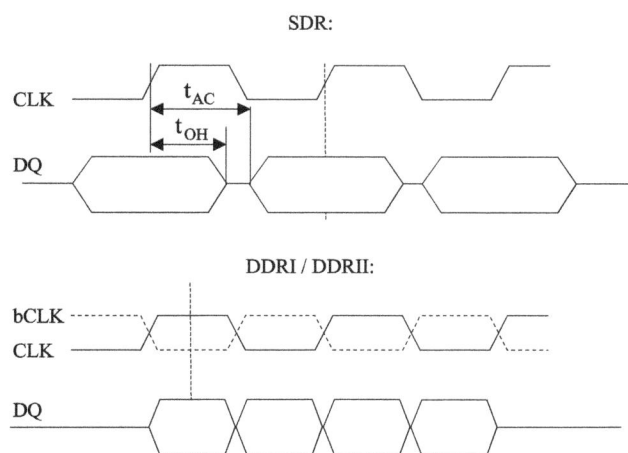

Fig. 5. The graph depicts the external clock inputs for SDR and DDRI/DDRII. The crossing of CLK and bCLK defines the point of time t1 much more fail-safe compared to the edge of a single ended clock.

to sustainable values for each architecture, the frequency of the data lines is differing strongly. Here, architectural differences between the different types are becoming relevant. The following section will touch upon these.

5.1 Commands and requests

The memory controller needs a way to initiate activity in the memory. The classical way is the simple command protocol

as implemented on the evolutionary SDR/DDR family of devices. All control and address information for one command in the DRAM is transmitted during a single clock cycle. This protocol requires typically 5 control- and 15 address-lines to encode all relevant commands (e.g. activate row, precharge row, read column, write column, ...). In this approach, the complexity of the address and control lines is limited because these lines are run in single-data-rate fashion, however, on the expense of a rather large number of lines. Another advantage is the small latency: complete decoding of the command can procede after only one clock cycle. RDRAMs follow a completely different approach. Here, e.g., a single set of eight lines for addresses and control may be used. Three of these lines transmit row oriented requests; five transmit column oriented requests. The three row-oriented lines handle row activation and row precharge. To be able to encode an e.g. 16bit wide address onto 3 lines, the information is time-multiplexed and transmitted in DDR-like fashion. On a first look this introduces multiple disadvantages. Four clock cycles are required to transmit all information required in one access, which increases access latency compared to the one-clock cycle approach of SDR/DDR devices. Furthermore, a capture-DLL needs to be implemented as the data-eye onto the shared control, data, and address lines (the bus) gets too narrow to be received with conventional means. As, however, the number of lines gets very small, the bus can be run at very high frequency while still keeping total system cost at a reasonable level. In addition, seperation of the bus into row-request and a column-request sections enables a very flexible protocol: row and column operations can be initiated independent from and parallel to each other. For example, it is possible to issue a read to bank 0 and an activate to bank 1 in parallel.

5.2 Module design

SDR and DDR systems are addressing multiple devices on a module in a parallel fashion. As an example, Fig. 7 shows the 64 bit data lines from the SDR/DDR controller separating into four 16 bit sets of lines running to four 16 bit wide memories in parallel. In RDRAM, devices are placed serially on the bus. At a given time, only one device can be actively reading or writing. To select one out of multiple devices, this approach requires an additional device identifier besides the conventional row or column address for selection within one memory. Here again, the distinction between wide, lower-frequency SDR/DDR data lines and the narrow,

Prefetch 2 (DDRI):   Chip ⋮ System

Prefetch 4 (DDRII):

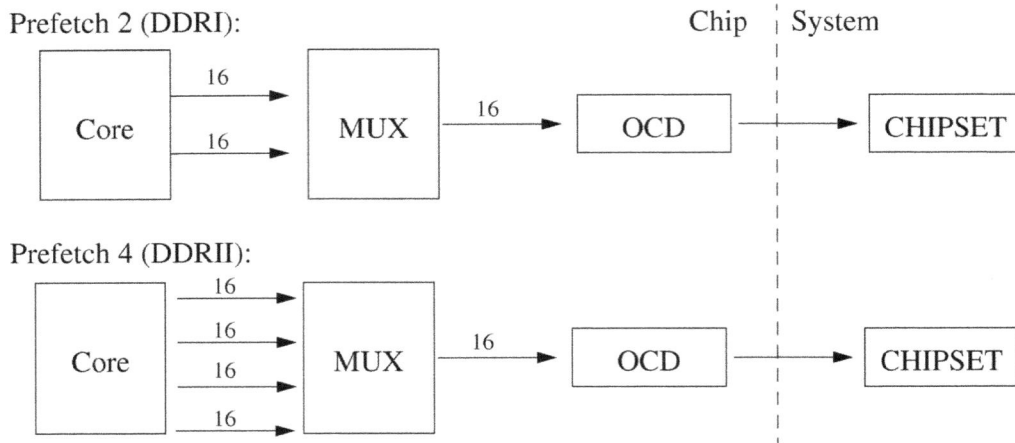

Fig. 6. Prefetch: Capturing of 2x resp. 4x the required data volume and time-multiplexing it onto the output leads to a simple multiplication of the external data-rate of the memory device.

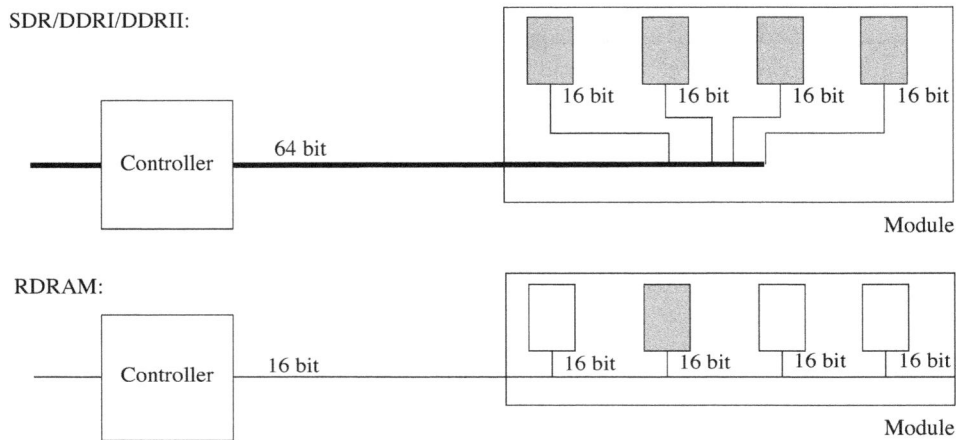

SDR/DDRI/DDRII:

RDRAM:

Fig. 7. SDR- and DDR-modules are operating in desktop-systems with a data-width of 64 bit and address typically 8 devices in parallel. The RDRAM-bus is significantly narrower (16 bit) and only one device is addressed. RDRAM bandwidth is increased by running the bus at a very high clock-frequency.

higher-frequency RDRAM bus gets obvious. On the other hand, RDRAMs are forced to operate at higher clock-rate because for a given clock-rate a DDR-system would be able to deliver four times the RDRAM peak bandwidth simply because of the 64 bit data-width.

5.3 Termination, clocking and strobing

As discussed before, SDR-systems are using a simple, single-ended external clock signal. All data lines are non-terminated and no specific additional means exist to improve signal quality and data valid window. Output drivers are simple voltage source push-pull type drivers. On chip-level, DDR and RDRAM implement DLLs for better alignment of the data output. However, differences do not stop here but continue on the architectural level. Figure 8 shows the clocking and output schemes realized in the different advanced DRAM types.

DDRI uses a mid-level termination of all lines. Addition-

ally, stub-resistors are introduced to limit reflections. A further, important change is the introduction of the data-strobe signal (DQS). DQS is a bi-directional synchronization signal, which is generated by the controller during a write access to the memory and by the memory during a read-access to the controller. In this way, DQS travels in the same direction as the data signal, itself. DQS is generated in a way that each edge (rising and falling) indicates valid data. As the flight time is the same for DQS and DQ, both controller and memory can use the DQS signal to capture incoming data. Different arrival times of the uni-directional always-on clock signal can thus be effectively cancelled out.

DDRII uses the very same DQS-scheme. Only termination is moved onto the chip itself (on-die-termination, ODT). ODT further improves signal quality and reduces total system cost as discrete termination resistors are no longer required. To support multi-module systems, all associated devices need to be able to turn-on or turn-off ODT on-demand.

(a)   (b)

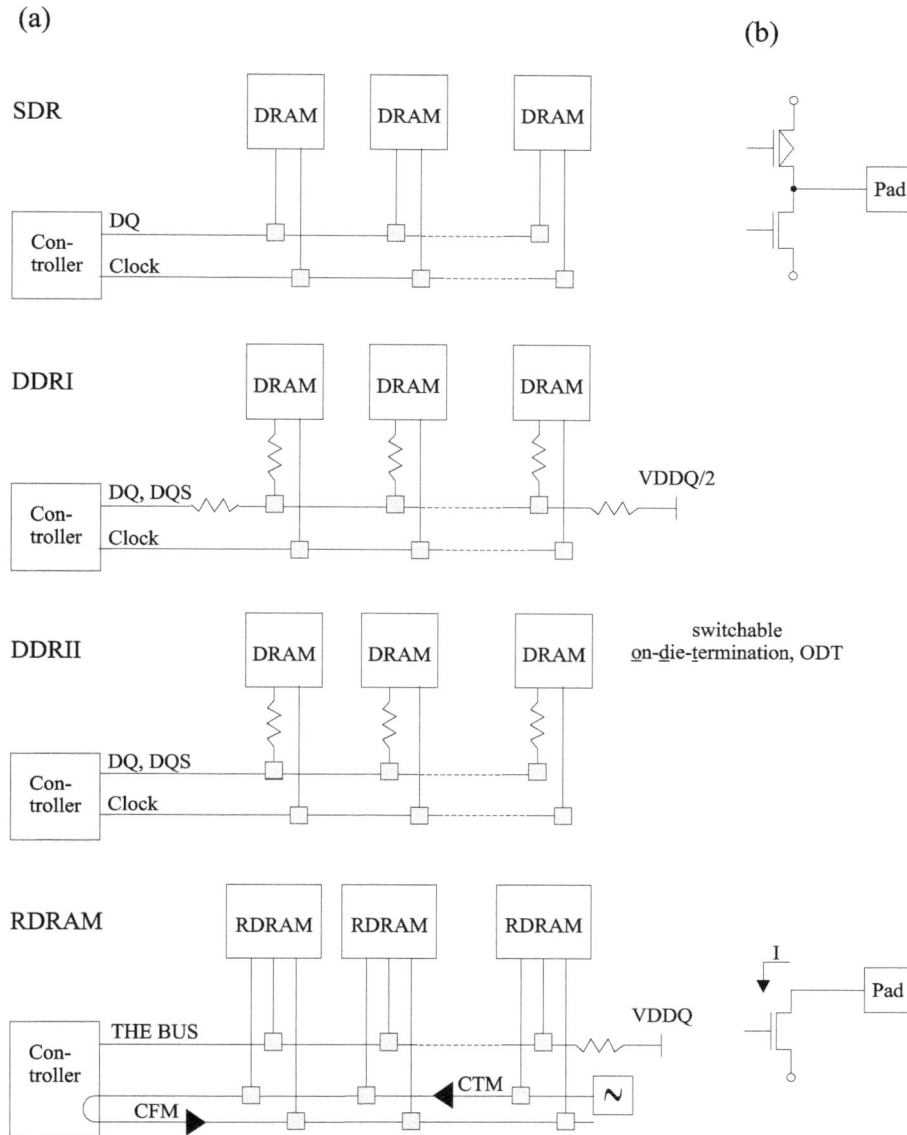

Fig. 8. Data line topologies for the different DRAM types (**a**): SDR, DDRI, DDRII, and RDRAM. (**b**) Simplified final driver stage.

RDRAMs employ a very different scheme. A clock is generated at the far end of the bus. The CTM-clock travels in the direction of the controller, gets mirrored-back at the controller and afterwards travels as CFM-clock on its way back to the far end of the bus. These two clocks can be used to synchronize write data to the memory with CTM, and read-data from the memory with CFM. In this scheme, travel direction of relevant clock and data is the same such that flight time differences are cancelled out. A high level termination is present only at the far-end of the bus such that the controller represents an open end to the bus. At this open end signal reflection occurs, which has to be accounted for in the RDRAM system design. Especially, RDRAM drivers need to be designed with a high differential output resistance such that no second reflection occurs when a reflected wave passes a data-driving memory. This can be guaranteed by designing

the output-driver as a current source, which makes its design very much different than the SDR/DDR voltage mode driver. Signaling levels are currents of 0 mA for a logical '0' and typically 30 mA for a logical '1'. Only on the data lines, the travelling current wave translates to a travelling voltage wave which can be received by memory respective controller.

6 Chip area

If the same technology and number of storage cells are assumed the architecture influences the chipsize. Here, the number of banks is important. Also a higher prefetch needs more datalines running from the banks to the multiplex unit. The number is increased by a factor equal to the prefetch. The 16 data lines from a SDR-SDRAM are doubled for

DDRI and quadrupled for DDRII. The package-oriented communication between a controller and an RDRAM requires a complex decoder logic block in the I/O part. The reduced supply voltages of emerging products adds to the area because of complicated on-chip pump and generator systems.

7   Outlook

It is possible to run the core of an SDRAM with frequencies of 133, 166 or even more than 200 MHz. Combining it with a high prefetch and a large number of parallel devices, data rates in the region of several gigabytes per second are possible. Otherwise to have a noticeable speed up in a computer system all components of such a system have to be balanced. With these bandwidth image processing and simulations are possible at high level. To enter new sophisticated applications a higher operating speed of computer systems is still desirable.

Acknowledgements. The Authors thanks H. Ruckerbauer, R. Klein, and E. Brass for the detailed discussions.

References

Itoh, K.: VLSI Memory Chip Design, Springer-Verlag, Berlin Heidelberg New York, 2001.

Infineon Technologies, SDRAM Datasheets, www.infineon.com/cgi/ecrm.dll/ecrm/scripts/prod_ov.jsp

24

Efficient MAP-algorithm implementation on programmable architectures

F. Kienle, H. Michel, F. Gilbert, and N. Wehn
University of Kaiserslautern, Germany

Abstract. Maximum-A-Posteriori (MAP) decoding algorithms are important HW/SW building blocks in advanced communication systems due to their ability to provide soft-output informations which can be efficiently exploited in iterative channel decoding schemes like Turbo-Codes. Multi-standards demand flexible implementations on programmable platforms.

In this paper we analyze a quantized turbo-decoder based on a Max-Log-MAP algorithm with Extrinsic Scaling Factor (ESF). Its communication performance approximate to a Turbo-Decoder with a Log-MAP algorithm and is less sensitive to quantization effects. We present Turbo-Decoder implementations on state-of-the-art DSPs and show that only a Max-Log-MAP implementation fulfills a throughput requirement of ~2 Mbit/s. The negligible overhead for the ESF implementation strengthen the use of Max-Log-MAP with ESF implementation on programmable platforms.

1 Introduction

Third generation's wireless communication systems comprise advanced signal processing algorithms that increase the computational requirements more than ten-fold over 2G's systems (Third Generation Partnership Project). Numerous existing and emerging standards require flexible implementations ("software radio") (Greifendorf et al., 2002). It is argued in Bickerstaff et al. (2000) that "the trends in decoding algorithms are moving from standard Viterbi towards more computationally-expansive algorithms like soft-output Viterbi algorithm (SOVA) and maximum a posteriori (MAP) algorithm. The implementation efficiency of these algorithms will become a differentiating factor for next generation wireless communications – particularly for those employing programmable DSP-devices."

Turbo-Codes, introduced by Berrou et al. (1993), have near Shannon-limit error correction capacity and are among the most advanced channel-coding algorithms and thus used in many communication standards like the 3GPP standard.

The important innovation was the reintroduction of iterative decoding schemes of convolutional codes by means of soft-output information exchange. The basic building block of a turbo-decoder is the component decoder which provides soft-output information. This information is a measure on the decoder confidence in its decoding decision. These component decoders are typically based on the MAP algorithm.

To reduce the MAP implementation complexity the algorithm is transformed into the logarithm-domain (Log-MAP) (Robertson et al., 1997) which reduces the operation strength but implicates the so-called max^* operation. This operation is composed of a maximum search and a correction term. Discarding the correction term results in the so called Max-Log-MAP algorithm, its implementation is less complex and thus faster but implies a loss in the communication performance of up to 0.3 dB.

The implementation of the max^* operation is especially very time consuming on standard DSP-processors. The additional assembler commands to add the correction term decreases the overall throughput of a Log-MAP implementation by a factor 2–3 compared to a Max-Log-MAP implementation. (the throughput degradation in a dedicated VLSI implementation is less worse). This trade-off between communication performance versus implementation complexity is a typical problem in the design of communication systems. Recently some very advanced DSPs like the TigerSharc from Anlog Devices have implemented a special max^* instruction to support an efficient Log-Map implementation.

In this paper we investigate different MAP-algorithms with respect to their communication performance and implementation complexity on state-of-the-art DSPs. In Vogt et al. (1999) an Extrinsic Scaling Factor (ESF) was proposed to improve the turbo-decoder performance with Max-Log-MAP component decoders. We observe the usage of ESF in a quantized turbo-decoder model and show that the Max-Log-MAP in combination with ESF is less sensitive to quantization effects than the Log-MAP algorithm. We present different turbo-decoder implementations on modern DSPs and state the enormous throughput difference between a Log-MAP and Max-Log-MAP implementation. The low implementation complexity of ESF and Max-Log-MAP and the good communication performance strengthen the use of this combination for turbo-decoder implementations on state-of-the-art DSPs.

The remainder of this paper is structured in two parts: Sect. 2 explains the system model and the MAP algorithm. In Sect. 2.2 we present turbo-decoder performance under quantization and SNR mismatch. Section 3 is devoted to Turbo-Decoder implementation on state-of-the-art DSP architectures: Starcore SC140 from Motorola/Lucent, ST120 from ST-Microelectronics and TigerSharc from AnalogDevices. Section 4 concludes this paper.

2   Turbo-System

Forward error correction is enabled by introducing parity bits. In Turbo-Codes, the original information (x^s), denoted as *systematic information*, is transmitted together with the parity information (x^{1p}, x^{2p}). In the Third Generation Partnership Project (3GPP), the encoder consists of two recursive systematic convolutional (RSC) encoders with constraint length $K = 4$. One RSC encoder works on the block of information in its original, one on an interleaved sequence, see Fig. 1. On the receiver side a corresponding component decoder for each of them exists. The MAP-Decoder has been recognized as the component decoder of choice as it is superior to the Soft-Output Viterbi Algorithm (SOVA) in terms of communications performance and implementation scalability, see Vogt et al. (1999).

The soft-output of each component decoder (Λ) is modified to reflect only its own confidence (z) in the received information bit of being sent either as "0" or "1". These confidences are exchanged between the decoders to bias their next estimations iteratively. During this exchange, the produced information is interleaved following the same scheme as in the encoder. The exchange continues until a stop criterion, see Worm et al. (2000b), is fulfilled. The last soft-output is not modified and becomes the soft-output of the Turbo-Decoder (Λ^2). Its sign represents the 0/1 decision and its magnitude the confidence of the Turbo-Decoder in it.

2.1   The MAP algorithm

Given the received samples of systematic and parity bits (*channel values*) for the whole block (y_0^N, where N is the block length), the MAP algorithm computes the probability for each bit to have been sent as $d_k = 0$ or $d_k = 1$. The logarithmic likelihood ratio (LLR) of these probabilities is the soft-output, denoted as:

$$\Lambda_k = \log \frac{\Pr\{d_k = 1 | y_0^N\}}{\Pr\{d_k = 0 | y_0^N\}}. \qquad (1)$$

Equation 1 can be expressed using three probabilities, which refer to the encoder states S_k^m, where $k \in \{0 \dots N\}$ and $m, m' \in \{1 \dots 8\}$:

The *branch metrics* $\gamma_{m,m'}^{k,k+1}(d_k)$ is the probability that a transition between S_k^m and $S_{k+1}^{m'}$ has taken place. It is derived

from the received signals, the a-priori information given by the previous decoder, the code structure and the assumption of $d_k = 0$ or $d_k = 1$, for details see Robertson et al. (1997). From these branch metrics the probability α_m^k that the encoder reached state S_m^k given the initial state and the received sequence y_0^k, is computed through a forward recursion:

$$\alpha_{m'}^k = \sum_m \alpha_m^{k-1} \cdot \gamma_{m,m'}^{k-1,k}.$$

Performing a backward recursion yields the probability $\beta_{m'}^{k+1}$ that the encoder has reached the (known) final state given the state $S_{m'}^{k+1}$ and the remainder of the received sequence y_{k+1}^N:

$$\beta_m^k = \sum_{m'} \beta_{m'}^{k+1} \cdot \gamma_{m,m'}^{k,k+1}$$

αs and βs are both called *state metrics*. Equation 1 can be rewritten as:

$$\Lambda_k = \log \frac{\sum_m \sum_m' \alpha_m^k \cdot \beta_{m,m'}^{k+1} \cdot \gamma_{m,m'}^{k,k+1}(d_k = 1)}{\sum_m \sum_m' \alpha_m^k \cdot \beta_{m'}^{k+1} \cdot \gamma_{m,m'}^{k,k+1}(d_k = 0)}. \qquad (2)$$

The original probability based formulation implies many multiplications and has thus been ported to the logarithmic domain resulting in the *Log-MAP Algorithm* (Robertson et al., 1997). Multiplications turn into additions and additions into the already mentioned *max** operation which is defined as:

$$max^*(\delta_1, \delta_2) = max(\delta_1, \delta_2) + ln(1 + e^{-|\delta_2 - \delta_1|}). \qquad (3)$$

This transformation does not decrease the communication performance. Arithmetic complexity can further be reduced by discarding this correction term which results in a 0.3 dB communication performance loss.

If we multiply the extrinsic information which is passed between the different constituent decoders with an appropriate scaling factor ESF the communication performance can be approximated to that of a Log-MAP decoder (J. Vogt, 2000). Simulations show that the optimal scaling factor is 0.7. For fixed-point implementation (which is a must on DSPs) an ESF=0.75 is used, which can be easily implemented by a shift operation.

2.2   Turbo-Decoder performance under quantization and SNR mismatch effects

SNR estimation is a difficult task in wireless communication systems. An imprecise SNR estimation can have strong influence on the communication performance of the decoding process. Worm et al. (2000a) have proven that Turbo-decoding based on a Max-Log-MAP algorithm is SNR independent, whereas the decoding performance with the Log-MAP algorithm depends on the accuracy of the SNR estimation. In this case the authors propose to work with SNR operating points L_{op}.

The estimated SNR values are used to scale the received channel input values. These values are interpreted as Log-Likelihood values and are calculated as follows:

$$\Lambda_k = \frac{4E_s}{N_0} a_k y_k \qquad (4)$$

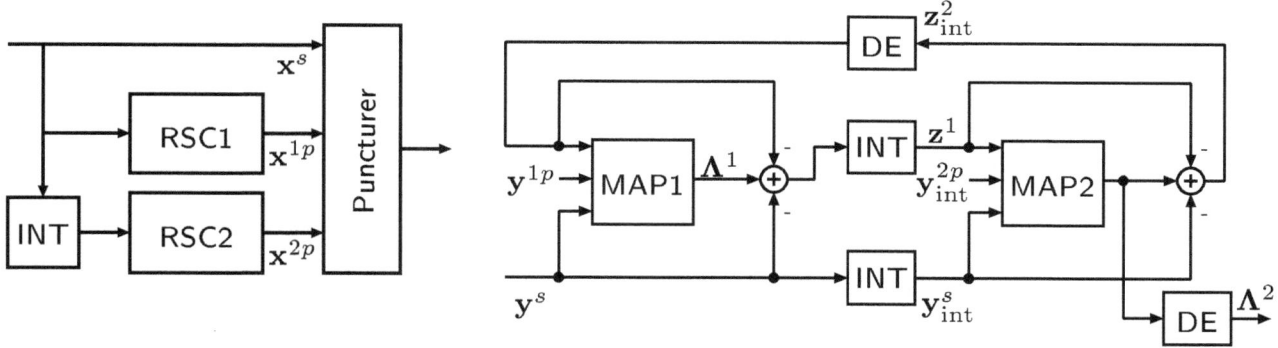

Fig. 1. Turbo-Encoder and Decoder

The factor $\frac{4E_s}{N_0}a_k$, denoted as L_c in the following, is commonly referred to as channel reliability factor (Hagenauer et al., 1996). If the channel characteristics does not change over time, it is sufficient to use a single SNR operating point which is constant, i.e. $L_{op} = L_c = \text{const}$.

The investigations in Worm et al. (2000a) were based on floating-point models - however VLSI- or DSP implementations require fixed-point representations with limited accuracy. The move from floating to fixed-point models (*quantization*) with minimized wordwidth requires a thorough exploration: the bitwidth has to be wide enough to cover the dynamic range and the number of fractional bits has to be large enough to ensure an appropriate processing accuracy (Michel and Wehn, 2001). To avoid an implementation of a multiplier for the channel reliability factor to evaluate Eq. (4), we approximate the SNR operating points with 2^x, $x \in \mathcal{Z}$. Thus the multiplication with L_{op} in Eq. (4) can be substituted by simple shift operations with a resulting SNR operating point granularity of 3 dB.

Figure 2 shows the communication performance of different MAP decoding algorithms. Bit Error Rates (BER) are simulated with an Additive White Gaussian Noise (AWGN) channel, the blocksize is 5114 bits (3GPP standard). Eight iterations are carried out.

In Fig. 2a we have compared the performance of a floating point Log-MAP algorithm with a fixed-point Log-MAP, a Max-Log-MAP and a Max-Log-MAP combined with ESF=0.75 scaling algorithm respectively. A time invariant stable operating point $L_{op} = 2$ is used, which is an optimum SNR operating point for an AWGN channel. The quantization (bitwidth 6, fractional part 2 bit) is accurate enough to prevent an error floor or any other larger quantization effects. Thus the fixed-point Log-MAP algorithm has a very small performance loss compared to the reference Log-MAP algorithm. The fixed-point Max-Log-MAP algorithm has a degradation of $\sim$0.3 dB due to its algorithmic simplification. But remarkable is that the Max-Log-MAP algorithm with ESF=0.75 scaling degrades only minimal compared to the fixed-point Log-MAP algorithm. Thus in the case of stable operating points we can use the fixed-point Max-Log-MAP algorithm with ESF scaling which has a similar communica-

tion performance as the Log-MAP algorithm.

In Fig. 2b we have investigated the influence of SNR mismatches for different fixed-point algorithms. The performance with an optimal SNR operating point ($L_{op} = 2$) is compared with the operating points $L_{op} = 1$ and $L_{op} = 4$ which correspond to a -3 dB and $+3$ dB SNR mismatch. Obviously the fixed-point Log-MAP algorithm is very sensitive with respect to the SNR operating point e.g. the performance degradation is 0.2 dB for $L_{op} = 4$ and even larger than $\sim$0.4 dB for $L_{op} = 1$. The performance of the Max-Log MAP algorithm with ESF scaling is by far not so sensitive to L_{op} variations. The degradation ranges between 0.05 and 0.1 dB for $L_{op} = 4$ and $L_{op} = 1$ respectively. In a floating-point model all the Max-Log-MAP graphs for the different operating-points would coincide, but due quantization different graphs result.

Under the consideration of quantization effects and SNR mismatches we recommend the use the Max-Log-MAP algorithm with ESF scaling in turbo-decoder implementations.

3 Turbo-Decoder Implementation on modern DSP architectures

Modern DSP architectures attempt to increase the signal processing performance by exploiting the inherent parallelism of many signal processing algorithms. This class of DSP architectures provides several independent ALU units along with wide and fast busses to the internal memories. To allow this increased degree of instruction level parallelism, parallel executed instructions for each active unit are grouped together to so-called very large instruction words (VLIW). Further, the processing units usually support the single-instruction/multiple-data approach (SIMD). This exploits sub-word parallelism (SWP), where several sub-words of a data word can be processed with the same operation. In the following we present state-of-the art DSP architectures and implementation results.

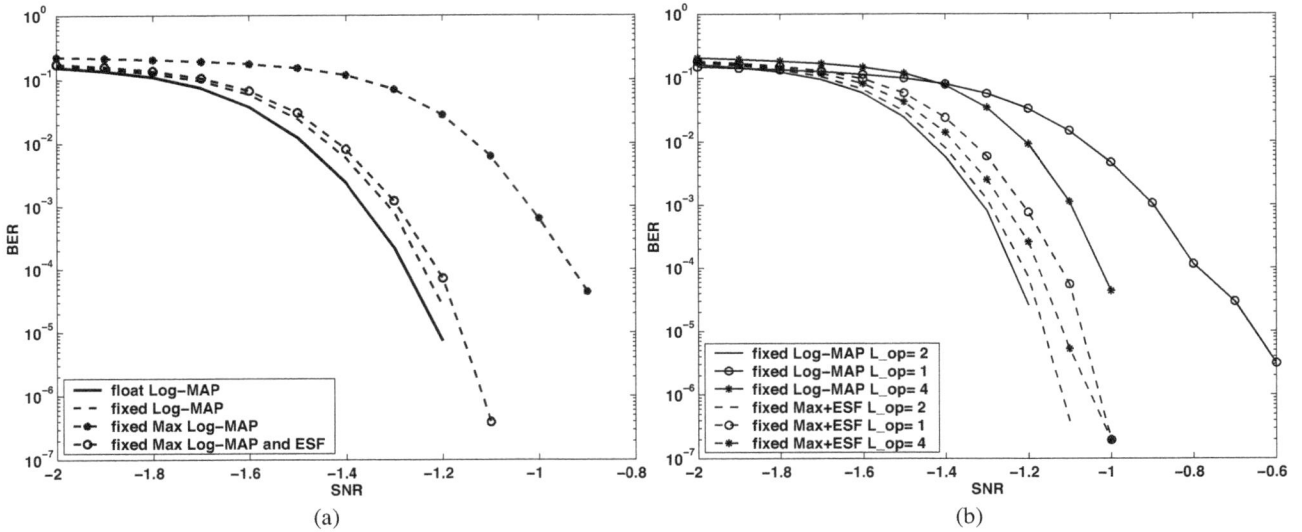

Fig. 2. BER of UMTS Turbo-Decoder: **(a)** Different MAP implementations 8 iterations, **(b)** Performance of different algorithms dependent on the SNR operating points.

3.1   Motorola/Lucent SC140

The **Starcore SC140** is jointly designed by Motorola and Lucent. A significant design issue is the variable length instruction set (VLES). Most instructions are 16-bit wide, but can be grouped into VLES packtets of 128 bit. We use the SC140 as an example for a multiple ALU VLIW DSP, supporting sub-word parallelism. Its architecture employs 4 ALU and 2 AGU units. In one clock cycle the SC140 is thus able to perform 4 ALU operations, each using 32 bit wide operands, and a 128 bit data transfer. A (Max-)Log-MAP implementation for decoding an 8-state Turbo-Code on this DSP should exploit the benefits of the sub-word parallelism by using 16-bit packed data types.

In Chass and Gubeskys (2000) the update of the path metrics of a Max-Log-MAP, comprising four butterflies, needs 3 clock cycles. This fully utilizes the architectural capabilities of the DSP. One Max-Log-MAP decoder can be realized in 16 cycles per bit. Assuming 5 iterations the resulting Turbo-decoding performance is 1875 kbit/s at 300 MHz. The component decoder with a Log-MAP implementation has a cycle count of approximately 50 cycles per bit, leading to a Turbo-decoding throughput of 600 kbit/s at 300 MHz (see Table 1).

This decrease in decoding performance results from the complexity of the max*-operation (see Eq. 3), which includes the steps: difference of parameters, absolute value, access to lookup-table (LUT), and adding the LUT result to maximum of parameters. The sequence of these steps needs 10 clock cycles in contrast to just one cycle needed for the plain maximum operation (Chass and Gubeskys, 2000).

3.2   STM ST120

The **ST120** is provided by ST-Microelectronics. It features two ALU units and supports three different instruction sets:

a 16-bit instruction set (GP16) for compact microcontroller code, a 32-bit instruction set (GP32) for higher performance and more complex instructions, and a third one for an increased level of instruction parallelism. In this 4×32-bit Score-boarded Long Instruction Word (SLIW) mode the processor is able to execute four GP32 instructions in one clock cycle. Following the SIMD approach, the processor supports 2×16-bit data packed into one 32-bit data word.

An optimized hand-coded assembly language implementation of an 8-state Turbo-Decoder has a performance of 37 cycles per bit and MAP, using packed data types and Max-Log-MAP algorithm. Assuming 5 iterations the Turbo-decoding performance results to 540 kbit/s at 200 MHz. For a Log-MAP implementation the throughput degrades to 200 kbit/s with 100 cycles per bit and MAP decoder.

3.3   TigerSharc

The TigerSharc DSP processor from Analog Devices is a high-performance architecture, which is targeted, e.g. for wireless infrastructure applications, such as cellular base stations. With its VLIW architecture, TigerSharc is capable to execute up to four intructions in a single cycle and combines hierarchically both types of data-level parallelism: SIMD and SWP.

The TigerSharc is the only processor with dedicated instruction support to implement the Log-MAP algorithm. A max* instruction is provided, which operates only on a set of enhanced communication registers. Transferring values between the ALU register file and these special registers causes significant data transfer overhead. The max* operation processes a (sub-word parallel) maximum selection of the input parameters and adds a respective correction term value from an integrated lookup-table to the maximum. Thus, full Log-MAP support is achieved without performance penalty.

Table 1. Turbo-Decoder throughput with (Max-)Log-MAP decoder on DSPs

Processor	Architecture	Clock freq.	cycles/(bit · MAP)	Throughput @ 5 iter.
		Max-Log-MAP		
ST120	VLIW, 2 ALU	200 MHz	37	540 kbit/s
SC140	VLIW, 4 ALU	300 MHz	16	1875 kbit/s
		Log-MAP		
ST120	VLIW, 2 ALU	200 MHz	≈ 100	≈ 200 kbit/s
SC140	VLIW, 4 ALU	300 MHz	50	600 kbit/s
ADI TS	VLIW, 2 ALU	180 MHz	27	666 kbit/s

Unfortunately, the integrated LUT is unsymmetrical to zero. Thus, the two parameters of the max* operation are not commutative any more, which complicates the validation of implementation's bit-true behavior versus a bit-true model written in a high-level language.

A single Log-MAP decoder requires 27 cycles per bit. Assuming 5 iterations the overall throughput of this Turbo-Decoder implementation on TigerSharc results to 666 kbit/s at 180 MHz.

3.4 Summary

Table 1 summarizes performance results of 3GPP compliant Turbo-Decoder implementations. The number of cycles per bit and MAP is three times higher for a Log-MAP than for a Max-Log-MAP implementation. The required Turbo-Decoder throughput for UMTS (up to 2 Mbit/s) can only be reached with the SC140 processor, using the Max-Log-MAP algorithm. The Turbo-Decoder with Log-MAP implementation achieves only a throughput of 666 kbit/s. For higher throughput requirements a multiprocessor architecture is mandatory. By using the Max-Log-MAP with ESF=0.75 the communication degradation can be almost avoided with an negligible implementation overhead (1 cycle/(bit· MAP)).

4 Conclusions

Turbo-Codes are part of the 3G cellular wireless standard. The complexity of the decoding algorithm and the throughput requirements pose great demands on the computational power of the signal processing devices. Current DSPs support the kernel operations of the Viterbi algorithm, however for the MAP algorithm, this support is lacking. Using VLIW DSPs one 3G data channel can be processed using the sub-optimal Max-Log-MAP algorithm. This validates the throughput of 1875 kbit/s for the Starcore SC140. Therefore we propose a Max-Log-MAP with Extrinsic Scaling Factor. Its communication performance is close to the performance of a Log-MAP implementation and is less sensible to SNR mismatch. The implementation complexity of the ESF Max-Log-MAP is nearly equal to the Max-Log-MAP and can be implemented on standard DSPs without specific instruction extensions.

Acknowledgements. This work has been supported by the Deutsche Forschungsgesellschaft (DFG) under grant We 2442/1-1 within the Schwerpunktprogramm "Grundlagen und Verfahren verlustarmer Informationsverarbeitung (VIVA)".

References

Berrou, C., Glavieux, A., and Thitimajshima, P.: Near Shannon Limit Error-Correcting Coding and Decoding: Turbo-Codes, in: Proc. 1993 International Conference on Communications (ICC '93), pp. 1064–1070, Geneva, Switzerland, 1993.

Bickerstaff, M., Hughes, G., Nicol, C., Xu, B., and Yan, R.-H.: DSP Systems for Next-Generation Mobile Wireless Infrastructure, in: Proc. ICASSP 2000, pp. 3710–3713, 2000.

Chass, A. and Gubeskys, A.: On Performance/Complexity Analysis and SW Implementation of Turbo Decoding, in: Proc. 2nd International Symposium on Turbo Codes & Related Topics, pp. 531–534, Brest, France, 2000.

Greifendorf, D., Stammen, J., and Jung, P.: The Evolution of Hardware Platforms for Mobile 'Software Defined Radio' Terminals, in: Proc. 2002 International Symposium on Personal, Indoor, and Mobile Radio Communications (PIMRC '02), 2002.

Hagenauer, J., Offer, E., and Papke, L.: Iterative Decoding of Binary Block and Convolutional Codes, IEEE Transactions on Information Theory, 42, 429–445, 1996.

J. Vogt, A. F.: Improving the Max-Log-MAP Turbo Decoder, IEEE Electronic Letters, 36, 2000.

Michel, H. and Wehn, N.: Turbo-Decoder Quantization for UMTS, IEEE Communications Letters, 5, 55–57, 2001.

Robertson, P., Hoeher, P., and Villebrun, E.: Optimal and Sub-Optimal Maximum a Posteriori Algorithms Suitable for Turbo Decoding, European Transactions on Telecommunications (ETT), 8, 119–125, 1997.

Third Generation Partnership Project, 3GPP home page, www. 3gpp.org.

Vogt, J., Koora, K., Finger, A., and Fettweis, G.: Comparison of Different Turbo Decoder Realizations for IMT-2000, in: Proc. 1999 Global Telecommunications Conference (Globecom '99), vol. 5, pp. 2704–2708, Rio de Janeiro, Brazil, 1999.

Worm, A., Hoeher, P., and Wehn, N.: Turbo-Decoding without SNR Estimation, IEEE Communications Letters, 4, 193–195, 2000a.

Worm, A., Michel, H., Gilbert, F., Kreiselmaier, G., Thul, M. J., and Wehn, N.: Advanced Implementation Issues of Turbo-Decoders, in: Proc. 2nd International Symposium on Turbo Codes & Related Topics, pp. 351–354, Brest, France, 2000b.

25

ESD compact-simulation: investigation of inverter failure

S. Drüen[1], K. Esmark[2], W. Stadler[2], H. Gossner[2], and D. Schmitt-Landsiedel[1]

[1]TU München, Germany
[2]Infineon Technologies, München, Germany

Abstract. An ESD failure occurring inside the core circuitry known as "inverter failure" will be presented and analysed in this paper. The compact model utilised for this investigation is shortly presented. It will be shown that not only properties of the failed structure are relevant, but also surrounding circuitry. So the gate of an inverter will be connected during the simulations in diverse ways to V_{DD} and V_{SS}. The different possibilities of influence of pre drivers can be appraised in this way. In order to achieve a detailed understanding of the individual failure, it is necessary to include ambient circuitry as well as parasitics like resistors and capacitances.

1 Introduction

Susceptibility to Electro Static Discharge (ESD) increases with the shrinking feature size of VLSI technology. The circuit elements which are connected directly to the pads are particularly endangered and have to be protected. ESD protection structures are present in every modern IC which wouldn't survive without these measures. ESD protection elements are placed in parallel to susceptible structures in order to divert the stress current and to clamp the node voltage to safe values. Nevertheless the protected elements have to be considered particularly as they still represent a possible path for the discharge current. With the growing complexity of IC's associated with an increasing ESD sensitivity the optimization of ESD structures requires a high effort. The behaviour of complex systems under ESD stress is difficult to predict even for ESD engineers and requires a huge level of expert knowledge. Circuit simulation helps to close this gap between huge circuit complexity and understanding of circuit behaviour during discharge. Even apparently simple structures can behave in different ways due to interaction with surrounding circuitry. It is important to find critical structures in circuit architectures even before production of IC's or construction of libraries and to accelerate search of causes

of failed elements. In this paper, a well known failure will be investigated which has also been addressed in the literature (Chaine et al., 1997; Krakauer et al., 1994; Puvvada and Duvvury, 1998). Despite correct protection structures circuit elements parallel to protections may be damaged by ESD. It has been observed that the nMOS of an inverter fails for a particular setup of an inverter. Especially nMOS of inverters with pMOS with a large channel width are susceptible. It will be shown that the surrounding circuitry plays an additional role and influences the failure of the inverter. The investigation of the inverter failure will be shown in Sect. 3. The compact model is presented in short terms in Sect. 2. A conclusion is given at the end of the paper.

2 Compact model

The behaviour of an ESD stressed MOSFET can be ascribed to the turn-on of the parasitic bipolar transistor, caused by avalanche multiplication at the drain diffusion space charge region. A combination of a standard MOS model with a bipolar model extension is already presented in Wolf et al. (1998). The MOS model describing the FET behaviour in the normal operating regime and additional lumped elements describing the FET behaviour during ESD are included in the compact model used to investigate circuitry under ESD stress. Beside two exemplary simulations, Soppa et al. (2002) describes the implementation of this model in a proprietary simulator. Basically, a few elements determine the high current behaviour of the FET which are illustrated in Fig. 1. The bipolar transistor is implemented using the Ebers-Moll model (Ebers and Moll, 1954). This model also includes capacitors, which are mainly necessary for the simulation of the transient behaviour.

The IV-characteristic of a grounded-gate-nMOSFET (ggn-MOS) can be divided into four regions shown in Fig. 2. Up to a certain voltage, the reverse current of the drain-bulk junction can be observed (region 1). Due to a high electric field in the space charge region, charge carriers are generated, the

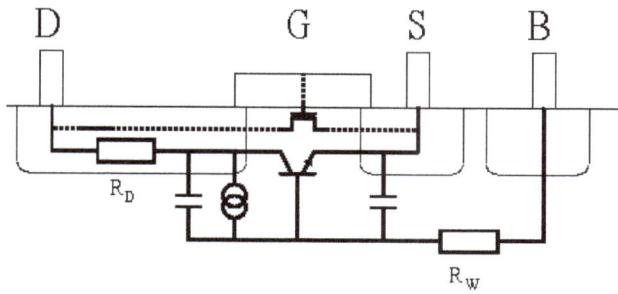

Fig. 1. Profile of nMOSFET with parasitic elements implemented in SQ3.

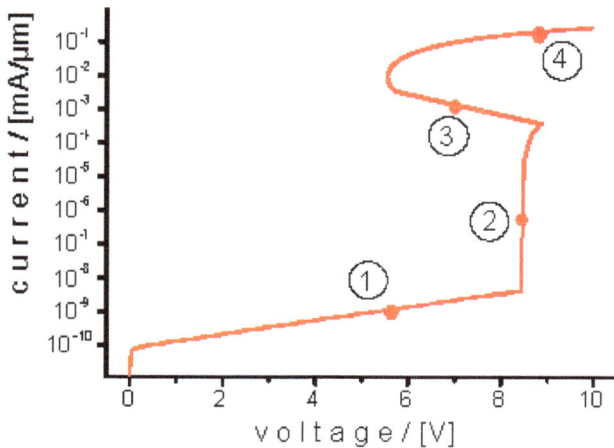

Fig. 2. High-current behaviour of ggnMOS: **1.** reverse current, **2.** avalanche generation, **3.** snapback, **4.** high current regime

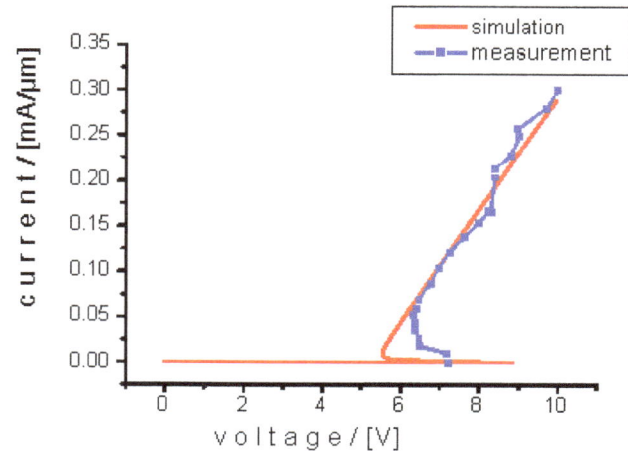

Fig. 3. Simulation and measurement of high current regime.

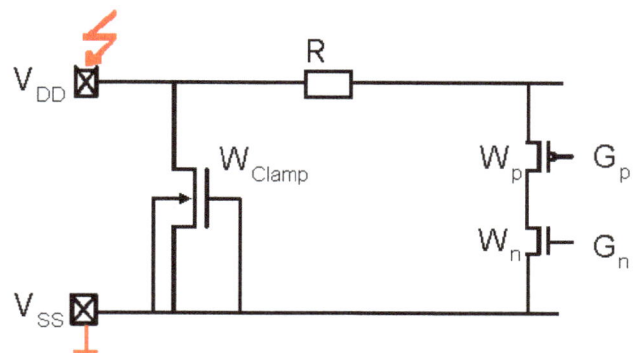

Fig. 4. Schematic of investigated inverter structure.

current (region 2) increases and causes a voltage drop across the well resistor R_W. If this voltage drop is high enough (~0.7 V) to forward bias the base emitter junction of the bipolar transistor, the voltage snaps back (snapback) to the hold voltage (region 3). When the bipolar transistor is activated, the high current behaviour (region 4) is determined by the diffusion resistor R_D. Simulation of snapback is mainly achieved by implementing the bipolar transistor, the current source I_{ava} and the resistor R_W.

Measurement and simulation of the high current regime are shown in Fig. 3. The measurement of the high current characteristic is done step by step with short pulses (Amerasekera and Duvvury, 2002) for diverse current levels.

3 Inverter failure

Many fails discussed in literature are caused by turn on of parasitic elements like pnpn structures (Amerasekera and Duvvury, 2002). In contrast failures where breakdown of parasitic elements is not observable are also mentioned in literature. Here the local distances of guardrings, diffusions etc. play a secondary role. Design properties like finger width of transistors are important and make it possible to investi-

gate the failure mechanisms by compact simulation. In new technologies it can be observed that nMOS of inverters with large pMOS are susceptible to ESD discharge. This chapter presents some basic examinations of the behaviour of an inverter parallel to an ESD protection element during ESD stress. The basic schematic of the investigated inverter structure is shown in Fig. 4.

The worst case will be investigated in the following simulations. A V_{DD} Pad is stressed positively (2 kV HBM) with a grounded VSS Pad, so that the ESD Protection element (Clamp) is operating in breakdown mode with a higher voltage drop than in its diode mode. In this stress mode the nMOS goes into breakdown and determines the triggering behaviour of the inverter path. The ESD pulse can take two paths from V_{DD} to V_{SS}: through the clamp and through the inverter. The high current characteristics of the inverter nMOS and the Clamp are shown in Fig. 5. The two possible paths have a low resistance in the high current regime, so that the current distribution depends strongly on the triggering behaviour of the paths.

With additional series resistors like a pMOS or bus resistances, the Clamp triggers first and shunts the main current. Different setups will be simulated to examine the behaviour

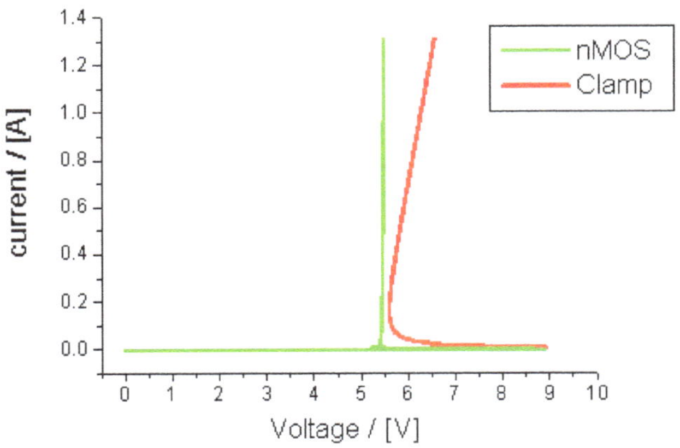

Fig. 5. High current characteristic of nMOS and Clamp.

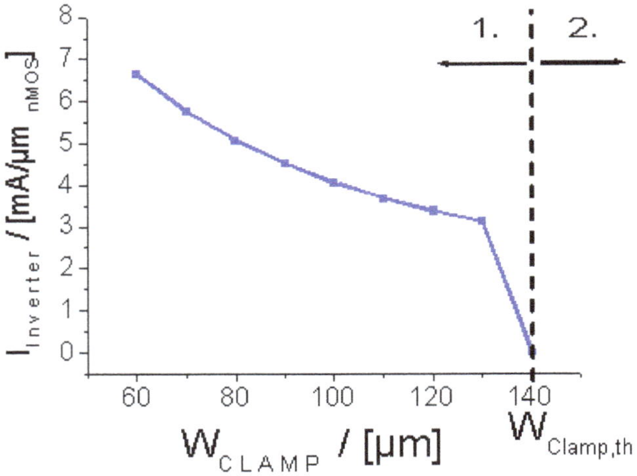

Fig. 6. Dependence of W_{Clamp} on inverter current.

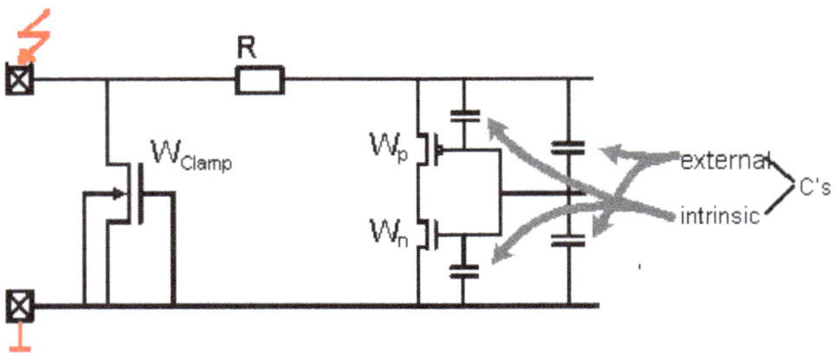

Fig. 7. Setup with intrinsic and external capacitors.

of an inverter during ESD stress. The gate potential of the inverter transistors can be torn to different levels during an ESD pulse. Therefore they are not connected in Fig. 4. For sake of simplicity the pre drivers being also responsible for the gate potential are not included to the following investigations.

In the first investigation the gates are connected to V_{SS}, so that the trigger mechanisms can be explained. The snapback of the inverter is invoked by avalanche multiplication, if the gate is connected to the bulk-node of the nMOS (grounded gate). Additional displacement currents also influence the triggering. The voltage drop at the inverter is essential for triggering of the nMOS. In other words, a good clamping behaviour of a protection device can be inhibited by additional voltage drop in the main current path due to resistors or diodes. If the voltage exceeds the trigger voltage, the nMOS triggers and current can flow through the inverter getting low ohmic. During ESD stress the voltage drop at the inverter and the inverter current are determined by the width of the clamp (W_{Clamp}). Figure 6 shows the influence of W_{Clamp} on the inverter-current $I_{Inverter}$. The threshold width $W_{Clamp,th}$ divides this graph into two regions:

– Region 1: current through inverter

– Region 2: no inverter current

A good clamping behaviour is achieved using a clamp with a high width and a low resistance in the high current regime. The trigger-threshold is not attained and thus no current flows in the inverter. Reduction of W_{Clamp} does not show an influence until W_{Clamp},th is reached. The current is increasing suddenly if W_{Clamp} is falling below $W_{Clamp,th}$. Further reduction of W_{Clamp} increases the current continuously. The width of the pMOS additionally influences the current through the inverter. Despite a good clamping behaviour of the power clamp, current can flow through an inverter if the resistance of the pMOS is low enough. Equivalent results are obtained if the pMOS is replaced with a resistor. In this case, variation of the resistor is equal to variation of the pMOS width. Tran-

sient effects play a role as well as the reduction of the series resistance in the inverter path.

Normally the potential of the gate during an ESD event is between V_{SS} and V_{DD}. In the following investigation the gate potential is determined by lumped capacitances. An investigation of the impact of the gate potential on the current through the inverter will be done in this chapter. As the gate voltage increases by capacitive coupling, the nMOS is in on-state before reaching the breakdown voltage. The gate potential will be determined by the capacitors connected from the gate to V_{DD} or V_{SS}. This setup corresponds to a circuit where the transistors connected to the gate are turned off. It should be recognized that potentials are not constant values, but adjusted during ESD pulse due to capacitors and resistors. Two setups are simulated in this investigation. In one case the gate potential is determined by the intrinsic diffusion-well capacitance of the pMOS, in the other case the gate potential is fixed by capacitors outside the transistor. These two setups differentiate whether external capacitors or intrinsic capacitors dominate the gate potential. The schematic with intrinsic and external capacitors is depicted in Fig. 7.

In addition to the ESD protection element, the width of the pMOS and the gate potential plays an important role, which is investigated in this chapter. The high current regime resistance of the inverter path with triggered nMOS is defined

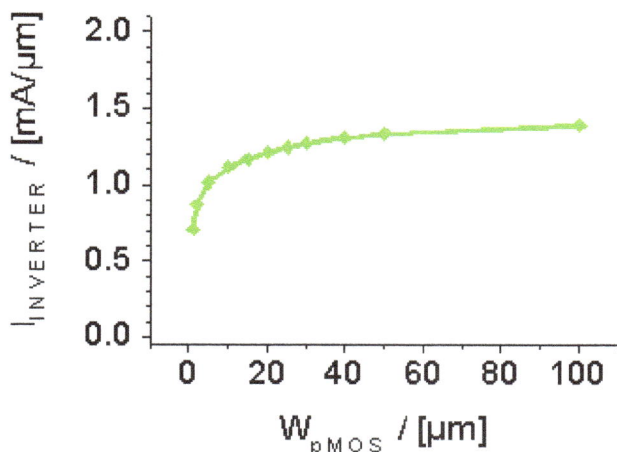

Fig. 8. Inverter current with intrinsic capacitors controlling the gate voltage.

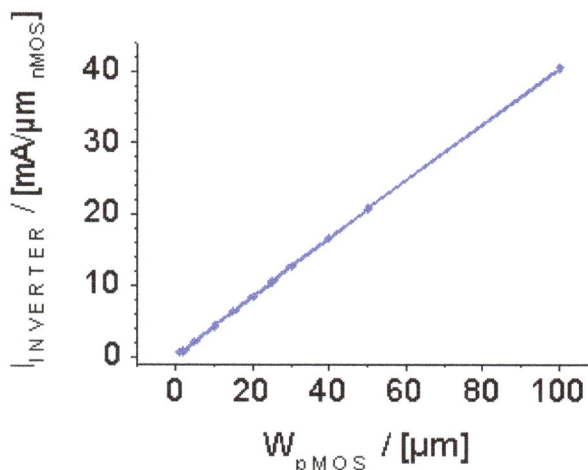

Fig. 9. Inverter current with external capacitors controlling the gate voltage.

by the width of the pMOS and by the gate potential of the pMOS. A high gate potential or a low pMOS width restrain the discharge current through the inverter during ESD stress. Figure 8 shows the influence of pMOS width on the current through the inverter. In these simulations the intrinsic capacitors determine the gate potential. The current doesn't show the expected linear increase with the width.

Two effects influence this behaviour. Increasing the width of the pMOS, the current should increase linearly. The deviation of the curve shown in Fig. 8 from the expected characteristic is caused by the increase of the gate-source capacitance with the width of the pMOS. Due to this capacitance the gate potential approaches the V_{DD} potential and the pMOS is driven into its blocking-state. In this case the gate voltage limits the current through the inverter. With large external capacitors at the gate of the inverter the gate voltage doesn't depend on the pMOS width anymore. The expected linear curve can be observed in Fig. 9.

This investigation demonstrates that the MOS parameters (pMOS width, gate voltages, attached capacitances) had a significant impact on the current through the inverter and with that on the failure of the susceptible nMOS. Assuming a failure current of approximately $2 \, mA/\mu m$, the nMOS would survive in the first case but would fail for a certain pMOS width using the second setup.

4 Conclusion

In addition to the width of powerclamp and pMOS the gate potential controlled by capacitors has a significant influence on the current through the inverter. It depends on the capacitors, which can be external or intrinsic, if the current through the inverter surmounts the failure current, and thus if the inverter fails. This is one possible answer to the question why only a few of many equal inverter structures are susceptible to ESD events. ESD performance doesn't only depend on

the design of the inverter, but also on the behaviour of the surrounding circuitry during ESD. In real circuitry, diverse factors can affect the current and thus the failure of the inverter. So adjacent circuitry as well as parasitics like capacitors and resistors have to be included to the investigation of ESD failures by circuit simulation.

Acknowledgements. The authors would like to thank Heinrich Wolf, FhG IZM-ATIS München, for his support with the parameter extraction and the valuable discussions concerning the compact model. The authors also thank Winfried Soppa, FH Osnabrueck, for his support and useful contributions regarding the compact model.

References

Amerasekera, A. and Duvvury, C: ESD in Silicon Integrated Circuits, Second Edition, John Wiley, Chichester, England, 2002.

Chaine, M., Smith, S, and Bui, A.: Unique ESD Failure Mechnisms During Negative To Vcc HBM Tests, EOS/ESD Symp. Proc. pp. 346, 1997.

Ebers, J. J. and Moll, J. L.: Large Signal Behavior of Junction Transistors." Proc. I.R.E., 42, pp.1761–1772, 1954.

Krakauer, D., Mistry, K., and Partovi, H.: Circuit Interactions During Electrostatic Discharge, EOS/ESD Symp. Proc. pp. 113, 1994.

Puvvada, V. and Duvvury, C.: A Simulation Study of HBM Failure in an Internal Clock Buffer and the Design Issues for Efficient Power Pin Protection Strategy, EOS/ESD Symp. Proc. pp. 104, 1998.

Soppa, W., Drueen, S., Wolf, H., Stadler, W., Esmark, K., and Schmitt-Lansiedel, D.: VHDL-AMS-Modellierung von Schutzstrukturen einer 0.18-μm-CMOS-Technologie zur Simulation von ESD-Stress, Tagungsband ANALOG'02, pp.401–406, 2002.

Wolf, H., Gieser, H., and Stadler, W.: Bipolar model extension for MOS transistors considering gate coupling effects in the HBM ESD domain, Proceedings of the EOS/ESD Symposium 1998, pp. 271–280, 1998.

26

Theory of circuit block switch-off

S. Henzler[1]**, J. Berthold**[2]**, G. Georgakos**[2]**, and D. Schmitt-Landsiedel**[1]

[1]Lehrstuhl für Technische Elektronik, Technische Universität München, Theresienstrasse 90, D-80290 Munich, Germany
[2]Corporate Logic, Infineon Technologies AG, Balanstrasse 73, D-81541 Munich, Germany

Abstract. Switching-off unused circuit blocks is a promising approach to supress static leakage currents in ultra deep sub-micron CMOS digital systems. Basic performance parameters of Circuit Block Switch-Off (CBSO) schemes are defined and their dependence on basic circuit parameters is estimated. Therefore the design trade-off between strong leakage suppression in idle mode and adequate dynamic performance in active mode can be supported by simple analytic investigations. Additionally, a guideline for the estimation of the minimum time for which a block deactivation is useful is derived.

1 Introduction

The static power dissipation of digital CMOS circuits in deep sub-micron technologies becomes one of the most challenging topics in digital system design. The exponentially increasing number of transistors per chip as well as the rising leakage currents per device make the standby power dissipation an appreciable portion of total power consumption. Various Circuit Block Switch-Off schemes (CBSO) have been proposed in order to to reduce the static power dissipation of circuit blocks which are currently not used. Therefore, an additional transistor is used to separate the considered circuit block from either the vdd or the vss potential. Thus the leakage current of this circuit block is reduced significantly during idle mode. The design of a Circuit Block Switch-Off scheme is driven by the trade-off between a strong leakage suppression, low area overhead due to the switch devices and a small performance degradation of the circuit in active mode. Additionally, the minimum time for which a block deactivation is useful has to be known in order to implement a reasonable power-down control logic. The fundamental approaches for answering these design questions are given in this paper. In the first part the static behaviour of a power switched circuit block is considered. Following, the minimum power-down time is derived and finally a compact simulation approach to investigate the dynamic delay degradation in active mode is given.

2 Power-down state

After a circuit block has been cut-off, it reaches a certain steady state. This state has to be known in order to estimate the remaining leakage current. Usually the total transistor width in the logic block is much larger than the width of the switch device. Hence the potential of the virtual power-rail(s) has to change until the leakage current of the logic block is equal to the leakage of the switch. Figure 1 shows an equivalent circuit which models an n-block switched circuit. The cross coupled inverters are dimensioned very large in order to describe a large logic block with half of the nodes at logic high level and half of the nodes at logic low level. This simple equivalent circuit has a low transistor count, thus even large logic blocks can be modeled efficiently. The cross coupled inverter structure neglects interface effects between a power switched circuit block and the surrounding circuitry. Thus other models have to be used if these effects are to be investigated. In order to find the final state after switching-off a circuit block, the potential of the virtual power rail has been ramped up. The current through the switch device as well as the total current in the logic is given in Fig. 2. The steady state of the idle system is determined by the intersection of the two current voltage characteristics of the switch and the logic. Depending on the transistor widths and threshold voltages, the potential of the virtual power rail ($vvss$) moves versus the non switched power rail (vdd). In typical n-block switched circuits the $vvss$ potential as well as the potential of all internal signal nodes charges up to a value slightly below the vdd potential. Thus all transistors operate in subthreshold region:

$$I_p = I_{0p} \exp\left(\frac{vdd - v_1 + v_{tp}}{\eta v_T}\right)\left[1 - \exp\left(\frac{v_1 - vdd}{v_T}\right)\right] \quad (1)$$

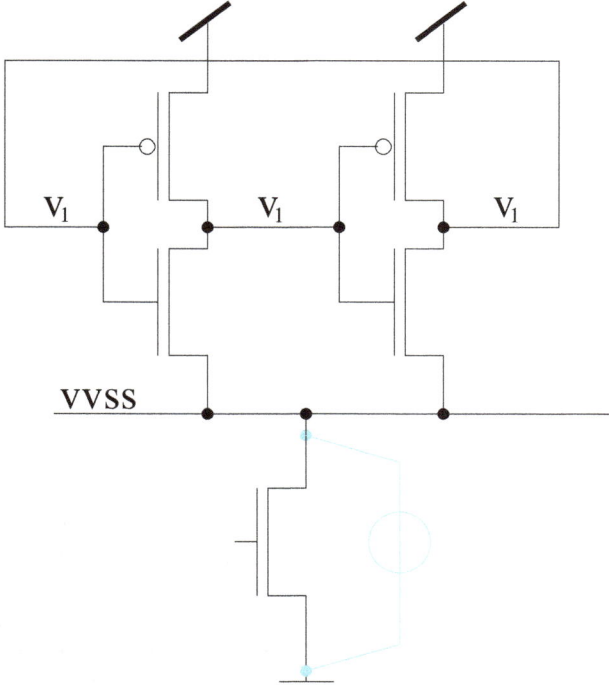

Fig. 1. Equivalent circuit to model a large power switched circuit block. Interface effects between the circuit block and the surrounding circuitry are neglected by this circuit. A voltage source between the virtual power rail and ground can be used to determine the current voltage characteristic of the circuit block and the switch device.

$$I_n = I_{0n} \exp\left(\frac{v_1 - vvss - v_{tn}}{\eta v_T}\right)\left[1 - \exp\left(-\frac{v_1 - vvss}{v_T}\right)\right] \quad (2)$$

$$I_s = I_{0s} \exp\left(\frac{-v_{ts}}{\eta v_T}\right)\left[1 - \exp\left(-\frac{vvss}{v_T}\right)\right] \quad (3)$$

Hence the equivalent circuit of Fig. 3 can be used to derive an analytic expression for the steady state. The total logic width is assumed to be much larger than the width of the switch device. Thus Fig. 2 states that the potential of the virtual power rail is slightly below vdd. Therefore the $vvss$ dependency of the subthreshold current in the switch device can be neglected and the switch threshold voltage v_{ts} is approximated by the switch threshold voltage at maximum drain-to-source voltage $V_{ds}^{switch} = vdd$. As the deviation of the $vvss$ potential with respect to vdd is small, the potential V_1 of the signal nodes is also slightly below vdd. Thus a perturbation approach can be used to describe these voltages:

$$V_1 = vdd - \Delta V_1 \qquad 0 < \Delta V_1 \ll vdd \quad (4)$$

$$vvss = V_1 - \Delta V_2 \qquad 0 < \Delta V_2 \ll vdd \quad (5)$$

solving the linear equation system 1-3 results in

$$\Delta V_1^{1,2} = \frac{1}{2}\eta v_T \left(-1 \pm \sqrt{\underbrace{1 + \frac{4}{\eta}\frac{I_{0s}}{I_{0p}}\exp\left(-\frac{v_{ts} + v_{tp}}{\eta v_T}\right)}_{>1}}\right) \quad (6)$$

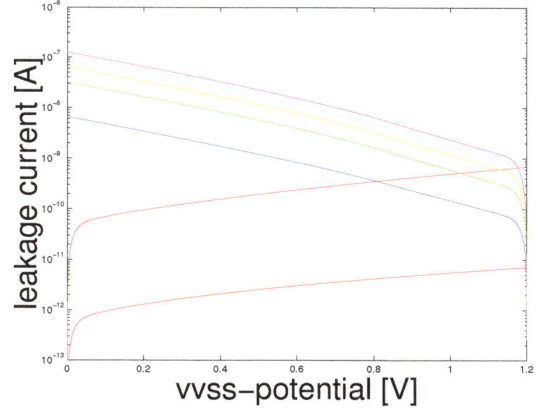

Fig. 2. Simulated leakage-voltage-characteristic of several logic circuits with different total transistor widths and subthreshold characteristic of two switch devices with low and high threshold voltage. The intersection of these curves determines the stationary state after block deactivation.

$$\Delta V_2^{1,2} = \frac{1}{2}\eta v_T \left(-1 \pm \sqrt{1 + \frac{4}{\eta}\frac{I_{0s}}{I_{0n}}\exp\left(\frac{v_{tn} - v_{ts}}{\eta v_T}\right)}\right) \quad (7)$$

The upper solution is physical meaningful and for $I_{0s} \ll I_{0p}$, I_{0n} Eq. (6) and Eq. (7) can be approximated by

$$\Delta V_1 = v_T \frac{I_{0s}}{I_{0p}}\exp\left(-\frac{v_{ts} + v_{tp}}{\eta V_T}\right) \quad (8)$$

$$\Delta V_2 = v_T \frac{I_{0s}}{I_{0n}}\exp\left(\frac{v_{tn} - v_{ts}}{\eta V_T}\right) \quad (9)$$

These equations state that the voltage difference of the signal- and $vvss$-nodes decreases exponentially with increasing switch threshold voltage and linear with decreasing width ratio $\frac{W_s}{W_{\log}}$.

3   Leakage reduction ratio

The target of any CBSO scheme is the reduction of the leakage current of unused circuit blocks. The quality of this leakage reduction can be quantified by the leakage reduction ratio (LRR):

$$LRR := \frac{I_L^{no\ switch}}{I_L^{switch\ off}} \quad (10)$$

$I_L^{no\ switch}$ is the total leakage current of the considered circuit block if no cut-off switch is used and $I_L^{switch\ off}$ is the remaining leakage current of this circuit block if a cut-off switch is added and turned off. Without cut-off switch, the leakage of an arbitrary static CMOS circuit can be calculated according to

$$I_L^{no\ switch} = \sum_{\{L_V\}}\left(W_P^{eff} I_p' e^{\frac{v_{tp}}{\eta v_T}} + W_n^{eff} j_T^n\right) +$$

$$\sum_{\{H_V\}}\left(W_n^{eff} I_n' e^{\frac{v_{tn}}{\eta v_T}} + W_p^{eff} j_T^p\right) \quad (11)$$

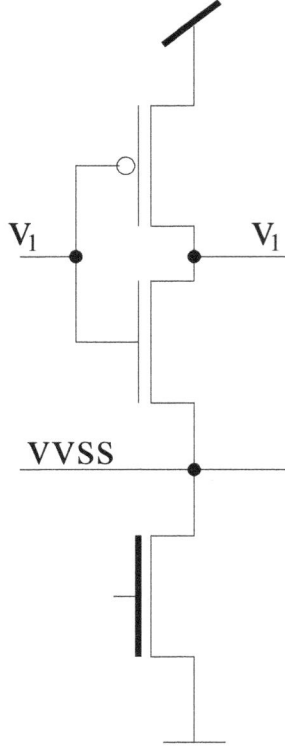

Fig. 3. Reduced equivalent circuit to determine the stationary cut-off state analytically.

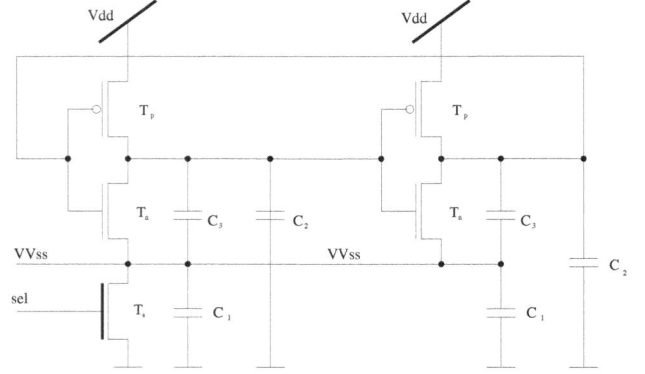

Fig. 4. Equivalent Circuit of a large logic block with internal capacities

$\{L_v\}$ describes the set of gates with a low output level and $\{H_v\}$ the disjoint set of gates with a high output level. I'_n and I'_p respectively describe the subthreshold leakage per transistor width and j_T^n and j_T^p refer to the gate tunneling currents per transistor width. (Assuming all transistors have the same channel length.) Averaged over all system states, the leakage current can be approximated by a common effective logic width W_L^{eff}:

$$I_L^{\text{no switch}} \approx \frac{1}{2} W_L^{eff}$$

$$\left(I'_p e^{\frac{v_{tp}}{n v_T}} + I'_n e^{-\frac{v_{tn}}{n v_T}} + j_T^n + j_T^p \right) \tag{12}$$

In an n-block switching scheme, the virtual ground potential charges up to a voltage slightly below vdd. Thus the residual leakage current can be estimated by

$$I_L^{\text{switch off}} = W_s \left(I'_s e^{-\frac{v_{ts}}{n v_T}} + j_T^s \right) \tag{13}$$

Insertion of Eqs. (12) and (13) into (10) results in the **LRR** of the n-block switching scheme.

4 Power-down process

When a circuit block is not used it can be turned off by a cut-off switch. As the circuit block includes inner capacities, the

leakage current does not collapse instantaneously. Instead, as shown in Fig. 5 there is a smooth transition between the active leakage and the residual leakage in idle mode. During this transition the $vvss$ rail as well as the internal signal nodes charge up to a potential usually slightly below vdd. Figure 4 shows the equivalent circuit introduced in Sect. 2 and its internal capacities. The capacities labeled by C_2 represent the capacity of each signal node versus the vdd-rail and the capacities labeled by C_3 represent the capacitance with respect to the virtual ground rail. Hence the collapsing supply current charges the C_1 and C_2 capacitances. The capacities C_3 however are discharged. Thus the intrinsic energy dissipation W_{intr} due to charging the inner capacities can be estimated by

$$W_{\text{intr}} = vdd^2 \left(\sum C_1 + \sum_{\{L_v\}} C_2 - \sum_{\{H_v\}} C_3 \right)$$

$$\approx vdd^2 \left(\sum C_1 + \frac{1}{2} \sum C_2 - \frac{1}{2} \sum C_3 \right)$$

$$= \frac{1}{2} vdd^2 \sum (2C_1 + C_2 - C_3) \tag{14}$$

When the circuit is turned on again, about half of the signal nodes keep a logic high level and hence the energy $W_{\text{boot, cap}} = \frac{1}{2} vdd^2 C_3$ is dissipated in order to charge up the C_3 capacitances associated to these nodes.

In order to estimate the minimum time T_{min} for which it is useful to turn off the circuit block, the collapse of the supply current after the switch is turned off has to be modeled properly. The model of this current has to be accurate but also simple enough to be used in analytic expressions. The leakage current after turning off the system can be decomposed according to

$$i_L (t > t_{\text{off}}) = i_L^{\text{off}} + i_{\text{inertia}}(t) \quad \lim_{t - t_{\text{off}} \to \infty} i_{\text{inertia}} \to 0 \tag{15}$$

into the stationarily remaining leakage current I_L^{off} and the so called inertial current i_{inertia}. The power dissipation due to the non instantaneously collapsing leakage current ist given

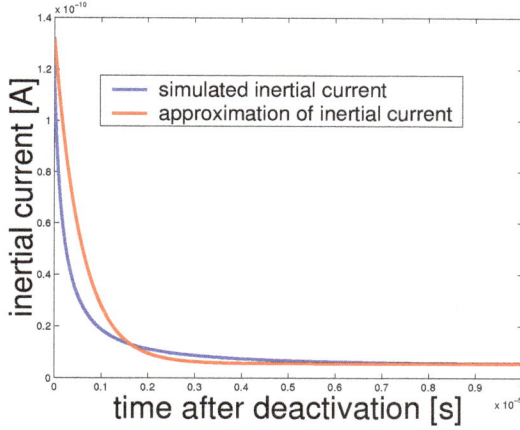

Fig. 5. Time dependence of inertial current and its approximation by an exponential function

by $\int_{t_{\text{off}}}^{\infty} i_{\text{inertia}}(t)\, dt$. If the leakage current was collapsing instantaneously the time which one has to wait in order to save the power penalty due to W_{intr} is given by

$$T_{\text{inertia}}^{eff} = \frac{\int_{t_{\text{off}}}^{\infty} i_{\text{inertia}}(t)\, dt}{\left(I_L^{\text{on}} - I_L^{\text{off}}\right)} = \frac{W_{\text{intr}}}{vdd\left(I_L^{\text{on}} - I_L^{\text{off}}\right)} \quad (16)$$

This time should be called effective inertial time. A simple approximation for the inertial current is the exponential function:

$$i_{\text{inertia}}^{\text{approx}} = \left(I_L^{\text{on}} - I_L^{\text{off}}\right) \exp\left(-\frac{t - t_{\text{off}}}{\tau}\right) \quad (17)$$

It is suggestive to determine the time constant τ such that the total energy dissipation due to the exponential function is equal to the real energy consumption. Equating this postulation results in $\tau = T_{\text{inertia}}^{eff}$.

5 Minimum powerdown time

Turning a circuit block off and on causes not only intrinsic energy dissipation W_{intr} but also extrinsic losses W_{extr}. These losses consist of losses in the switch device, in the switch driver and in the power-down-logic. Additionally there are losses during the power-up process (power-up glitches, $W_{\text{boot, cap}}$). Summing up all these energy dissipations results in the total energy penalty due to the power switching: $W_{\text{tot}} = W_{\text{intr}} + W_{\text{extr}}$.

The power saving due to the power switching at time $t > t_{\text{off}}$ can be calculated by

$$P_{\text{save}}(t) = \left(\Delta I_L - i_{\text{inertia}}(t)\right) vdd \quad \Delta I_L := I_L^{\text{on}} - I_L^{\text{off}} \quad (18)$$

If the circuit block is switched on again at time $t_{\text{off}} + T$ the energy penalty and the saved energy are given by

$$W_{\text{loss}} = vdd \int_{t_{\text{off}}}^{T + t_{\text{off}}} i_{\text{inertia}}(t)\, dt + W_{\text{extr}} \quad (19)$$

$$W_{\text{saved}} = vdd \int_{t_{\text{off}}}^{T + t_{\text{off}}} \Delta I_L\, dt \quad (20)$$

Equating these two expressions results in the so called energy equivalent powerdown time T_{eeq} which describes the minimum time for which it is useful to switch the circuit block off under power considerations:

$$\frac{W_{\text{extr}}}{vdd} = \Delta I_L \left[T_{eeq} + T_{\text{inertia}} \left(\exp\left(-\frac{T_{eeq}}{T_{\text{inertia}}}\right) - 1 \right) \right] \quad (21)$$

If the extrinsic losses are large, $T_{eeq} \gg T_{\text{inertia}}$ and therefore

$$T_{eeq} \approx \frac{W_{\text{extr}}}{vdd\,\Delta I_L} + T_{\text{inertia}} \quad (22)$$

Although this time is the main contributor to the minimum powerdown time T_{min} there are additional terms due to the signal propagation delay in the power-down-logic and the switch driver. Additionally, there is a certain settling time $T_{\text{power on}}^{CB}$ before the circuit block can be used after block activation. The minimum power-down time is the sum of all these contributors:

$$T_{\text{min}} = 2T_{\text{tot}}^{\text{switch}} + T_{eeq} + T_{\text{power on}}^{CB}. \quad (23)$$

6 Dynamic behaviour of CBSO-systems

As the cut-off device in a CBSO-system has a finite on-resistance, the supply current through the logic block causes a voltage drop across the switch. Hence the effective supply voltage is reduced and the signal propagation delay in the circuit increases. The voltage-delay dependence of static CMOS gates can be derived by analytic MOSFET models like the alpha-power-law:

$$d \propto \frac{vdd}{(vdd - v_{th})^{\alpha}} \quad (24)$$

In this equation vdd is the nominal supply voltage, v_{th} is the threshold voltage and α is a technology dependent coefficient with values between 1 and 2. As the whole supply current of a power switched circuit flows across the cut-off switch, the delay degradation depends on the switching activity of all gates assigned to the switch. Therefore not only the time critical paths but the ensemble of all gates which switch their outputs while the signal propagates through the critical paths determine the overall delay of the circuit. Furthermore, the voltage-delay characteristic Eq. (24) is nonlinear and hence it is not a trivial task to determine the optimum dimensioning of the switch device. A large switch transistor reduces the delay degradation but suffers from large area consumption and poor leakage supression. In order to estimate the delay degradation for a given circuit if a certain cut-off switch is added, the supply current profile of the circuit without cut-off switch is assumed to be known. The addition of a cut-off switch causes delay degradation but does not affect the logic function of the circuit. Hence the switching events are the same whether a cut-off switch is inserted or not and the current profiles of the two cases are similar.

The supply current of a static CMOS gate is given by the sum of the static leakage current, the short circuit current during the output transition and the dynamic current which charges the load capacitance. As the effective supply voltage is reduced due to the voltage drop across the switch device, the charge which is necessary to charge the load capacity is reduced. The voltage swing of the virtual ground line in a n-block-switching scheme is small in order to keep delay degradation low. Hence the charge on the nonlinear load capacitance of an arbitrary gate can be estimated by linear taylor approximation: $Q(v)=c(v)v$ describes the dependence of the charge on this capacity on the voltage. In the vicinity of vdd, $c(v)$ is assumed to be a weak function of v. If the effective supply voltage is reduced by $vvss$, the charge necessary to charge up the load capacity can be expressed by

$$Q(vdd - vvss)=Q(vdd)-\frac{dQ}{dv}(vdd)vvss$$
$$=c(vdd)vdd-c(vdd)vvss-\frac{dc}{dv}(vdd)vdd\,vvss$$
$$\approx c(vdd)\left[vdd - vvss\right]. \tag{25}$$

A similar approach can be done for the charge that is injected into the virtual power rail due to short circuit and leakage currents. Hence the supply current of a logic circuit can be decomposed according to

$$i(t) = \frac{dq}{dt} = \zeta(t)vdd(t) \tag{26}$$

into an effective conductivity $\zeta(t)$ and the momentary supply voltage. If there is no cut-off switch the supply voltage is constant and ζ is given by $\zeta(t)=\frac{i(t)}{vdd}$. Using this expression the charge which is injected into the ground node during the time interval $[t; t+dt]$ can be expressed by $dq=\zeta(t)dt\,vdd$. Assuming that the supply current profile of a large circuit block is given and assuming further that this current profile results from many discrete switching events, this current profile can be partitioned into small current pulses of width $d\tau$. The distinct current pulses are shaped in the following way in order to estimate the current profile of the circuit with cut-off switch: The charge which is injected into the virtual ground node is reduced due to the reduced effective supply voltage. Hence ζ has to be multiplied by this effective supply voltage $vdd-vvss(t)$. Moreover the time interval during that this charge is injected is stretched with respect to the time interval if there was no switch device. Assume $i(\tau)$ is the current profile of a given circuit block without cut-off switch for a certain input transition. The dynamic of this circuit is described in the time variable τ. The current profile $i(t)$ of the circuit block with cut-off switch has to be determined by shaping the given profile. The time variable t describes the dynamic of the power switched circuit block. The charge which is injected into the virtual ground node during the time interval $[t; t + dt]$ is given by

$$dq(t) = \zeta(\tau(t)) \cdot [vdd - vvss(t)]\,d\tau(t) \tag{27}$$

The relation between the two infinitesimal time elements dt and $d\tau$ is given by the normalized supply voltage dependent delay degradation $\delta(v)$ of a CMOS gate:

$$dt = \delta(vdd - vvss(t))\,d\tau \tag{28}$$

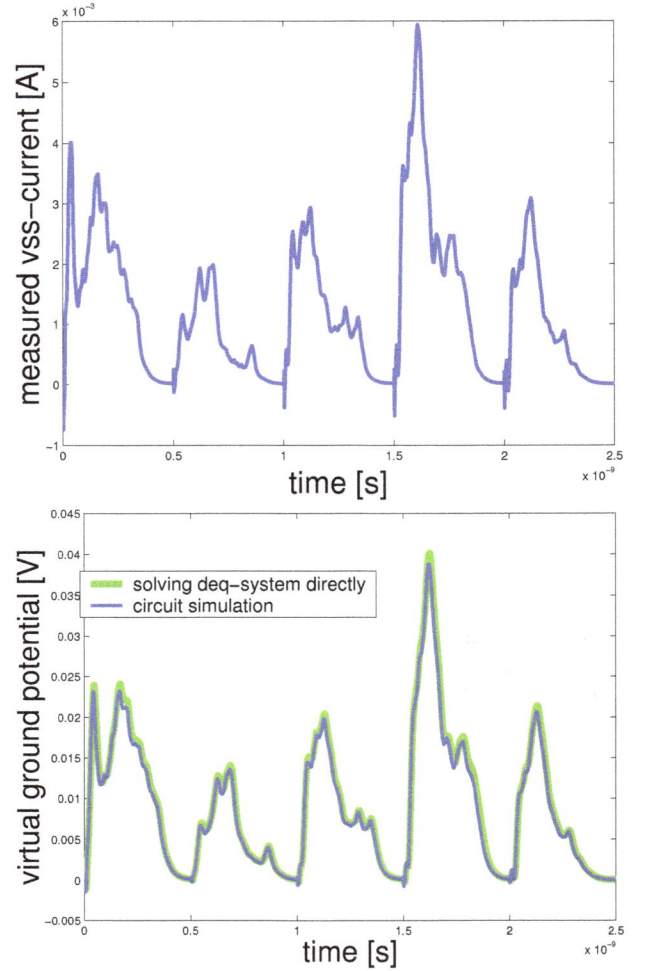

Fig. 6. Example for a current supply-current-profile and the resulting potential of the virtual ground line. The congruence of the curve determined by analog circuit simulation and the direct solving of the derived deq-system is very good.

Combining Eqs. (27) and (28) results in an expression for the current profile in the t domain which can be inserted into the Kirchhoff current equation of the virtual ground node:

$$\frac{dQ}{d\,vvss}v\dot{v}ss(t) + I(vvss(t))$$
$$= \zeta(\tau(t))\delta^{-1}(vdd - vvss(t))[vdd - vvss(t)] \tag{29}$$
$$dt = \delta(vvss)d\tau \tag{30}$$

Solving this system of ordinary differential equations results in an estimation of the current profile of the circuit with current switch. The nonlinear functions $q_c=Q(vvss)$ and $i_R=I(vvss)$ describe the charge on the $vvss$-capacitance and the current through the switch device respectively. These two functions have to be determined once by analog simulation and then many different current profiles can be evaluated by the fast current-profile-shaping algorithm. Figure 6 shows a given current profile of a 32−bit Kogge-Stone adder for five input transistions. The resulting $vvss$-potential after the insertion of a certain cut-off switch is given in the lower plot. The congruence of the voltage profile acquired by analog circuit simulation and by solving the deq-system respectively is very good.

Thus many different input transistions, switch sizes or switch device types can be examined by this fast compact simulation methodology.

7 Conclusion

In this paper we have described analytically the bahaviour of a power switched circuit block during and after the block deactivation. The minimum power down time has been defined and the main contributors have been derived. Finally a fast analytic methodology to estimate the influence of a cut-off switch on the dynamic performance of a circuit has been proposed.

References

Henzler, S., Koban, M., Berthold, J., Georgakos, G., and Schmitt-Landsiedel, D.: Design Aspects and Technological Scaling Limits of ZigZag Circuit Block Switch-Off Schemes, IFIP International Conference on Very Large Scale Integration of System-on-Chip, 2003.

Mutoh, S., Douseki, T., Matsuya, Y., Aoki, T., Shigematsu, S., and Yamada, J.: 1-V Power Supply High-Speed Digital Circuit Technology with Multithreshold-Voltage CMOS, IEEE Journal of Solid-State Circuits, 30, 8, 847–854, 1995.

Kawaguchi, H., Nose, K., and Sakurai, T.: A Super Cut-Off CMOS (SCCMOS) Scheme for 0.5-V Supply Voltage with Picoampere Stand-By Current, IEEE Journal of Solid-State Circuits, 35, 10, 1498–1501, 2000.

Min, K., Kawaguchi, H., Sakurai, T.: Zigzag Super Cut-off CMOS (ZSCCMOS) Block Activation with Self-Adaptive Voltage Level Controller: An Alternative to Clock-Gating Scheme in Leakage Dominant Era, IEEE Solid State Circuits Conference, 2003.

Meer, P. R. and Staveren, A.: New standby-current reduction technique for deep sub-micron VLSI CMOS circuits: Smart Series Switch, European Solid State Circuits Conference, 2002.

27

Resonant 90 degree shifter generator for 4-phase trapezoidal adiabatic logic

A. Bargagli-Stoffi[1], E. Amirante[1], J. Fischer[1], G. Iannaccone[2], and D. Schmitt-Landsiedel[1]

[1]Institute for Technical Electronics, Technical University Munich, Theresienstrasse 90, D-80290 Munich, Germany
[2]Dipartimento di Ingegneria dell'Informazione, Università degli Studi di Pisa, Via Diotisalvi 2, I-56122 Pisa, Italy

Abstract. Many adiabatic logic families make use of multi phase trapezoidal or sinusoidal power clocks to recover the energy stored in the load capacitances. A key aspect for the evaluation of the performance of adiabatic logic is then the study of a system that includes the power clock generator. A four-phase trapezoidal power clock generator, according to the requirements of the most promising architectures, namely the ECRL and PFAL, has been designed and simulated. The proposed circuit, realized with a double-well 0.25 μm CMOS technology and external inductors, is a resonant generator designed to oscillate at a frequency of 7 MHz, which is within the optimum frequency range for adiabatic circuits realized with this CMOS technology. The generator has been simulated with the equivalent load of fifty 1-bit adders and the operating behavior of a 4-bit adder has been evaluated. The key aspects of a generator for adiabatic logic are its power consumption and the phase relationships between its output signals. The proposed generator has a conversion efficiency higher than 80%, and it is robust with respect to variations of technology parameters. The four power supplies exhibit the correct relationship of phase also in the presence of no equally distributed loads.

1 Introduction

Power consumption is a crucial requirement of present and future circuits and systems since the increasing demand of portable electrical applications makes the tradeoff of computing power versus battery life time more critical. Furthermore, the number of gates per chip area is constantly increasing, while the gate switching energy does not decrease at the same rate, so power consumption rises and heat removal becomes more difficult and expensive. Then, to limit the power consumption, alternative solutions at each level of abstraction are proposed.

In static CMOS, any circuit can be described as a load capacitor and two switches that connect it alternately to the power supply or to ground, and hence abruptly modify its voltage. If the constant power supply is replaced by a time variable power clock, with a rise time longer than the time constant of the charging path (Athas et al., 1994), the switching operation can be accomplished with ideal vanishing dissipation. Furthermore, when the capacitor is connected to the decreasing power clock, its charge flows back to the power source. Circuits that fulfill the low power requirements recovering the energy through time variable power supplies (power clocks) are called "adiabatic". In literature, choosing different power clocks and methods to control the switching, many multi-phase adiabatic families have been proposed by exemplifying the operating principle with simple circuits and emphasizing the energy saving with respect to standard CMOS. Although these studies illustrate important aspects of the logic circuits, they usually are not comprehensive since the energy dissipated in the DC/AC converter that generates the multi-phase clock is often not taken into account. In some papers the converter is not considered ideal, but the designed solutions have not very high efficiency (Moon and Jeong, 1998) or the generated signals differ from the optimal waveform for the adiabatic circuits (Maksimovic and Oklobdzija, 1995; Moon and Jeong, 1996, 1998).

We believe that a key aspect in the evaluation of the potential of adiabatic logic families is the performance of a complete system, including the power clock generator and the interface with conventional CMOS logic. In this paper, a high efficiency power supply generator is presented, which can be used for logic circuits implemented using three adiabatic families, namely ECRL (Moon and Jeong, 1996, 1998), PFAL (Vetuli et al., 1996; Blotti et al., 2000), and 2N-2N2P (Kramer et al., 1995; Liu and Lau, 1998). These families require, as power supplies, four trapezoidal waveforms equally spaced in phase.

The paper is organized as follows: in Sect. 2 some oscillators proposed in literature are overviewed, illustrating advantages and drawbacks of each architecture. In Sect. 3 the new oscillator and its amplitude and frequency expressions are introduced. In Sect. 4 the simulation results, such as conversion efficiency and sensibility to parameter variations, are reported. In Sect. 5 an alternative solution is evaluated.

2 Overview of adiabatic oscillators

The main requirements for a DC/AC converter for adiabatic circuits are the capability to recover the energy stored in the load capacitances, and a high power-conversion efficiency, defined as the ratio of the load power to the total DC supply power. Oscillators based on LC resonant circuits can meet these requirements, and therefore ensure that the complete adiabatic system provides significant power saving with respect to its standard CMOS counterpart. In the few schemes appeared in the literature, the adiabatic logic is usually represented by its equivalent load, i.e. a resistance R_e and a capacitance C_e. In the 1N single-phase power clock generator (Maksimovic and Oklobdzija, 1995) (see Fig. 1) an inductance L connects the DC power supply to the logic, that has a nMOSFET in parallel. The oscillator generates only one sinusoidal phase φ_0 and needs a square waveform control signal ctr. The output φ_0 has the same frequency as the control signal, and its amplitude is determined by the ctr duty cycle. When the output signal frequency is close to the resonant frequency, the generator has a conversion efficiency around 70%.

In Fig. 2 is reported the 2N2P two-phase power clock generator (Maksimovic and Oklobdzija, 1995; Moon and Jeong, 1996, 1998; Ye and Roy, 2001), which provides two sinusoidal clocks and requires only one inductor. If control signals are external, the frequency is easily enforced and with some additional circuitry two oscillators can be merged into a 4-phase generator. In this case, the conversion efficiency is around 40–50% (Moon and Jeong, 1996).

Because the adiabatic loads driven by the oscillator outputs are different and time variable (Kim et al., 2001), the circuit cannot be completely balanced with external capacitances, therefore a phase error up to 5% may occur in the output waveforms. If outputs are used as internal control signals, i.e. if $c_1 = \varphi_1$ and $c_2 = \varphi_2$, the circuit complexity and occupied area are reduced, but also the conversion efficiency decreases (Liu and Lau, 1998).

3 Shifting oscillator

The oscillator proposed in this paper is realized as a ring of four low-power 90-degree shifters, as shown in Fig. 3. Each shifter is realized with a CMOS inverter and an LC resonant circuit, so that the energy is transferred between reactive elements while the DC power supply only delivers the energy dissipated on the resistance and on the diodes.

Each output of the oscillator drives a stage of the adiabatic circuit, represented in Fig. 3 by its equivalent load, i.e. a resistance R_e and a capacitance C_e. The inductors are external to the chip. The output amplitude regulation between

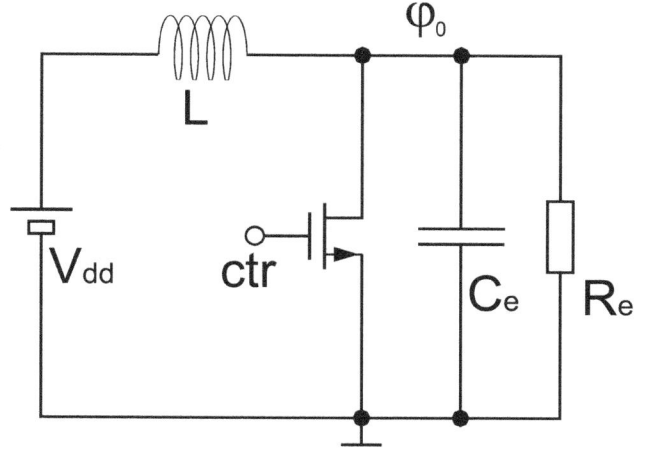

Fig. 1. The 1N generator.

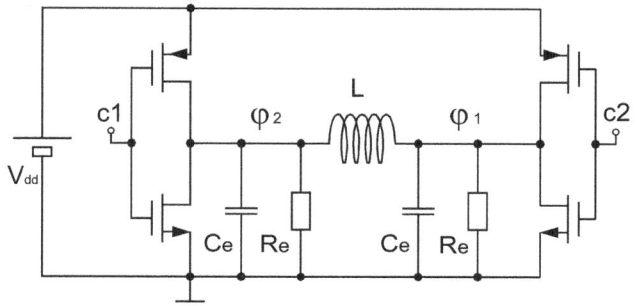

Fig. 2. The 2N2P generator.

0 V and V_{dd} is achieved with Schottky diodes, whose low V_γ ensures moderate energy dissipation when diodes are in conduction. The proper configuration of the reactive elements provides the required phase delay. Therefore, when the circuit oscillates, the four outputs have the same frequency and a quarter-period shift, as required by the considered adiabatic architectures.

To properly dimension the elements of the oscillator, let us consider its analytical transfer function. The ring transfer function is the product of four identical single-stage transfer functions, therefore Barkhausen criterion is met when a single-stage transfer function has a gain equal to 1 and a phase delay equal to $k\pi/2$. In a small signal approximation, the expression of a single-stage transfer function is:

$$\frac{\varphi_1(s)}{\varphi_0(s)} = \frac{-R_e \, r_{ds} \, (g_{mp} + g_{mn})}{r_{ds} + R_e + (L + r_{ds} R_e C_e)s + R_e C_e L s^2} \quad (1)$$

where r_{ds} is the small signal output resistance of the MOSFETs, and g_{mp} and g_{mn} are the pMOSFET and nMOSFET transconductances, respectively. By enforcing a phase of $3\pi/2$, we obtain the following expressions for the frequency and for the gain value:

$$f_0 = \frac{1}{2\pi} \sqrt{\frac{r_{ds} + R_e}{R_e C_e L}} \simeq \frac{1}{2\pi} \sqrt{\frac{1}{C_e L}} \quad (2)$$

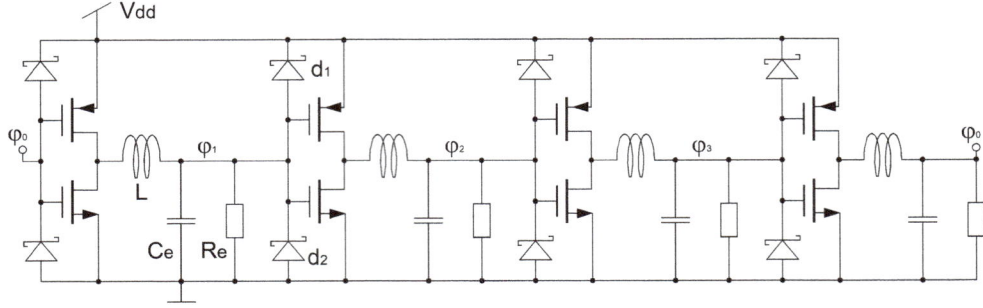

Fig. 3. Schematic of the 4-phase shifter oscillator.

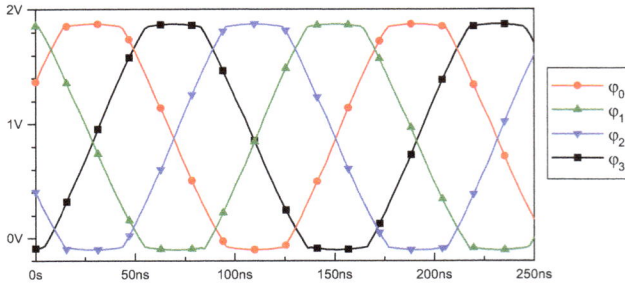

Fig. 4. Waveforms of the 4-phase power clock generator.

$$\left| \frac{\varphi_1}{\varphi_0} \right| = \frac{R_e\, r_{ds}\,(g_{mp} + g_{mn})}{L + r_{ds}\, R_e\, C_e} \sqrt{\frac{R_e\, C_e}{(r_{ds} + R_e)\, L}} \qquad (3)$$

In our simulation, a standard double-well 0.25 μm CMOS technology and a DC supply voltage of 1.8 V are considered. R_e and C_e are chosen as the equivalent impedance seen by the power supply of fifty 1-bit adders, so R_e is 2 $M\Omega$ and C_e is 500 fF. Since the considered adiabatic families with this technology process have the optimal frequency range between 1 and 10 MHz (Amirante et al., 2001), the clock frequency is chosen equal to 7 MHz. From Eq. (2), the inductance value is derived to be equal to 1 mH. Since the transfer function is based on a simplified circuit that does not consider the effect of diodes, the gain value must be larger than 1 to ensure oscillation. Nevertheless high gain values lead to square waveforms and to longer period of diode conduction, therefore the optimum gain is found between 1.5 and 3. To reduce dissipation on the channel resistance, the pMOSFET W/L ratio is 10 times larger than of the nMOSFET W/L.

4 Simulation results

The circuit functionality has been simulated with PSpice. The conversion efficiency η, defined as the ratio of the energy dissipated on the load to the total energy delivered by the DC supply, is one of the most important parameters of the generator, since energy dissipation is the main concern of adiabatic architectures. The oscillator, driving the equivalent load of the adiabatic system (R_e, C_e), shows a conversion efficiency of 85%. The time relationship among the power clock phases is another important specification, since between the input signals and the power clock a 90 degree delay must be present . The phase error is defined as the distance in degrees between the real waveform and the theoretical one. In the nominal conditions the proposed oscillator has a maximum phase error of 0.9 degrees.

The threshold voltages of the n- and p-MOSFETs are $V_{Tn} = 0.44$ V and $V_{Tp} = -0.43$ V, respectively. Increased efficiency can be obtained with higher threshold voltage devices. For this reason, also the possibility to develop the circuit with high V_T MOSFETs has been considered, and the simulation results with $V_{Tn} = -V_{Tp} = 0.9$ V are presented.

To evaluate not only the conversion efficiency of the oscillator, but also to evaluate the quality of the generated power clocks, a more complete adiabatic system is simulated. The system is made up of the proposed oscillator, of the synchronizers for the external inputs (Fischer et al., 2002), and of a 4-bit pipelined Ripple Carry Adder realized in ECRL logic. With this load the oscillator presents a conversion efficiency of 82.3% with the high V_T MOSFETs, and of 75.6% with the normal V_T MOSFETs. In both cases the output waveforms are close to be trapezoidal and exhibit the correct relationship of phase (Fig. 4).

The power conversion efficiency has been evaluated when variations of the value of a single reactive parameter occur. Since the inductances are external elements characterized by a tolerance range around the nominal value, our simulations take in account the worst case of technological parameter shift, whereas the capacitive load on each phase is determined by the logic function that the phase itself drives, and its value is usually a function of time, therefore the simulated capacitor conditions must predict the behavior of the circuit in case of the worst operating variations.

The simulation results reported in Table 1 are obtained modifying the value of one reactive element in a single shifter stage (whose modified value is reported in brackets), and leaving unaltered the values of all other reactive parameters.

Using high V_T devices, a 30% variation of one capacitance value causes a reduction of the conversion efficiency of less than 2.3%, while the frequency variation is less than 3%. With a 30% variation of the value of a single inductance, the efficiency is reduced by less then 1%, and the frequency variation amounts to 5%. The phase error of the oscillator

Table 1. Simulation results of the oscillator driving the adiabatic logic gate implemented with high (left) and low (right) V_T MOSFETs

V_T [V]	L [mH]	C [fF]	η [%]	f [MHz]	V_T [V]	L [mH]	C [fF]	η [%]	f [MHz]
0.9	1	500	82.3	7.13	0.4	1	500	75.6	7.14
0.9	1	500 (650)	80.0	6.94	0.4	1	500 (650)	74.0	6.94
0.9	1	500 (350)	81.9	7.35	0.4	1	500 (350)	75.2	7.40
0.9	1 (1.3)	500	81.4	6.93	0.4	1 (1.3)	500	72.9	7.96
0.9	1 (0.7)	500	81.6	7.48	0.4	1 (0.7)	500	74.0	7.54

with the nominal load presents is only 1 degree, while its worst case performance in presence of parameter variations amounts to 5.8 degrees. The oscillator with high V_T devices is therefore not only characterized by higher efficiency value than the solution with low V_T devices, but also by better robustness against parameter variations.

5 Alternative solution without Schottky diodes

Since Schottky diodes are not usually available in a standard CMOS process, alternative solutions have been investigated, in particular, the possibility to replace the diodes with MOS-FETs. In this alternative design, the diodes *d1* and *d2* (Fig. 3) are replaced by a pMOSFET and a nMOSFET, respectively, with their gates connected to their sources. The conversion efficiency of this solution simulated with an equivalent load amounts to 76%, instead of the 85% obtained with the Schottky diodes. Moreover, this implementation also exhibits other drawbacks: the clipping of the output waveforms is not accurate, therefore the power-clocks reach voltages lower than 0 V and higher than V_{dd} (-0.7 V; 2.6 V), and the waveforms are almost sinusoidal rather than trapezoidal. To reduce these effects, the diodes are replaced by low V_T MOSFETs ($V_T = 0.2$ V) whose sources are connected to voltage references ($V_{ref,p} = 1.3$ V; $V_{ref,n} = 0.5$ V). In this case, the efficiency amounts to 76%, and the dynamic range of the power supplies is comparable to the results obtained with the implementation using Schottky diodes (-0.3 V; 2.1 V).

6 Conclusions

A high conversion efficiency oscillator capable to generate the 4-phase trapezoidal power-clocks required by many adiabatic logic families has been presented. The generator utilizes Schottky diodes and shows operation at 7 MHz, which is within the optimum frequency range for the considered 0.25 μm CMOS technology. To evaluate the performance of a complete adiabatic system, the oscillator has been simulated driving a pipelined 4-bit adder. The generator produces almost trapezoidal output signals, without the need for any auxiliary control circuit, and it has a conversion efficiency higher than 80%. The robustness to technological and operational parameter variations has been characterized. In case of a 30% variation of a reactive element value, the conversion efficiency decreases only by 2%, while the frequency

variation amounts to less than 5%. In addition, the generator exhibits a low energy dissipation even if the load capacitances are not equally distributed on the phases. To avoid the use of Schottky diodes, an alternative solution using low V_T MOSFETs has been discussed, together with its cost in terms of circuit complexity. In this case, the conversion efficiency amounts to 76%. The proposed oscillator, with its high efficiency and its almost ideal trapezoidal waveforms, gives to the adiabatic logic the possibility to compete with static CMOS logic in low power applications.

References

Amirante, E., Bargagli-Stoffi, A., Fischer, J., Iannaccone, G., and Schmitt-Landsiedel, D.: Variations of the Power Dissipation in Adiabatic Logic Gates, Proc. 11th Int. Work. on Power And Timing Modeling, Optimization and Simulation, Yverdon-les-Bains, Switzerland, Sep., pp. 9.1.1–9.1.10, 2001.

Athas, W. C., Svensson, L., and Koller, J. G.: Low-power digital systems based on adiabatic-switching principles, IEEE Transactions on VLSI Systems. 2, 398–407, 1994.

Blotti, A., Di Pascoli, S., and Saletti, R.: Sample Model for positive feedback adiabatic logic power consumption estimation, Electronics Letters, Vol. 36, No. 2, pp. 116–118, Jan. 2000.

Fischer, J., Amirante, E., Bargagli-Stoffi, A., and Schmitt-Landsiedel, D.: Adiabatische Schaltungen: Wandler für statische CMOS-Eingangssignale, U.R.S.I. Kleinheubacher Tagung 2002, Miltenberg, Deutschland, 30 September–2 October, 2002.

Kim, S., Ziesler, C. H., and Papaefthymiou, M. C.: A True Single-Phase 8-bit Adiabatic Multiplier, DAC2001, Las Vegas, Nevada, USA, 758–763, 2001.

Kramer, A. and Denker, J. S.: 2nd order adiabatic computation with 2N-2P and 2N-2N2P logic circuits, Proc. Intern. Symp. Low Power Design, 191–196, 1995.

Liu, F. and Lau, K. T.: Improved structure for efficient charge recovery logic, Electronics Letters, 34 (18), 1731–1732, 1998.

Maksimovic, D. and Oklobdzija, V. J.: Integrated power clock generators for low energy logic, 26th Annual IEEE Power Electronics Specialists Conf., Atlanta, GA, June 1995, 61–67, 1995.

Moon, Y. and Jeong, D.-K.: An efficient charge recovery logic circuit, IEEE J. Solid-State Circuits, 31, 514–522, 1996.

Moon, Y. and Jeong, D.-K.: A 32×32-bit adiabatic register file with supply clock generator, IEEE J. Solid-State Circuits, 33 (5), 696–701, 1998.

Vetuli, A., Di Pascoli, S., and Reyneri, L. M.: Positive feedback in adiabatic logic, Electronics Letters, 32, 20, 1867ff, 1996.

Ye, Y. and Roy, K.: QSERL: Quasi-Static Energy Recovery Logic, IEEE J. Solid-State Circuits, 36 (2), 239–248, 2001.

28

Beam-Splitting in plasmonic multimode waveguides based on the self-imaging effect

A. Edelmann, S. Helfert, and J. Jahns

FernUniversität in Hagen, Universitätsstr. 27/PRG, 58084 Hagen, Germany

Correspondence to: A. Edelmann (andre.edelmann@fernuni-hagen.de)

Abstract. A plasmonic 1×2 beamsplitter based on the self-imaging effect is analysed. The simulations were performed by a 3-D full-vectorial numerical calculation. We discuss the coupling efficiency depending on structural parameters and on the attenuation.

1 Introduction

The field of plasmonics deals with electromagnetic waves which propagate at the interface between a dielectric and metallic medium. These electromagnetic waves are called surface plasmon polaritons (SPPs). The SPPs couple with the free electrons of the metal excited by an evanescent field, resulting e.g. from a prism or grating. The incident light has to be TM polarised to obtain an electric component normal to the interface. The SPPs have the same frequency with the incident light but due to the larger k-vektor a smaller wavelength. A profound knowledge of the physical understanding of SPPs is given, e.g., in Raether (1988). Plasmonics offer the potential towards higher integration densities for optical circuits by combining the high capacity in photonics and the miniaturisation technologies in electronics. One of the challenges in plasmonics refers to the attenuation of the propagating surface waves caused by the metal. Here propagation lengths of several tens of micrometers are achievable. The attenuation depends on various parameters, e.g., the wavelength or the dielectric media and can be generally reduced by accepting a higher extension of the field in the dielectric. More detailed information about the potential of plasmonics are specified in Ozbay (2006); Maier and Atwater (2005).

Beam splitters are an important functionality for integrated optical circuits. Several concepts in plasmonics have been developed, e.g., Y-shaped splitters (Han et al., 2010), T-shaped splitters (Veronis and Fan, 2005) and splitters relates to the multimode interference (MMI) (Han and He , 2007; Yuan et al., 2009). Here, we focus on beam splitters re-

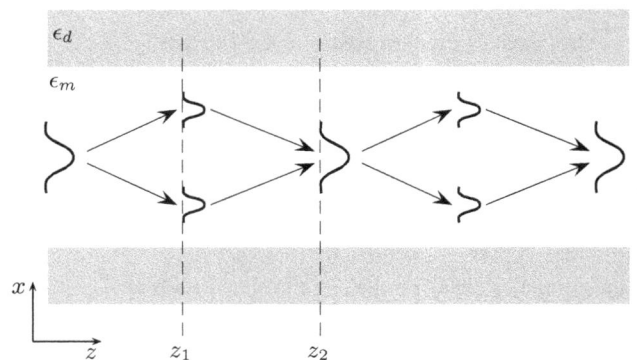

Fig. 1. Self-imaging effect in a multimode waveguide. At position z_1 the field is split and at the position z_2 the input field is reproduced.

garding the MMI. These are based on the self-imaging phenomenon. Our intention is to find out detailed information about the characteristic regarding the beam splitter by varying several structural parameters.

This article is organised as follows: in Sect. 2 we show how the beam splitting can be achieved in multimode waveguides; in Sect. 3 we describe the numerical algorithm and the model of the structure; in Sect. 4 we present and discuss our findings and in Sect. 5 we conclude our work.

2 Self-imaging based beam splitting

The self-imaging effect in dielectric and plasmonic waveguides leads to periodic repetition of the injected field profile (Soldano and Pennings, 1995; Edelmann et al., 2010). Fig. 1 shows the principle. One can find a 2-D plasmonic waveguide with the metallic core at the permittivity ϵ_m in the center surrounded by a dielectric media with the permittivity ϵ_d. As indicated in the figure, we inject a field with narrow shape (no eigenmode) on the left into the waveguide. The input field propagates in z-direction. As shown, we find doubled self-images of the field at the position z_1 and an image of the input field distribution at the position z_2. The self-imaging phenomenon occurs because of interference of the excited modes in the waveguide. Note: in Fig. 1 we have shown the self-imaging effect for the special case of a symmetric field profile which is injected at the center of the waveguide. The periodicity of the self-imaging is determined by $z_2 = 3/8\lambda_0/\Delta\bar{\beta}$. Here, $\Delta\bar{\beta}$ means the difference in normalized propagation constants of the two lowest-order modes. More specified information can be found in (Soldano and Pennings, 1995).

Based on the self-imaging effect in multimode waveguides, one can realise a beam splitters by out-coupling the doubled images via two separated waveguides at the position z_1. Approaches for that kind of beam splitting in plasmonic waveguides are studied in e.g., Han and He (2007); Yuan et al. (2009).

3 Numerical modelling of the structure

Our results are gained with full vectorial calculations using the Method of Lines (MoL). Here we restrict ourselves to a brief summary, detailed information can be found in e.g., Pregla and Pascher (1989); Pregla (2008). The MoL is an eigenmode propagation method. The cross-section of the considered structure is discretized with finite differences (FD). The eigenmodes are obtained by combining the FD expressions in an operator matrix. Then the eigenvalues and eigenvectors of this operator matrix represent the propagation constant and the field distribution of the eigenmodes. The further calculations are done analytically.

Figure 2a shows the cross-section of the plasmonic waveguide which is used in our studies. One can find the metallic layer in the center surrounded by dielectric medium. The computation window has the dimensions $w_x = 4.15\,\mu m$ and $w_y = 0.875\,\mu m$ with the discretisation distances $h_x = 0.05\,\mu m$ and $h_y = 0.0125\,\mu m$. The calculations are carried out at the wavelength $\lambda = 0.633\,\mu m$. For the metallic layer width we choose $w_m = 3\,\mu m$ and for the thickness $t_m = 0.125\,\mu m$. For the metal we use gold with the permittivity $\epsilon_m = -14.2035 - j0.7621$ (obtain from the Drude-model (Palik, 1985) with the time dependence according to $e^{+j\omega t}$) and for the dielectric we choose $\epsilon_d = 4$. The calculated field profile of the first fundamental mode is also plotted in Fig. 2a. As is known, the maxima of the field are located at the interfaces in y-direction and exponentially decay into the dielectric media and the metal. The longitudinal section at the metallic/dielectric interface of the analysed beam splitter is shown in Fig. 2b. The geometrical parameters of the structure are: width of the input waveguide $w_i = 1.1\,\mu m$; length of the input waveguide $z_i = 1.0\,\mu m$; width of the middle multimode waveguide $w_m = 3.0\,\mu m$; length of the middle wave-

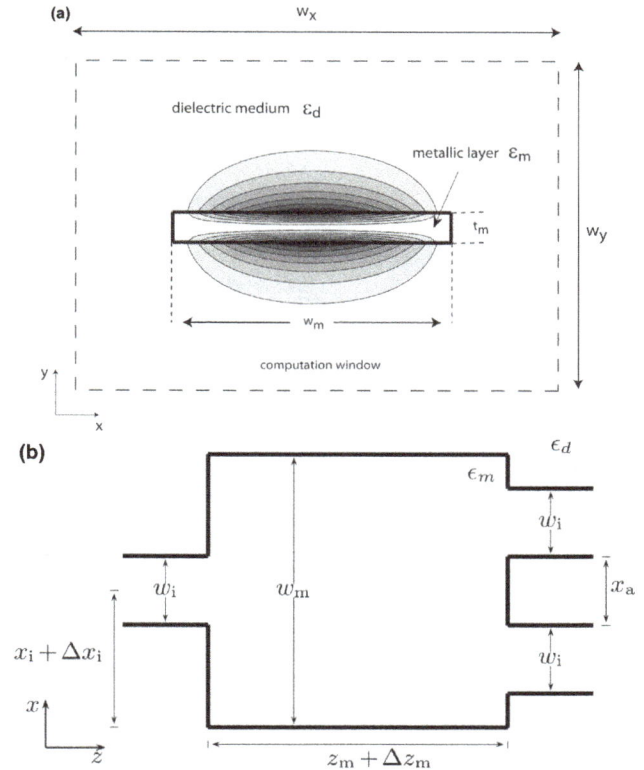

Fig. 2. (a) Cross-section of the computed structure and field profile of the first fundamental mode. **(b)** Longitudinal section of the analysed beam splitter.

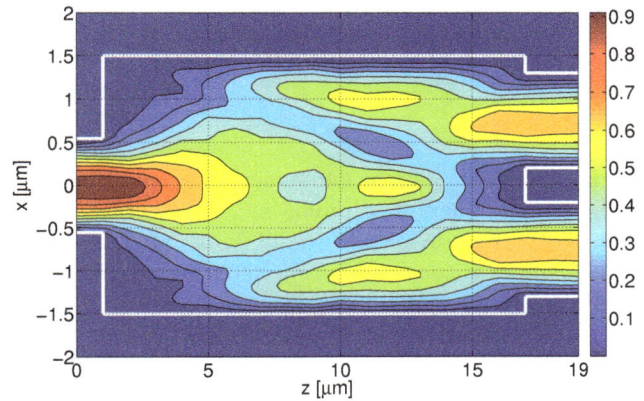

Fig. 3. Calculated field profile of the plasmonic beam splitter shown in Fig. 2 (losses are neglected).

guide $z_m = 16\,\mu m$. The output waveguides have the same dimensions as the input waveguide. The output waveguides are symmetrically placed around the centre of the structure with distance between them of $x_a = 0.4\,\mu m$.

In the following we examine the influence of a few geometrical parameters.

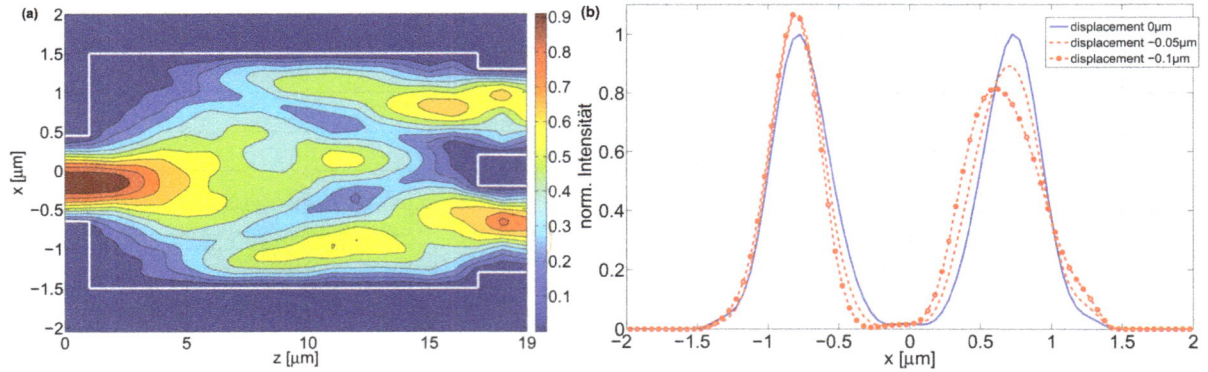

Fig. 4. (a) Calculated field (3-D) at the metallic/dielectric interface. The input waveguide is shifted by $\Delta x_i = -0.1\,\mu$m to the central position x_i. **(b)** Related intensity profile at the output waveguide for for the shifts $\Delta x_i = -0.05\,\mu$m and $\Delta x_i = -0.1\,\mu$m.

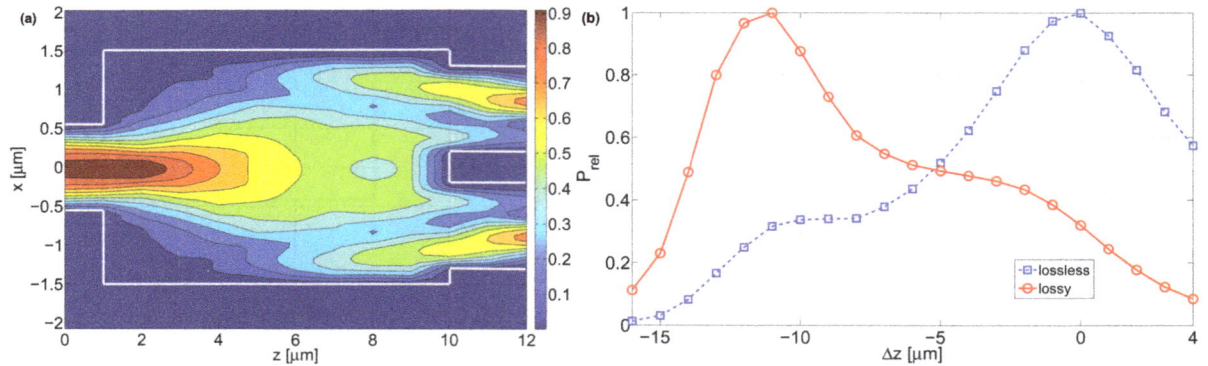

Fig. 5. Calculated field (3-D) at the metallic/dielectric interface. The middle waveguide is shorted by $\Delta z_m = -6\,\mu$m relative to the optimal position z_m (losses are neglected). **(b)** Peak power in the output waveguides normalised to peak power at the optimal coupling position.

4 Results

In this section we present the numerical results. Figure 3 shows the calculated field profile in the dielectric/metallic interface for the structure shown in Fig. 2. The fundamental mode of the input waveguide is injected into the middle multimode waveguide. Within the middle waveguide the field is symmetrically split due to the self-imaging effect into two separate images which couple into the output waveguides. Below we discuss the influence of the position of the input waveguide in dependance of Δx_i and the influence of the length of the middle waveguide by varying Δz_m.

4.1 Position of the input waveguide

Figure 4a shows the field distribution when the input waveguide is shifted at $\Delta x_i = -0.1\,\mu$m. The field distribution in the middle waveguide and in the output waveguides is no longer symmetric. In Fig. 4b the intensity profile in the output section is plotted. The continuous curve relates to the unshifted case. The dashed lines relate to a shift in negative di-

rection with $\Delta x_i = -0.05\,\mu$m and $\Delta x_i = -0.1\,\mu$m. If we increase the displacement of the input waveguide the intensity for upper output waveguide decreases, whereas the intensity in the lower output waveguide remains almost constant. The source of that characteristic results from the unsymmetric interference of the excited modes at the out-coupling position. Note: Due to the symmetric structure we find analogue results if we shift the input waveguide in positive direction.

4.2 Length of the middle waveguide

Next we analysed the influence of the length of the middle waveguide z_m. In Fig. 5a the field distribution for $\Delta z_m = -6\,\mu$m (i.e. the length of the middle waveguide is reduced) is plotted. If we compare this figure with Fig. 3 we find significant deviations of the field distribution due to the worse coupling position. To quantify the relation between the coupling efficiency and the length of the center waveguide we determine the maximum of the field intensity in the output section (as shown in Fig. 4b) as function of the deviation of the center waveguide Δz_m. The result can be seen in Fig. 5b. The dashed line shows the behaviour for neglected losses whereas the continuous line takes the losses of the metal into account. Assuming lossless metal, the maximum intensity is located at $\Delta z_m = 0$ and decays by shifting Δz_m in positive or negative direction. At $\Delta z_m = 0$ the best coupling position appears, where we have the doubling image of the input signal. If we take the losses into account the maximum of the intensity is given for a shift of $\Delta z_m = -11 \, \mu$m. Here the coupling of the field to the output waveguides is inferior to $\Delta z_m = 0$ but due to the reduced attenuated amplitude of the field we obtain in sum a higher intensity in the output waveguides.

5 Conclusions

We have analysed a plasmonic beam splitter based on the self-imaging effect. The field distribution of the beam splitter was calculated by 3-D full vectorial calculations using the MoL. Particular we studied the influence of the lateral position of the input waveguide and the length of the middle waveguide on the coupling efficiency. We found that the ratio of the field at the output waveguides can be modified by geometrical parameters (symmetrically as well as asymmetrically). In particular we have considered the losses of the metal regarding the coupling efficiency. It is found that the highest coupling efficiency does not occur at the location of the doubled self-imaging, but at an earlier position, due to the propagation losses.

Acknowledgements. We thank Christopher Alain Jones for reading this manuscript.

References

Edelmann, A. G., Helfert, S. F., and Jahns, J.: Analysis of the self-imaging effect in plasmonic multimode waveguides, Appl. Opt., 49, A1–A10, 2010.

Han, Z. and He, S.: Multimode interference effect in plasmonic sub-wavelength waveguides and ultra-compact power splitter, Opt. Comm., 278, 199–203, 2007.

Maier, S. A., Atwater, H. A.: Plasmonics: Localization and guiding of electromagnetic energy in metal/dielectric stuctures, J. Appl. Phys., 98, 011101:1-10, 2005.

Han, Z., Elezzabi, A. Y., and Van, V.: Wideband Y-splitter and aperture-assisted coupler based on sub-diffraction confined plasmonic slot waveguides, Appl. Phys. Lett., 96, 131106:1-3, 2010.

Ozbay, E.: Plasmonics: Merging photonics and electronics at nanoscale dimensions, Science, 311, 189–193, 2006.

Palik, E. D.: Handbook of Optical Constants of Solids, Academic Press, London, UK, 1985.

Pregla, R. and Pascher, W.: The method of lines, in: Numerical Techniques for Microwave and Millimeter Wave Passive Structures, edited by: Itoh, T., J. Wiley Publ., New York, USA, 381–446, 1989.

Pregla, R.: Analysis of electromagnetic Fields and Waves: The Method of Lines, John Wiley & Sons, Ltd., West Sussex, England, 2008.

Raether, H.: Surface plasmons on smooth and rough surfaces and on gratings, Springer, Berlin, Germany, 1988.

Soldano, L. B. and Pennings, E. C. M.: Optical multi-mode interference device based on self-imaging: principles and applications, J. Lightwave Technol., 13, 615–627, 1995.

Veronis, G. and Fan, S.: Bends and splitters in metal-dielectric-metal subwavelength plasmonic waveguides, Appl. Phys. Lett., 87, 131102:1-3, 2005.

Yuan, G., Wang, P., Lu, Y., and Ming, H.: Multimode interference splitter based on dielectric-loaded surface plasmon polariton waveguides, Opt. Express, 19, 12594–12600, 2009.

29

Network modelling with Brune's synthesis

F. Mukhtar[1], Y. Kuznetsov[2], and P. Russer[1]

[1]Institute for Nanoelectronics, Technische Universität München, Germany
[2]Theoretical Radio Engineering Department, Moscow Aviation Institute, Russia

Abstract. Network modelling of general, lossy or lossless, one-port and symmetric two-port passive electromagnetic structures in systematic manner is presented. Rational function representation of the numerical data of Z- or Y-parameters is obtained with the use of Vector Fitting procedure. A systematic strategy for obtaining equivalent lumped element circuit from the rational function, applying Brune's circuit synthesis, is also presented.

1 Introduction

In microwave circuit design compact models for distributed circuits are a valuable tool for efficient modeling of complex circuits. Lumped element circuit models are not only compact but also allow to represent the physical properties, including energy- and power properties (Russer, 2006; Felsen, 2009; Russer, 2010). For linear lossless reciprocal multi-port structures Foster circuit representations are canonical circuit representations which easily can be synthesized in a systematic way. One can also try to apply the Foster method synthesis to lossy structures. This works in principle, however yields in the general case lumped element equivalent circuits exhibiting negative elements.

Otto Brune published in 1931 a method for the synthesis of a finite one-port network whose driving–point impedance is a prescribed function of frequency (Brune, 1931; Guillemin, 1957, 343–358 pp.). This method allows to synthesize lumped element equivalent circuits with a minimum number of elements which are positive if the function describing the one-port impedance is positive real (PR). The circuit elements are inductors, capacitors and resistors only. The interconnect structure in general contains ideal transformers. In (Russer, 2010) also the modeling of lossy structures on the basis of Brune's equivalent circuit realization has been discussed.

Correspondence to: F. Mukhtar
(mukhtar.farooq@mytum.de)

In this work we describe a systematic strategy for Brune's synthesis of one-ports described by positive real impedance functions. This method is also applied to symmetric two-ports. In the case of symmetric two-ports Bartlett's theorem (Guillemin, 1957, p. 196) is applied to represent the symmetric two-port by an equivalent circuit consisting of four one-ports. Subsequently Brune's synthesis is applied.

2 Procedure

The synthesis of network model is carried out in four steps:

Step 1: Data acquisition
Z or Y-parameters data is acquired through numerical full-wave simulation of the structure. Care has to be taken to ensure that the data does not violate passivity conditions, i.e. data matrix for each frequency point is positive definite. Thus if $\mathbf{R} = \Re[\mathbf{Z}]$, then for all $i = 1, 2, \cdots, N$, where N is the order of matrix, $R_{ii} \geq 0$ and the determinant with all principle minors of $\mathbf{R}$ should also be positive for $\mathbf{Z}$ to be positive real (Guillemin, 1957, Ch.1). For the achievement of P.R. data, taking finer mesh size of structure for simulation is helpful.

Step 2: Decomposition to one-ports
This step concerns only symmetric two-ports. Using the following equations and topology in Fig.1, one can decompose symmetric two-port circuit to two one-port circuits (Guillemin, 1957, Ch.6).

$$Z_a = Z_{11} - Z_{12}$$
$$Z_b = Z_{11} + Z_{12} \tag{1}$$

Step 3: System identification
Vector Fitting(V.F.) Method (Gustavsen, 1999, 2006; Deschrijver, 2008) is used to find poles and residues for curve fitting. It should be kept in mind that V.F. method is a mathematical tool providing the poles giving best fit. The poles may or may not represent the dynamical properties of the system.

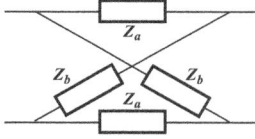

Fig. 1. Lattice structure for symmetric two port circuit.

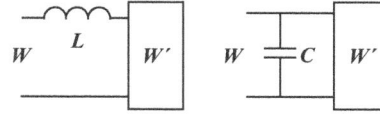

Fig. 2. Possible network relizations for cases 1 and 2.

Step 4: Network synthesis

Once a positive real rational function is obtained, the equivalent lumped element circuit can be synthesized by many methods and algorithms. Authors have chosen Brune's process (Brune, 1931). An implementation of this algorithm is accomplished in MATLAB using its Symbolic Math Toolbox for high precision calculations. The reasons for choosing Brune's method and the implementation are described in next section.

3 Brune's circuit synthesis

3.1 Advantages of Brune's synthesis

A detailed comparison between different synthesis methods are given in (Darlington, 1955). Here a brief comparision between Brune's synthesis and Darlington's synthesis is given.

1. In Brune's process, minimum reactive elements suffice, which is not the case in Darlington's synthesis.

2. Brune's extraction process defines its own topology. In Darlington's method one has to choose the circuit topology which is extremely difficult for generalized n-port.

3. Brune's synthesis is stage by stage extraction and each cycle decreases the degree of function by 2, while Darlington's method is NOT stage by stage process, rather one has to take the whole function and compare it to a predefined loss-less two-port.

4. In literature, the extension of Brune's method to n-port is available (Tellegen, 1953), while there is no extension of Darlington's method to n-port.

3.2 Some properties of P.R. functions

A detailed discussion on properties of P.R. functions is given in (Guillemin, 1957, Ch 1). Some of these properties, related to this work, will be discussed here briefly. Typical rational function, having same degree of polynomials in the numerator and the denominator, is given as

$$W = \frac{a_n s^n + a_{n-1} s^{n-1} + a_{n-2} s^{n-2} + \cdots + a_0}{b_n s^n + b_{n-1} s^{n-1} + b_{n-2} s^{n-2} + \cdots + b_0} \quad (2)$$

where $W \in \{Z, Y\}$. For W to be a P.R. function, the coefficients $a_n, a_{n-1}, \cdots, a_0$ and $b_n, b_{n-1}, \cdots, b_0$ should be positive

real numbers and non can be zero except a_n, b_n, a_0, b_0 under certain conditions. It should also be noted that the difference in degrees of denominator and numerator cannot exceed 1. All poles and zeros of a P.R. function lie on left half plane or on the imaginary axis of complex frequency s. If a pole or a zero lie on imaginary axis, it can be separated from the rational function without disturbing its P.R. character. Also, $s = \infty$ is considered as located on imaginary axis.

3.3 Cases of P.R. function

With all considerations described above, given P.R. rational function can only occur within seven cases and in each case a small part of function can be separated leaving behind a reduced P.R. rational function. First six cases occur in pairs while seventh one is the default case which is called Brune's process or cycle.

3.3.1 Case 1

$b_n = 0 \Rightarrow W \to \infty$ as $s \to \infty$ i.e. there is a pole at infinity. From implementation point of view, one has to divide the numerator with denominator. Computer software tools are designed to give first order and constant terms as quotient, leaving remainder one degree lower than the divisor. A modification can be done to obtain only one term as quotient and have remainder of same degree as that of divisor: multiply divisor (denominator) by s before division and divide by it later. These steps are shown below.

$$\begin{aligned}
W &= \frac{a_n s^n + a_{n-1} s^{n-1} + a_{n-2} s^{n-2} + \cdots + a_0}{b_{n-1} s^{n-1} + b_{n-2} s^{n-2} + \cdots + b_0} \\
&= s \left[\frac{a_n s^n + a_{n-1} s^{n-1} + a_{n-2} s^{n-2} + \cdots + a_0}{b_{n-1} s^n + b_{n-2} s^{n-1} + \cdots + b_0 s} \right] \\
&= s \left[\frac{a'_{n-1} s^{n-1} + a'_{n-2} s^{n-2} + \cdots + a'_0}{b_{n-1} s^n + b_{n-2} s^{n-1} + \cdots + b_0 s} + A \right] \\
&= \frac{a'_{n-1} s^{n-1} + a'_{n-2} s^{n-2} + \cdots + a'_0}{b_{n-1} s^{n-1} + b_{n-2} s^{n-2} + \cdots + b_0} + As = W' + As \quad (3)
\end{aligned}$$

If $W = Z$ i.e. an impedance then A is an inductor in series with rest of the circuit and if $W = Y$ i.e. an admittance then A is a capacitor in parallel with rest of circuit.

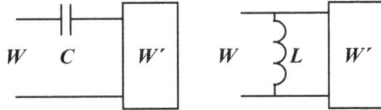

Fig. 3. Possible network realizations for cases 3 and 4.

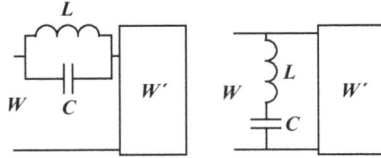

Fig. 4. Network realizations of cases 5 and 6.

3.3.2 Case 2

$a_n = 0 \Rightarrow W \to 0$ as $s \to \infty$ i.e a zero at infinity. We can take the reciprocal of W and proceed as in case 1, above. We can write

$$W = \left[\frac{1}{W'} + As \right]^{-1} \tag{4}$$

Here the role of A is reversed. For $W = Z$, A is a parallel capacitor and for $W = Y$, A is a series inductor. Figure 2 shows the possible networks for cases 1 and 2.

3.3.3 Case 3

$b_0 = 0 \Rightarrow W \to \infty$ as $s \to 0$ i.e. a pole at $s = 0$. Again, from implementation point of view, we shall substitute s with $\frac{1}{\tilde{s}}$ transforming the function into a form suitable for case 1. Equations below show the steps.

$$W(s) = \frac{a_n s^n + a_{n-1} s^{n-1} + \cdots + a_0}{b_n s^n + b_{n-1} s^{n-1} + \cdots + b_1 s}$$

$$W\left(\tilde{s} = \frac{1}{s}\right) = \frac{a_n + a_{n-1}\tilde{s} + \cdots + a_0 \tilde{s}^n}{b_n + b_{n-1}\tilde{s} + \cdots + b_1 \tilde{s}^{n-1}}$$

$$= \frac{a'_n + a'_{n-1}\tilde{s} + \cdots + a'_1 \tilde{s}^{n-1}}{b_n + b_{n-1}\tilde{s} + \cdots + b_1 \tilde{s}^{n-1}} + A\tilde{s}$$

$$W(s) = \frac{a'_n s^{n-1} + a'_{n-1} s^{n-2} + \cdots + a'_1}{b_n s^{n-1} + b_{n-1} s^{n-2} + \cdots + b_1} + \frac{A}{s} \tag{5}$$

If $W = Z$ then $\frac{1}{A}$ is the value of a series capacitor and if $W = Y$ then it is the value of parallel inductor.

3.3.4 Case 4

$a_0 = 0 \Rightarrow W \to 0$ as $s \to 0$ i.e. a zero at $s = 0$. It is the reciprocal case of case 3. In equation form

$$W = \left[\frac{1}{W'} + \frac{A}{s} \right]^{-1} \tag{6}$$

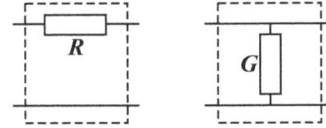

Fig. 5. Resistive part of Brune's Process.

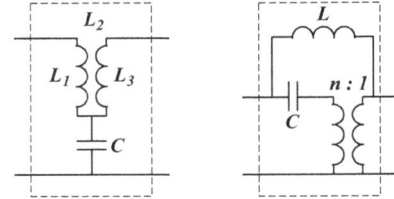

Fig. 6. Reactive part of Brune's Process.

The role of A is reversed. $\frac{1}{A}$ represents a parallel inductor or series capacitor for $W = Z$ and $W = Y$, respectively. Figure 3 shows the possible realizations for cases 3 and 4.

3.3.5 Case 5

When a pole pair occurs on the imaginary axis at finite frequency ω_p we can write

$$W = \frac{1}{s^2 + \omega_p^2} \frac{P}{Q}$$

$$W' = \frac{1}{s^2 + \omega_p^2} \frac{P}{Q} - \frac{2ks}{s^2 + \omega_p^2}$$

$$= \frac{1}{s^2 + \omega_p^2} \left[\frac{P - 2ks}{Q} \right] = W' + \frac{1}{As + \frac{B}{s}} \tag{7}$$

The value of residue k is calculated such that the polynomial $P - 2ks$ has the factor $s^2 + \omega_p^2$, thus cancelling it out in the remainder function W'.

3.3.6 Case 6

This is the reciprocal of above case having zero pair on imaginary axis at finite frequency ω_p

$$W = \left[\frac{1}{W'} + \frac{1}{As + \frac{B}{s}} \right]^{-1} \tag{8}$$

Possible network realizations for cases 5 and 6 are shown in Fig. 4

3.3.7 Case 7

When a P.R. rational function does not occur in any of the above six cases then all of its poles and zeros are strictly in left half plane. This default extraction procedure is called Brune's process and involves two steps. First step is to find the global minima of the real part of the function on imaginary axis and to subtract the value from the function. This

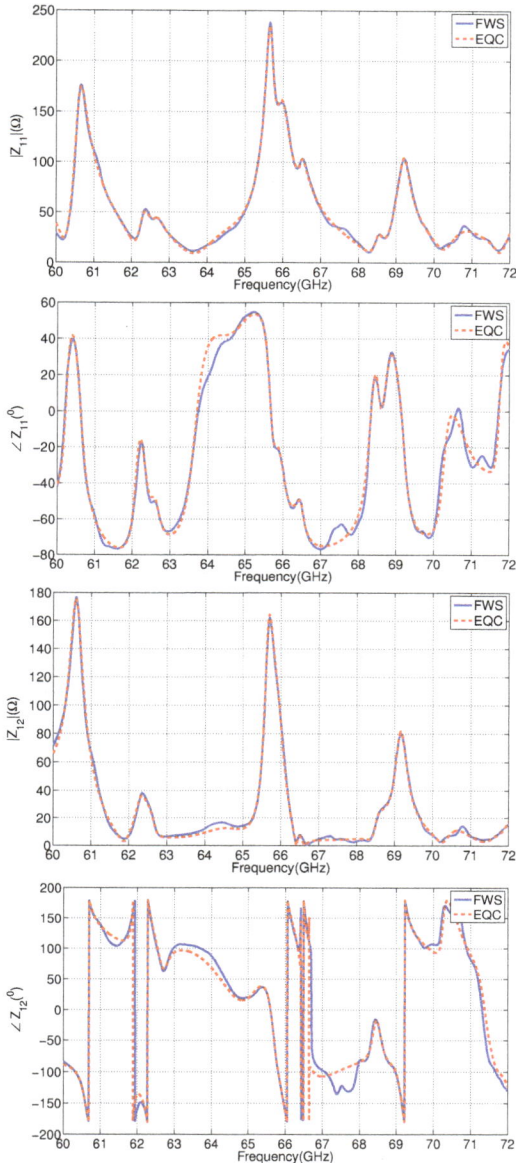

Fig. 7. Magnitude $|Z_{11}|$, $|Z_{12}|$ and phase $\angle Z_{11}$, $\angle Z_{12}$ of the wireless transmission link two-port from full-wave simulation (FWS) and equivalent circuit (EQC).

results in a series resistance (when $W = Z$) or a parallel conductance (when $W = Y$). Figure 5 shows the possible circuits.

Second step involves extraction of imaginary part of function at the above found minima, thus creating a zero pair on imaginary axis of complex frequency plane. The zero on imaginary axis can be extracted easily as discussed in above cases. The extracted imaginary part and the zero pair combine to give one of circuits shown in Fig. 6. A detailed discription and examples of Brune's process is given in (Brune, 1931) and in (Guillemin, 1957, Ch.9).

3.4 Numerical example

The procedure was applied to a structure discribed in (Mukhtar, 2010). These are two integrated antennas on one chip with a distance of 2 mm. Z-paramters were obtained from full wave simulation and modelled. Fig. 7 shows the comparision of Z-parameters from simulation and from network model.

4 Conclusions

A lumped element equivalent circuit model of general symmetric passive two-port microwave structure is presented and verified with an example.

Acknowledgements. This work was supported by Deutsche Forschungsgemeinschaft (DFG).

References

Brune, O.: Synthesis of a Finite Two-Terminal Network Whose Driving-Point Impedance is a Prescribed Function of Frequency, J. Math. Phys. Camb., 10, 191–236, 1931.

Guillemin, E. A.: Synthesis of Passive Networks, John Wiley & Sons, Inc., New York, 1957.

Tellegen, B. D. H.: Synthesis of the 2n-Poles by Networks Containing the Minimum Number of Elements, J. Math. Phys. Camb., 32, 1–18, 1953.

Darlington, S.: A Survey of Network Realization Techniques, Ire. T. Cir. Theo., 2, 291–297, 1955.

Gustavsen, B. and Semlyen A.: Rational Approximation of Frequency Domain Responses by Vector Fitting, Ieee. T. Power. Deliver., 14, 1052–1061, 1999.

Gustavsen B.: Improving the pole relocating properties of vector fitting, IIeee. T. Power. Deliver., 21, 1587–1592, 2006.

Deschrijver, D., Mrozowski, M., Dhaene, T., and De Zutter, D.: Macromodeling of Multiport Systems Using a Fast Implementation of the Vector Fitting Method, Ieee Microw Wirel Co, 18, 383–385, 2008.

Mukhtar, F., Yordanov, H., and Russer, P.: Network Model of On-Chip Antennas, Kleinheubacher Tagung 2010, Miltenberg, 4–6 October 2010, KH2010-D-1494, 2010.

Russer, P.: Electromagnetics, Microwave Circuit and Antenna Design for Communications Engineering, 2nd ed., Artech House, Boston, 2006.

Felsen, L. B., Mongiardo, M., and Russer, P.: Electromagnetic Field Computation by Network Methods, Springer, Berlin, 2009.

Russer, J. A., Kuznetsov, Y., and Russer, P.: Discrete-time Network and Steady State Equation Methods Applied to Computational Electromagnetics, Mikrotalasna Revija (Microwave Review), 2–14, 2010.

30

Minimizing interference in automotive radar using digital beamforming

C. Fischer[1], M. Goppelt[2], H.-L. Blöcher[1], and J. Dickmann[1]

[1]Daimler AG, Wilhelm-Runge-Straße 11, 89013 Ulm, Germany
[2]University of Ulm, Institute of Microwave Techniques, Albert-Einstein-Allee 41, 89069 Ulm, Germany

Abstract. Millimetre wave radar is an essential part of automotive safety functions. A high interference tolerance, especially with other radar sensors, is vital. This paper gives an overview of the motivation, the boundary conditions and related activities in the MOSARIM project funded by the European Union and concerned with interference mitigation in automotive radars. Current and planned activities considering Digital Beamforming (DBF) as a method for interference mitigation are presented.

1 Introduction: future automotive scenarios

A global trend or "megatrend" to be expected in the midterm future is increasing urbanization and thus the growth of megacities. In the process of this development, in addition to the increasing number of people living in cities, there will also be more cars in the streets (an illustrative example is given in Fig. 1). Recent studies also expect a growing number of cars in emerging countries (cf. IEA, 2008). Automotive safety is an essential requirement and vital necessity. There is another implication for the increasing number of cars equipped with safety functions based on radar technology: As the car density and absolute number of cars using radar increase, the chance of two radar sensors interacting in some way also increases. In anticipation of this development, the EU Seventh Framework Programme MOSARIM (see MOSARIM Webpage, 2010) was initiated by a number of OEMs, Tier 1 and scientific institutes. This project has three major goals:

- to identify and analyze the mechanisms of interference between automotive radar sensors

- and subsequently to derive guidelines to minimize the probability of interference in future sensor generations.

The project commenced in January 2010 and will end in December 2012. A workshop will be held where the final results will be presented to the public.

This paper is structured as follows: First a short estimation of the future development of radar equipped cars is presented. Then, Digital Beamforming is briefly presented and the methods considered for interference mitigation in Wirth (2001) are described. Finally, the simulation environment is introduced and results presented. The paper concludes with a summary and an outlook.

2 Digital beamforming

Digital Beamforming (DBF) is a signal processing method with growing importance. As semiconductor technology advances, more and more parts of a radar system can actually be computed in software rather than realized in hardware. Key components for DBF are the analog-digital converters (ADC) and the digital signal processing element realized in an FPGA, DSP or an ASIC. Compared with currently widely used automotive radar sensors, no moving parts are necessary to measure the incident angle of a signal. Complicated high frequency hardware like RF phase shifters may also be omitted because of the early digitization of the signals. This makes DBF an important candidate for cost reduction while at the same time increasing flexibility and reliability.

Correspondence to: C. Fischer
(christoph.c.fischer@daimler.com)

Fig. 1. Inner city traffic scenarios with a large number of radar sensors like this will be an increasing challenge for radar sensors and implemented interference mitigation techniques in the future.

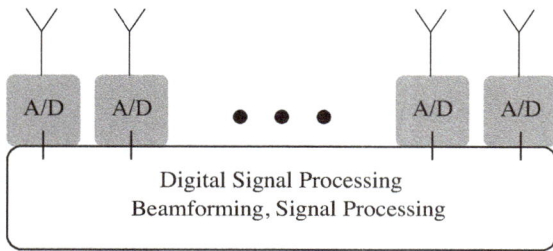

Fig. 2. The basic principle of Digital Beamforming: The received signals will be sampled close to the antenna and processed in parallel.

The basic principle of DBF is simple (see Fig. 2): A fixed number of antennas N at different positions receive in parallel. This can be regarded as spatial sampling that results in a number of signals:

$$s_i(t) \qquad \text{with} \quad i = 1 \dots N. \tag{1}$$

After reception and downconversion the signals are sampled in time by the ADCs:

$$x_i[k] = s_i(kT) \qquad \text{with} \quad T = \frac{1}{f_{\text{Sample}}}, k \in \mathbb{N}. \tag{2}$$

Being in the digital domain, the signals $x_i[k]$ can be processed with great flexibility. The best known application in DBF is the estimation of the incident angle (or direction) of arrival (AoA or DoA) of a target. This can be done for a linear antenna array simply by computing a discrete Fourier tranform (DFT) across the individual elements of the array. Because of this, the term *angular spectrum* is often used. In matrix notation this can be expressed simply by

$$\mathbf{x}_a[k] = \mathbf{A}\mathbf{x}[k], \tag{3}$$

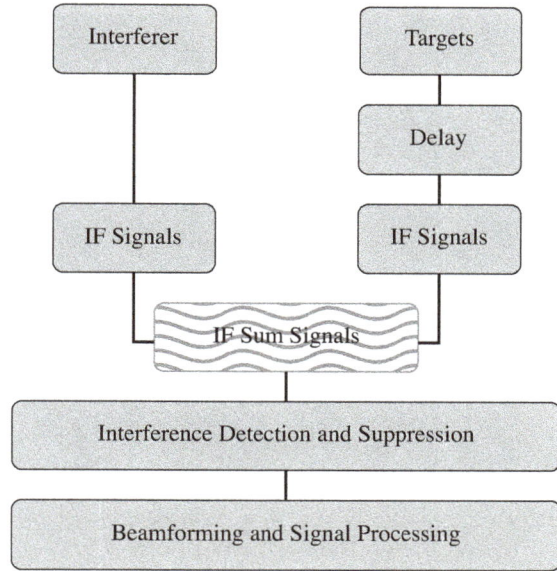

Fig. 3. Simplified block diagram of the implemented simulation environment.

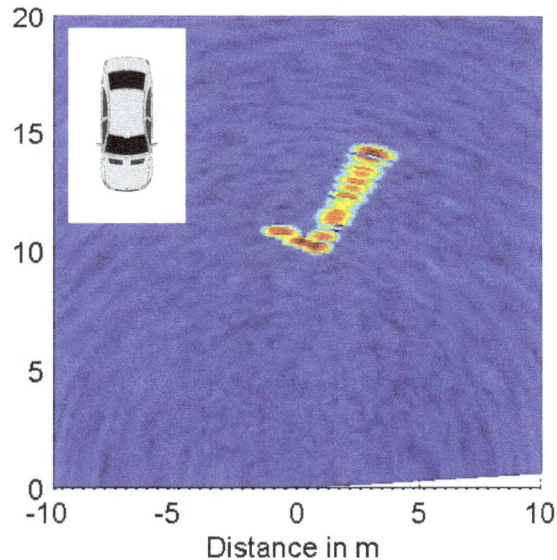

Fig. 4. An example of a simulation result: The basic simulation scenario without any interference but with additive white Gaussian noise (red = high amplitude, blue = low amplitude).

with $\mathbf{A}$ being the transformation matrix. Using this formalism it is possible to stack different linear processing steps into this matrix notation. This form is also advantageous for calculating transformations other than the DFT, for which the FFT already provides a fast algorithm. For matrix

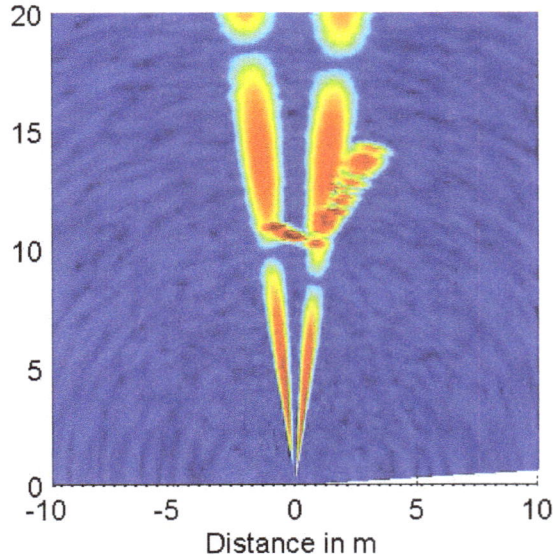

Fig. 5. Simulation result: The scenario with interference and additive white Gaussian noise but without any mitigation technique applied (red = high amplitude, blue = low amplitude).

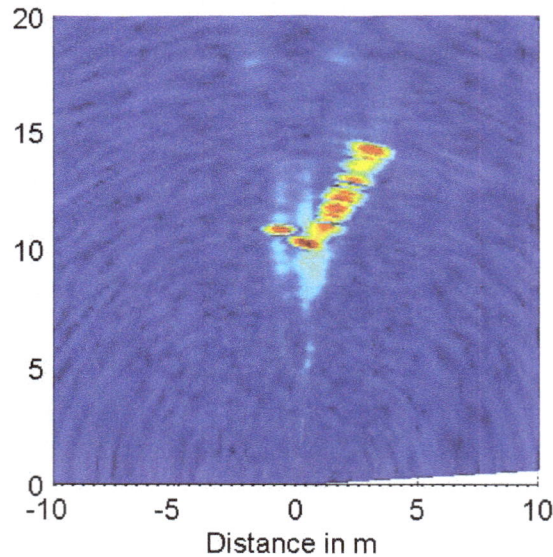

Fig. 7. Simulation result: The scenario with interference and additive white Gaussian noise (red = high amplitude, blue = low amplitude). The interference is canceled with LSMI.

3   Interference identification and suppression

Goppelt et al. (2010) showed that interference in FMCW radar systems can have severe impact on the IF signals. Although no influence on the object level has been observed so far, the signal to noise ratio degrades. To reduce this effect before the target acquisition step in the radar sensor, the interference should be suppressed as early as possible in the signal processing chain. In this paper, the focus is on short time interference with high amplitude relative to the wanted signals. Due to the architecture of FMCW signals, this kind of interference is most likely to occur in real driving situations, for example, when two radar equipped cars drive towards each other on a straight road. Considering ramp ups followed by ramp downs, which are typical in standard FMCW systems, there has to be a point in time where these ramps intersect and where the above mentioned interference pattern occurs.

The identification of the interference is the first step to its mitigation. In the case considered here, the use of the spiky time domain signal characteristics is obvious. The energy of the signals

$$e_i[k] = x_i[k]^2 \tag{4}$$

convolved with a moving average kernel is therefore considered. When the average energy in the considered timeframe rises above a given limit, defined by the scaled mean energy of the complete signal, an interference is declared. Since a DBF system with several receive channels is being consid-

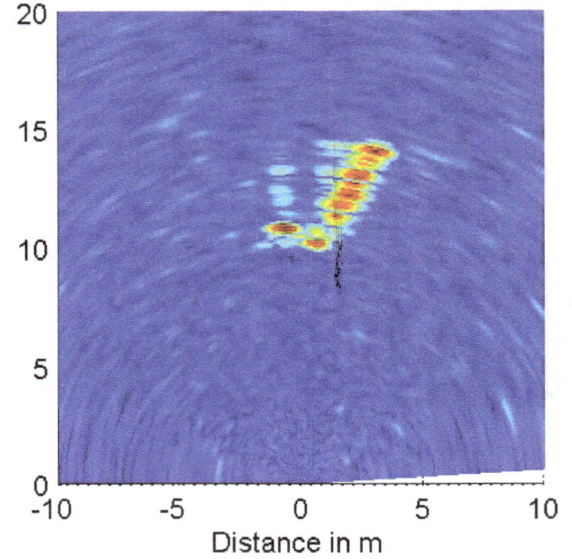

Fig. 6. Simulation result: The scenario with interference and additive white Gaussian noise (red = high amplitude, blue = low amplitude). The samples affected by the interference are zeroed.

multiplication, fast algorithms exist that can be used for implementation on the desired hardware platform, such as an FPGA, for example.

ered, this procedure is applied to all channels separately. After the detection step, the length of the interference over all channels is defined by the first and the last detection.

From these identified interference sequences $y_i[k]$, the covariance matrix $\mathbf{Q}$ must now be determined. It needs to be estimated based on S samples or snapshots of the received signals:

$$\hat{\mathbf{Q}} = \frac{1}{S} \sum_{s=1}^{S} \mathbf{y}_s \cdot \mathbf{y}_s^* = \frac{1}{S} \mathbf{YY}^*. \tag{5}$$

A noise matrix is added to reduce the influence of fluctuations of small eigenvalues:

$$\hat{\mathbf{Q}} = \frac{1}{S} \mathbf{YY}^* + \alpha \cdot \mathbf{I}. \tag{6}$$

This also ensures the covariance matrix to be non-singular and therefore invertible. After this preparation, the interference is suppressed by computing

$$\hat{\mathbf{x}}[k] = \mathbf{A}^* \hat{\mathbf{Q}}^{-1} \mathbf{x}[k]. \tag{7}$$

The reconstructed angular spectrum $\hat{x}_i[k]$ can then be further processed. This method is known as loaded sample matrix inversion (LSMI). One point to be mentioned is that a small part of the desired signal is also removed due to the very simple algorithm for estimating the covariance matrix. The interference suppression is not only a function of the length of the identified interference, but also of the interference to signal ratio. Note that if the interference is too weak, the suppression makes no more sense since the interference power is spread across the whole range profile and therefore has little influence on reasonably strong targets. Quantitative comparisons on that will be the subject of further research.

4 Simulation environment and first results

A simulation chain has been implemented in MATLAB® to simulate the effects of different interference scenarios on a DBF radar. This simulation can resemble an arbitrary FMCW radar with multiple receiving antennas. For now the individual blocks are assumed to be completely linear, but some quantization effects are already implemented. The block diagram in Fig. 3 shows the basic elements. In the first step, the targets are defined and the propagation delay to each of the the receiving antennas is calculated. The resulting intermediate frequency (IF) is calculated for each target from these delays and a velocity property of the targets and then combined with the interference signals generated separately. The result of this accumulation is a sampled multichannel IF signal. For the simulations shown in this paper, a value of 8192 samples per ramp has been considered, but this value can be changed to any value necessary for an application. Although we consider static scenarios here, it is also possible to simulate dynamic scenarios.

The scenario used for the interference tests is displayed in Fig. 4. The L-shape was selected as the target because it can be interpreted as the reflection pattern of a car. In the simulation, it was represented by a number of point scatterers that were positioned accordingly. The interferer was placed in front of the L-shape. The scenario can therefore be considered as a car equipped with a similar sensor to the simulated one, driving towards the victim at a certain angle.

In Figs. 5 to 7, some first results of the mentioned algorithm are displayed. For comparison, what happens when the affected bins are simply set to zero (see Fig. 6) was also simulated. It can clearly be seen that this procedure results in significant distortion in the range direction, whereas the LSMI displayed in Fig. 7 results in a slight distortion in the angular direction. Further techniques that use extrapolation or intrapolation to reconstruct the part of the signal affected by interference already exist. These will also be implemented in the setup and tested for practical use. In subsequent steps, the implemented algorithms will need to be applied to real data from an automotive sensor to show their usefulness in practice.

5 Conclusions

In the future, interference mitigation will become an important issue in automotive radar. Using a DBF simulator the applicability of well-known interference mitigation techniques could be displayed. Analysis will continue with the use of data from real sensors and the implementation of more sophisticated techniques. In the long term these techniques will be implemented on real hardware to undergo real world tests. The complexity of the algorithms and the possibility for realization on the FPGA of the existing platform will therefore also be an issue.

Acknowledgements. The research leading to these results has received funding from the European Community's Seventh Framework Programme (FP7/2007-2011) under grant agreement no 248 231 – MOSARIM Project.

References

Goppelt, M., Blöcher, H.-L., and Menzel, W.: Automotive radar –investigation of mutual interference mechanisms, Adv. Radio Sci., 8, 55–60, doi:10.5194/ars-8-55-2010, 2010.

IEA (International Energy Agency): World Energy Outlook, 2008.

MOSARIM Webpage: http://www.mosarim.eu, last access: 17 November 2010.

Wirth, W.-D.: Radar techniques using array antennas, IET radar, sonar, navigation and avionics series 10, edited by: Steward, N. and Griffith, H. The institution of engineering and technology, London, 2001

31

Combined lumped element network and transmission line synthesis for passive microwave structures

J. A. Russer[1]**, F. Mukhtar**[1]**, A. Baev**[2]**, Y. Kuznetsov**[2]**, and P. Russer**[1]

[1]Lehrstuhl für Nanoelektronik, Technische Universität München, Arcisstrasse 21, Munich, Germany
[2]Theoretical Radio Engineering Dept., Moscow Aviation Inst.,Volokolamskoe shosse, 4, GSP-3, Moscow, 125993, Russia

Abstract. Compact circuit models of electromagnetic structures are a valuable tool for embedding distributed circuits into complex circuits and systems. However, electromagnetic structures with large internal propagation delay are described by impedance functions with a large number of frequency poles in a given frequency interval and therefore yielding equivalent circuit models with a high number of lumped circuit elements. The number of circuit elements can be reduced considerably if in addition to capacitors, inductors, resistors and ideal transformers also delay lines are included. In this contribution a systematic procedure for the generation of combined lumped element/delay line equivalent circuit models on the basis of numerical data is described. The numerical data are obtained by numerical full-wave modeling of the electromagnetic structure. The simulation results are decomposed into two parts representing a lumped elements model and a delay line model. The extraction of the model parameters is performed by application of the system identification procedure to the scattering transfer function. Examples for the modeling of electromagnetic structures are presented.

1 Introduction

The importance for applying circuit-theoretic multi-port concepts in the modeling of wireless communication links already has been pointed out in Ivrlac and Nossek (2010). Different from simple information theoretic models multi-port models allow to consider the energy flow and the coupling of antenna elements. The use two-port models of antennas already has been discussed in literature (Rogers et al., 2003; Boryssenko and Schaubert, 2007; Aberle, 2008). These models yielded an improved description of antennas, however,

Correspondence to: J. A. Russer
(jrusser@tum.de)

due to the limited number of lumped elements the antenna could be modeled in a narrow frequency band only.

Distributed circuits can be modeled also in a broad frequency band with arbitrary accuracy using lumped element network models. A general way to establish network models is based on modal analysis and similar techniques (Russer, 2006; Felsen et al., 2009; Russer et al., 2010). In Russer et al. (2010) also the modeling of lossy structures on the basis of Brune's equivalent circuit realization has been discussed. However, electromagnetic structures with large internal propagation delay are described by impedance functions with a large number of frequency poles in a given frequency interval and therefore yielding equivalent circuit models with a high number of lumped circuit elements. The number of circuit elements can be reduced considerably if in addition to capacitors, inductors, resistors and ideal transformers also delay lines are included. A combined lumped elements and delay line approach already has been presented in Shevgunov et al. (2008).

In this contribution a systematic procedure for the generation of a combined lumped element/delay line equivalent circuit model for a wireless transmission link on the basis of numerical data is described. The numerical data are obtained by numerical full-wave modeling of the electromagnetic structure. The simulation results are decomposed into two parts representing a lumped elements model and a delay line model. The extraction of the model parameters is performed by application of the system identification procedure to the scattering transfer function. Examples for the modeling of electromagnetic structures are presented.

2 Model of a wireless link

Let us consider the easiest case of a wireless transmission link consisting of two antennas connected in the far-field over free-space. Figure 1 shows this arrangement schematically. In **Z** representation the voltage-current relations are given by

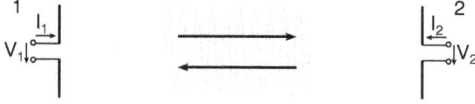

Fig. 1. Wireless transmission link with two antennas.

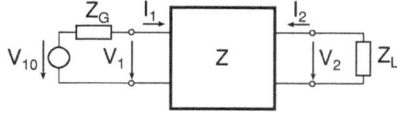

Fig. 2. Two-port representation of the wireless transmission link with source and load.

$$\begin{bmatrix} V_1 \\ V_2 \end{bmatrix} = \begin{bmatrix} Z_{11} & Z_{12} \\ Z_{21} & Z_{22} \end{bmatrix} \begin{bmatrix} I_1 \\ I_2 \end{bmatrix}. \tag{1}$$

Figure 2 shows the block diagram of the wireless link for the case where antenna 1 is the transmitting antenna and antenna 2 is the receiving antenna. The transmitter has the open source voltage V_{10} and the source impedance Z_G. The receiving antenna is terminated with the load impedance Z_L. Under these conditions the input and output currents I_1 and I_2, respectively are given by

$$I_1 = \frac{Z_{22} + Z_L}{(Z_{11} + Z_G)(Z_{22} + Z_L) - Z_{12}Z_{21}} V_{10}, \tag{2}$$

$$I_2 = \frac{Z_{21}}{(Z_{11} + Z_G)(Z_{22} + Z_L) - Z_{12}Z_{21}} V_{10}. \tag{3}$$

Figure 3 shows the schematic two-port representation of the link with source and load where Z_{12} and Z_{21} are represented by current controlled voltage sources. The input impedance Z_{in} of the wireless link two-port $\mathbf{Z}$ terminated by Z_L is given by

$$Z_{in} = Z_{11} - \frac{Z_{12}Z_{21}}{Z_{22} + Z_L}. \tag{4}$$

The available power P_a of the transmitter is given by

$$P_a = \frac{|V_{10}|^2}{8\Re\{Z_G\}}. \tag{5}$$

The power delivered to the load R_L is

$$P_L = \frac{1}{2}\Re\{Z_L\}|I_2|^2. \tag{6}$$

The transducer power gain is defined as

$$G_t = \frac{P_L}{P_a} = \frac{4|Z_{21}|^2\Re\{Z_G\}\Re\{Z_L\}}{|(Z_{11} + Z_G)(Z_{22} + Z_L) - Z_{12}Z_{21}|^2}. \tag{7}$$

In a wireless transmission link the reaction of the receiving antenna on the transmitting antenna usually is negligible, i.e. $|(Z_{11} + Z_G)Z_{22}| \gg |Z_{12}Z_{21}|$. In this case we can approximate the transducer power gain by the unilateral power gain.

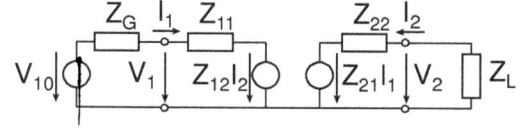

Fig. 3. Two-port representation of the wireless transmission link with source and load where Z_{12} and Z_{21} are represented by current controlled voltage sources.

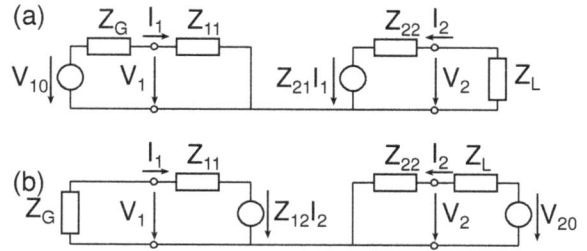

Fig. 4. Two-port representation of the unilateralized wireless transmission link with source and load (**a**) with antenna 1 as the transmitting antenna, (**b**) with antenna 2 as the transmitting antenna.

Depending on the direction of transmission we can replace the reciprocal wireless link models by the non-reciprocal models shown in Fig. 4. In this case the power gain G_t is approximated by the unilateral power gain

$$G_{tu} = \frac{4|Z_{21}|^2\Re\{Z_G\}\Re\{Z_L\}}{|(Z_{11} + Z_G)(Z_{22} + Z_L)|^2}. \tag{8}$$

We note that in this approximation the antenna feed impedances of antenna 1 and antenna 2 are given by Z_{11} and Z_{22} respectively.

The unilateral power gain exhibits its maximum value $G_{tu,max}$ for power matching at the input and output, i.e. $Z_G = Z_{11}^*$, $Z_L = Z_{22}^*$ and is given by

$$G_{tu,max} = \frac{|Z_{21}|^2}{4\Re\{Z_{11}\}\Re\{Z_{22}\}}. \tag{9}$$

3 The two-port antenna model

As long as the coupling of the antenna with the other antenna is neglected the antenna can be considered as a passive one-port. If the antenna is coupled with one or more other antennas over the far-field only, the driving point impedance Z_{in} of this antenna one-port will not be influenced by the other antennas. In the following we assume the antenna itself to be lossless. The real part of the antenna driving point impedance in that case is only due to the active power radiated by the antenna. Z_{in} is a positive real impedance function. A one-port with a positive real impedance function can be represented by a reactive reciprocal two-port $\mathbf{Z_R}$ terminated by a real resistor R_r (Guillemin, 1957). Figure 5 shows the block

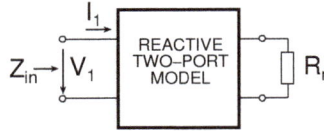

Fig. 5. Two-port antenna model.

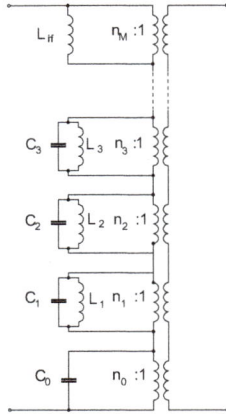

Fig. 6. Foster realization of the reactive two-port.

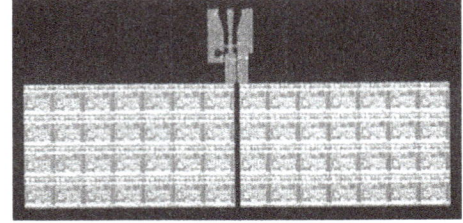

Fig. 7. Photograph of the open slot antenna (Yordanov and Russer, 2010).

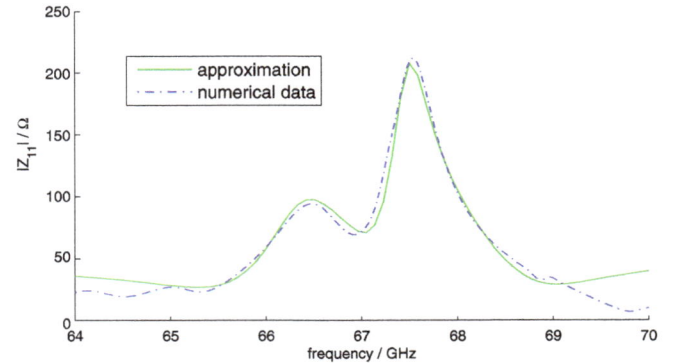

Fig. 8. $|Z_{11}|$ of the wireless transmission link.

diagram of the two-port antenna model. The determination of the positive rational impedance functions of the two-port was performed by system identification methods (Kuznetsov et al., 2004; Russer et al., 2010).

An approximation of the lossless reciprocal two-port based on a finite number of poles of its two-port reactance matrix $\boldsymbol{Z_R}$ can be realized with a finite number of capacitors, inductors and ideal transformers. Figure 6 shows the canonical Foster realization of the reactive two-port (Russer, 2006; Felsen et al., 2009; Russer et al., 2010).

The complete impedance matrix describing a circuit with a finite number of poles is obtained by parallel connecting the circuits describing the individual poles. In the canonical Foster representation, the impedance matrix $\boldsymbol{Z}(\omega)$ is given by

$$\boldsymbol{Z_\lambda}(\omega) = \frac{1}{j\omega C_0}\boldsymbol{B_0} + \sum_{\lambda=1}^{N}\frac{1}{j\omega C_\lambda}\frac{\omega^2}{\omega^2-\omega_\lambda^2}\boldsymbol{B_\lambda} + j\omega L_\infty \boldsymbol{B_\infty}.$$
(10)

The frequency independent rank 1 matrices $\boldsymbol{B_\lambda}$ are given by

$$\boldsymbol{B_\lambda} = \begin{bmatrix} n_\lambda^2 & n_\lambda \\ n_\lambda & 1 \end{bmatrix},$$
(11)

where the n_λ are the turns ratios of the transformers.

We have applied the methods described in this paper to model communication links between monolithic integrated on-chip antennas. Figure 7 shows a photograph of an open-chip antenna with CMOS circuits under the antenna electrode (Yordanov and Russer, 2010). The antennas have been measured on-wafer and diced.

Figure 8 shows the comparison of the $|Z_{11}|$ data obtained from numerical simulation with the data computed from the two-port model according to Fig. 5 with the reactive Foster two-port model shown in Fig. 6. The numerical full-wave simulations have been performed using CST software. Based on four pairs of poles and two single poles at zero and infinity the two-port model provides an accurate model of Z_{11} for the frequency band from 65 GHz to 69 GHz. The frequency range may be extended by increasing the number of poles.

4   The wireless link model

To extend the two-port antenna model to the full wireless link model we start from the assumption that the losses in the antenna are mainly radiation losses. In this case we can assume that the reactive two-port in Fig. 6 models the near-field of the antenna whereas the real resistor R_r models the energy dissipation in the far-field. Considering dispersion-free single-path propagation between the two antennas, we can model the free-space propagation of the electromagnetic wave by a dispersion-free lossy transmission line (TL). The length of this transmission line corresponds to the distance between the antennas and the attenuation follows from Friis transmission formula (Russer, 2006).

This yields the model of the complete wireless transmission link shown in Fig. 9. The scattering matrix of the lossy

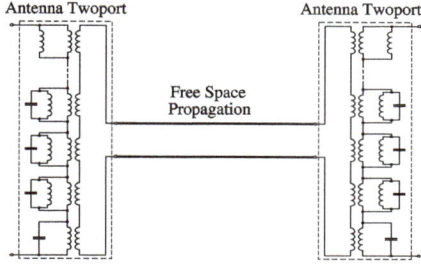

Fig. 9. Model of the complete wireless transmission link.

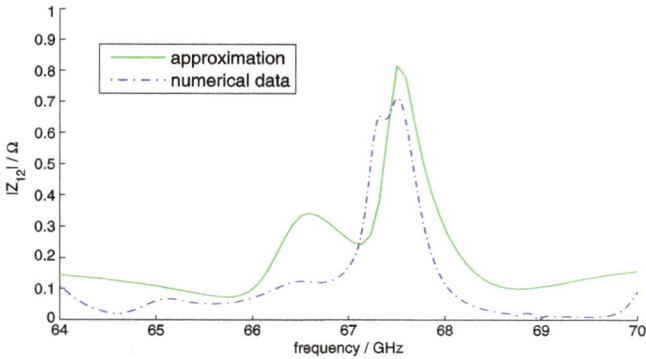

Fig. 10. $|Z_{12}|$ of the wireless transmission link.

transmission line with wave impedance R_r is given by

$$S_L = a \frac{\exp(-jkr)}{kr} \begin{bmatrix} 0 & 1 \\ 1 & 0 \end{bmatrix}, \tag{12}$$

where a represents the attenuation according to the Friis formula and r is the distance between the antennas. The corresponding Z-matrix is given by

$$\mathbf{Z}_L = Z_0 \begin{bmatrix} \dfrac{1 + \frac{a^2 \exp(-2jkr)}{(kr)^2}}{1 - \frac{a^2 \exp(-2jkr)}{(kr)^2}} & \dfrac{\frac{a \exp(-jkr)}{kr}}{1 - \frac{a^2 \exp(-2jkr)}{(kr)^2}} \\ \dfrac{\frac{a \exp(-jkr)}{kr}}{1 - \frac{a^2 \exp(-2jkr)}{(kr)^2}} & \dfrac{1 + \frac{a^2 \exp(-2jkr)}{(kr)^2}}{1 - \frac{a^2 \exp(-2jkr)}{(kr)^2}} \end{bmatrix}. \tag{13}$$

For $\frac{a \exp(-jkr)}{kr} \ll 1$ this can be approximated by

$$\mathbf{Z}_L \approx Z_0 \begin{bmatrix} 1 & \frac{a \exp(-jkr)}{kr} \\ \frac{a \exp(-jkr)}{kr} & 1 \end{bmatrix}. \tag{14}$$

Figure 10 shows the comparison of the numerical data of $|Z_{12}|$ obtained from the full-wave simulation of the wireless transmission link with the data computed from the model shown in Fig. 9. The stronger deviation between these data, compared with the case of $|Z_{11}|$ may be due to losses in the antenna and will be subject to further investigations.

5 Conclusions

We have presented an accurate method for generation of compact circuit models of wireless transmission links, where the antenna structure and the near-field regions are modeled by reactive lumped element networks and the far-field propagation by dispersion-free lossy transmission lines. Further investigations will extend this work towards multi-antenna systems and multi-path propagation. This should provide a framework for physical modeling of MIMO systems.

Acknowledgements. This work was supported by the Deutsche Forschungsgemeinschaft.

References

Aberle, J.: Two-Port Representation of an Antenna With Application to Non-Foster Matching Networks, Antennas and Propagation, IEEE Transactions on, 56, 1218–1222, 2008.

Boryssenko, A. O. and Schaubert, D. H.: On optimal port loading conditions for maximizing product of energy gain and bandwidth in broadband antenna links, IEEE Transactions on Antennas and Propagation, 55, 3668–3676, 2007.

Felsen, L. B., Mongiardo, M., and Russer, P.: Electromagnetic Field Computation by Network Methods, Springer, Berlin, Germany, 2009.

Guillemin, E. A.: Synthesis of Passive Networks, Wiley, New York, 1957.

Ivrlac, M. T. and Nossek, J. A.: Toward a Circuit Theory of Communication, Circuits and Systems I: Regular Papers, IEEE Transactions on, 57, 1663–1683, 2010.

Kuznetsov, Y., Baev, A., Coccetti, F., and Russer, P.: The Ultra Wideband Transfer Function Representation of Complex Three-Dimensional Electromagnetic Structures, in: 2004 International Symposium on Signals, Systems, and Electronics ISSSE'04, August 10-13, Linz, Austria, 2004.

Rogers, S. D., Aberle, J. T., and Auckland, D. T.: Two-port model of an antenna for use in characterizing wireless communications systems, obtained using efficiency measurements, IEEE Antennas and Propagation Magazine, 45, 115–118, 2003.

Russer, P.: Electromagnetics, Microwave Circuit and Antenna Design for Communications Engineering, Artech House, Boston, second edn., 2006.

Russer, J. A., Kuznetsov, Y., and Russer, P.: Discrete-time network and state equation methods applied to computational electromagnetics, Mikrotalasna Revija (Microwave Review), pp. 2–14, 2010.

Shevgunov, T., Baev, A., Kuznetsov, Y., and Russer, P.: Lumped element network synthesis for one-port passive microwave structures, in: Microwaves, Radar and Wireless Communications, 2008. MIKON 2008. 17th International Conference on, pp. 1–4, 2008.

Yordanov, H. and Russer, P.: Area-Efficient Integrated Antennas for Inter-Chip Communication, in: Proceedings of the 40th European Microwave Conference, EuMC 2010, Paris, France, 2010.

32

A curvature-corrected CMOS bandgap reference

O. Mitrea[1], C. Popa[2], A. M. Manolescu[2], and M. Glesner[1]

[1]Darmstadt University of Technology, Karlstr. 15, 64283 Darmstadt, Germany
[2]University Politehnica of Bucharest, Bd. Iuliu Maniu 1–3, Bucharest, Romania

Abstract. This paper presents a CMOS bandgap reference that employs a curvature correction technique for compensating the nonlinear voltage temperature dependence of a diode connected BJT. The proposed circuit cancels the first and the second order terms in the $V_{BE}(T)$ expansion by using the current of an autopolarized Widlar source and a small correction current generated by a MOSFET biased in weak inversion. The voltage reference has been fabricated in a $0.35\,\mu$m 3Metal/2Poly CMOS technology and the chip area is approximately $70\,\mu$m $\times$ $110\,\mu$m. The measured temperature coefficient is about 10.5 ppm/K over a temperature range of 10–90°C while the power consumption is less than 1.4 mW.

1 Introduction

Voltage references are widely used in applications such as A/D and D/A converters, acquisition data systems or smart sensors. As the precision of these circuits increases, the requirements for the reference stability with temperature, supply and process variations have also increased.

Since introduced by Widlar (1971), the bandgap reference (BGR) has been extensively used to obtain a constant value of the reference voltage with respect to temperature variations. The basic idea of a first-order compensated BGR is to cancel the negative temperature dependency of the V_{BE} by adding a correction factor proportional to the thermal voltage V_t which has a positive variation with temperature. The temperature coefficient (TCR) that can be obtained with this approach is generally greater than 30 ppm/K, being acceptable only for applications that do not require a very good accuracy (Ferro et. al., 1989; Tham and Nagaraj, 1995; Vermaas et al., 1998; Banba et al., 1999). This TCR limitation is determined by the fact that V_t is a fully linear function of T while V_{BE} is a complex function of T that contains higher order terms (Filanovsky and Chan, 1996) – see also Eq. (1).

A first way to improve the TCR of a bandgap reference is to correct the nonlinear temperature dependence of the base-emitter voltage by a suitable polarization of the bipolar transistor. In Filanovsky and Chan (1996), a polarization at a PTAT^3 + PTAT^4 collector current reduces the temperature coefficient to 4–8 ppm/K. The curvature-correction technique from Popa (2001), based on the polarization of the bipolar transistor at a PTAT^n current, reports a TCR of 20 ppm/K without trimming and chip temperature stabilization. Also these methods are conceptually simple, generating PTAT^n currents requires complex circuits which consume significant chip area and power.

Another possibility to improve the temperature dependence of a BGR is to add a correction voltage to the basic reference voltage (Salminen and Halonen, 1992), or a correction current to the PTAT current (Gunawan et al., 1993; Lee et al., 1994). The voltage reference presented in Salminen and Halonen (1992) has a relatively large temperature coefficient, about 30 ppm/K for a limited temperature range, due to the MOS parameters mismatching. The current compensation technique (Gunawan et al., 1993; Lee et al., 1994) decreases the temperature coefficient to less than 10 ppm/K but requires a large silicon area and it is not compatible with the CMOS technology.

This paper proposes a TCR improvement technique based on the compensation of the base-emitter nonlinearity with a current with an opposite temperature dependence, which will cancel the first and second-order harmonics in $V_{BE}(T)$. The main problem is to find a suitable implementation of a current generator with superior-order harmonics in its temperature expansion. In Sect. 2 the proposed circuit for implementing this technique is introduced and analyzed. Section 3 presents the experimental results: the measured temperature coefficient is 10.5 ppm/K for a temperature range of 10–90°C. The compatibility with CMOS technology is fulfilled because only MOS and pnp transistors have been used.

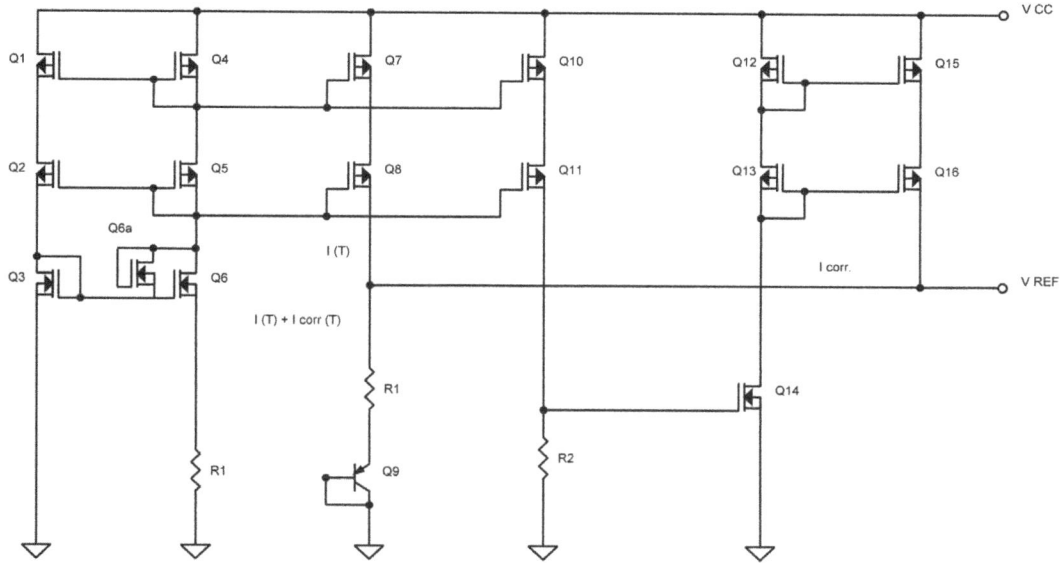

Fig. 1. Weak inversion curvature-corrected bandgap reference.

2 Circuit design

The temperature characteristic of the base-emitter voltage of a BJT can be expressed like Filanovsky and Chan (1996):

$$V_{BE}(T) = E_G(T) + \frac{T}{T_0}[V_{BE}(T_0) - E_G(T_0)] +$$

$$+ \frac{KT}{q} \ln\left[\frac{I_C(T)}{I_C(T_0)}\right] - \eta \frac{KT}{q} \ln\frac{T}{T_0} \qquad (1)$$

where T_0 is the reference temperature, E_G is the silicon bandgap voltage, $I_C(T)$ is the collector current of the bipolar transistor (considering a general temperature dependency) and η is a constant specific for each technology having typical values around 4. According to Lee et al. (1994), the temperature dependence of the silicon bandgap voltage can be very accurately modeled (with an error smaller than 0.2 mV) by the following empirical equation:

$$E_G(T) = a - bT - cT^2 \qquad (2)$$

$$a = 1.1785\,V$$

with $b = 9.025 \times 10^{-5}\,V/K$ for $150\,K \leq T \leq 300\,K$

$$c = 3.05 \times 10^{-7}\,V/K^2$$

$$a = 1.20595\,V$$

and $b = 2.7325 \times 10^{-4}\,V/K$ for $300\,K < T \leq 400\,K$.

$$c = 0$$

In order to improve the TCR we propose a CMOS implementation of a correction technique that exploits the exponential characteristic of a MOS transistor working in weak inversion – see Fig. 1.

The bipolar transistor Q9 and the resistor R1 form the core of the circuit: the reference voltage is the sum of V_{BE} and the voltage drop across R1. The autopolarized Widlar current source Q1-Q6 produces a PTAT current that is further mirrored through Q7-Q8 for biasing the BJT. Q6a is a start-up transistor that drives the circuit out of the degenerated bias point when the supply is turned on. Q10, Q11 and R2 generate the gate voltage of Q14 and should be dimensioned such that Q14 is biased in weak inversion ($V_{GS} < V_{TN}$). The current of Q14 is multiplied by Q12-Q16 and further injected in R1 to cancel the second order harmonics in $V_{BE}(T)$. The main PTAT current has the following expression (Razavi, 2001):

$$I(T) = \frac{2}{KN(W/L)_3 R1^2} \frac{(W/L)_7}{(W/L)_4} \left(1 - \sqrt{\frac{(W/L)_3}{(W/L)_6}}\right)^2 \qquad (3)$$

Because the bipolar transistor is biased at a collector current $I(T)$ (we can neglect I_{corr}), the base-emitter voltage will have the following general expansion around the reference temperature T_0:

$$V_{BE}(T) = V_{BE}(T_0) + \sum_{k=1}^{\infty} a_k(T - T_0)^k \qquad (4)$$

where a_k are constant coefficients of the expansion, with the following expressions $a_1 = [V_{BE}(T_0) - E_G(T_0)]/T_0 - b - 2cT_0 - (\eta-1)K/q$, $a_2 = (\eta-1)K/2qT_0 - c$, etc. The first term in Eq. (4) is a constant term. The next two terms (the linear and quadratic terms) will be cancelled by adding convenient correction factors to the base-emitter voltage. The influence of the superior-order terms ($a_3, a_4, ...$) will be further neglected.

The basic idea for improving the temperature behavior of the bandgap reference is to add a very small correction current, I_{corr} to the PTAT current $I(T)$, in order to compensate the first two-order terms from the base-emitter voltage expansion. Because this correction current has to be small and

Table 1. Measurement results

T (°C)	14	22	30	38	46	54	62	70	78	86	94
Vref (V)	1.1833	1.183	1.1834	1.1835	1.1834	1.1835	1.1831	1.183	1.1825	1.1828	1.1835

must contain at least second-order terms in its temperature expression, the proposed circuit for obtaining this current is based on a MOS transistor (Q14) working in weak inversion. Thus, the correction current is given by:

$$I_{\text{corr}}(T) = A \exp\left[\frac{I(T)R2 - V_{TN}}{nV_t}\right] \tag{5}$$

where $A = (W/L)_{14} I_{D0} \frac{(W/L)_{15}}{(W/L)_{12}}$ and $I(T) \gg I_{corr}(T)$. In this case, the reference voltage is:

$$V_{REF}(T) = V_{BE}(T) + [I(T) + I_{\text{corr}}(T)]R1 =$$
$$= V_{BE}(T) + V_{R1}(T) \tag{6}$$

Using the Taylor's series, the correction voltage given by the voltage drop across $R1$ can be expressed like:

$$V_{R1}(T) = V_{R1}(T_0) + \sum_{k=1}^{\infty} b_k (T - T_0)^k \tag{7}$$

where b_k are constant coefficients of the expansion, with the following expressions $b_1 = K_w + A\frac{V_{14}}{nV_{t0}}\frac{1}{T_0}\exp\left(\frac{V_{14}}{nV_{t0}}\right)$, $b_2 = -A\frac{V_{14}}{nV_{t0}}\frac{1}{T_0^2}\left(2 + \frac{V_{14}}{nV_{t0}}\right)\exp\left(\frac{V_{14}}{nV_{t0}}\right)$, etc. and $V_{14} = I(T_0)R2 - V_{TN}$. Finally, the output voltage is:

$$V_{REF}(T) = V_{REF}(T_0) + \sum_{k=1}^{\infty} a_k (T - T_0)^k +$$
$$+ \sum_{k=1}^{\infty} b_k (T - T_0)^k \tag{8}$$

In order to cancel the linear and the quadratic term from Eq. (8), the design conditions are:

$$a_1 + b_1 = 0 \quad \text{(to cancel the linear term)}$$
$$a_2 + b_2 = 0 \quad \text{(to cancel the quadratic term)} \tag{9}$$

The remaining nonlinearities of the BGR will be now given only by the superior-order terms (greater than two) from the reference voltage expansion. Other sources of errors could be transistors mismatches, the neglect of the correction current I_{corr} when Eq. (4) was deduced, or the finite value of the current gain of the pnp transistor.

3 Experimental results

The circuit was dimensioned using the previous design relations and Spectre® simulations. The following values have been employed for the transistors widths: W1–5 = 2 μm, W7–8,10–11 = 8 μm, W14–16 = 1 μm, W6,12–13 = 4 μm. All MOS devices have the minimum length $L = 0.3\,\mu$m.

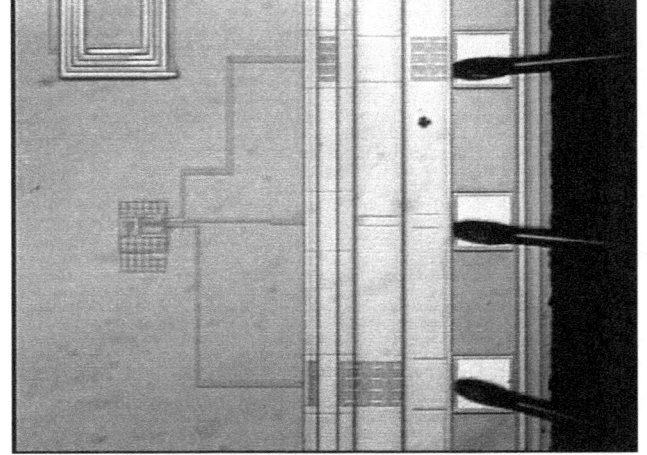

Fig. 2. Chip micrograph.

The values used for resistances are: R1 = 1.9 kΩ and R2 = 2.5 kΩ. The reference consumes approximately 0.4 mA from a supply voltage of 3.3 V.

The circuit was laid out and fabricated using the 0.35 μm CMOS AMS technology, available through the Europractice program – see the chip micrograph in Fig. 2. Special care was taken for avoiding mismatching between paired MOS devices and resistors. The silicon occupied area was about 70 μm × 110 μm, much smaller than that reported in other curvature-corrected voltage references with comparable performances (290 μm × 150 μm in Tham and Nagaraj (1995), 380 μm × 190 μm in Lee et al. (1994) and 600 μm × 220 μm in Gupta and Black (1996)).

Measurements and simulation results of the curvature-compensated voltage reference are presented in Table 1 and Fig. 3. The achieved temperature coefficient is about 10.5 ppm/K (without thermal stabilization of the chip) for a temperature range of 10–90°C.

4 Conclusions

A curvature-correction technique based on the compensation of the base-emitter voltage nonlinearity, using as correction the drain current of a MOS transistor working in weak inversion, was presented. The cancellation of the first and second-order term from the polynomial expansion of the base-emitter voltage allows the reduction of the temperature coefficient of the bandgap reference to about 10 ppm/K, for an extended temperature range. In addition, this circuit offers the possibility to cancel other superior-order terms which af-

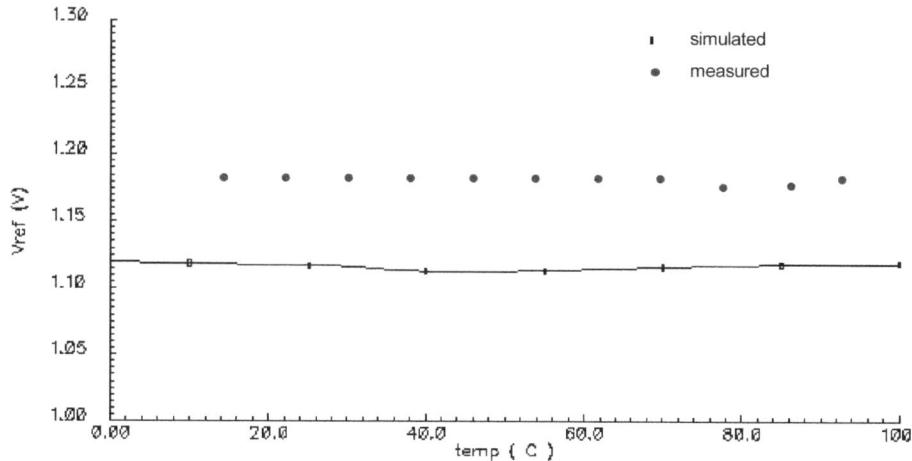

Fig. 3. Measurement and simulation results.

fect the TCR: the design condition for canceling the kth − order harmonic is $a_k + b_k = 0$.

References

Widlar, R. J.: New Developments in IC Voltage Regulators, IEEE Journal of Solid-State Circuits, 6, 2–7, 1971.

Ferro, M., Salerno, F., and Castello, R.: A Floating CMOS Bandgap Voltage Reference for Differential Applications, IEEE Journal of Solid-State Circuits, 24, 3, 690–697, 1989.

Tham, K. M. and Nagaraj, K.: A Low Supply Voltage High PSRR Voltage Reference in CMOS Process, IEEE Journal of Solid-State Circuits, 30, 5, 586–590, 1995.

Vermaas, L. L. G., De Mori, C. R. T., Moreno, R. L., Pereira, A. M., and Charry, R. E.: A Bandgap Voltage Reference Using Digital CMOS Process, IEEE International Conference on Electronics, Circuit and Systems, 2, 303–306, 1998.

Banba, H., Shiga, H., Umezawa, A., Miyaba, T., Tanzawa, T., Atsumi, S., and Sakui, K.: A CMOS Bandgap Reference Circuit with Sub-1-V Operation, IEEE Journal of Solid-State Circuits, 670–674, 1999.

Filanovsky, I. M. and Chan, Y. F.: BiCMOS Cascaded Bandgap Voltage Reference, IEEE 39th Midwest Symposium on Circuits and Systems, 943–946, 1996.

Popa, C.: Curvature-compensated Bandgap Reference, The 13th International Conference on Control System and Computer Science, University "Politehnica" of Bucharest, 540–543, 2001.

Salminen, O. and Halonen, K.: The Higher Order Temperature Compensation of Bandgap Voltage References, IEEE International Symposium on Circuits and Systems (ISCAS) 1992, 3, 1388–1391, 1992.

Gunawan, M., Meijer, G. C. M., Fonderie, J., and Huijsing, J. H.: A Curvature-corrected Low-voltage Bandgap Reference, IEEE Journal of Solid-State Circuits, 28, 6, 667–670, 1993.

Lee, I., Kim, G., and Kim, W.: Exponential Curvature-compensated BiCMOS Bandgap References, IEEE Journal of Solid-State Circuits, 1396–1403, 1994.

Razavi, B.: Design of Analog CMOS Integrated Circuits, McGraw Hill, 2001.

Gupta, S. and Black, W.: A 3 V to 5 V CMOS Bandgap Voltage Reference with Novel Trimming, IEEE 39th Midwest Symposium on Circuits and Systems, 969–972, 1996.

33

First harmonic injection locking of 24-GHz-oscillators

M. R. Kühn and E. M. Biebl

Technische Universität München, Fachgebiet Höchstfrequenztechnik, Arcisstr. 21, 80333 München, Germany

Abstract. An increasing number of applications is proposed for the 24 GHz ISM-band, like automotive radar systems and short-range communication links. These applications demand for oscillators providing moderate output power of a few mW and moderate frequency stability of about 0.5%.

The maximum oscillation frequency of low-cost off-the-shelf transistors is too low for stable operation of a fundamental 24 GHz oscillator. Thus, we designed a 24 GHz first harmonic oscillator, where the power generated at the fundamental frequency (12 GHz) is reflected resulting in effective generation of output power at the first harmonic. We measured a radiated power from an integrated planar antenna of more than 1 mW. Though this oscillator provides superior frequency stability compared to fundamental oscillators, for some applications additional stabilization is required.

As a low-cost measure, injection locking can be used to phase lock oscillators that provide sufficient stability in free running mode. Due to our harmonic oscillator concept injection locking has to be achieved at the first harmonic, since only the antenna is accessible for signal injection. We designed, fabricated and characterized a harmonic oscillator using the antenna as a port for injection locking. The locking range was measured versus various parameters. In addition, phase-noise improvement was investigated. A theoretical approach for the mechanism of first harmonic injection locking is presented.

1 Introduction

Injection locking is a low-cost measure to phase lock oscillators that provide sufficient stability in free running mode. In our case, injection locking has to be achieved at the first harmonic, since only the antenna is accessible for signal injection. Thus, the coupling circuit is integrated in the planar antenna structure. We investigated the locking range, i.e. the acceptable deviation between free-running oscilla-

tion frequency and frequency of the synchronization source versus different parameters like output power of the oscillator and power of the synchronization source. Moreover, phase-noise improvement due to injection locking was investigated. The mechanism of first harmonic injection locking and the resulting dependencies will be discussed.

2 Oscillator design

The oscillator design is based on the one port approach. The fundamental wave is at 12 GHz and reflected at the filter, i.e. there is no power delivered at the fundamental frequency. The oscillating conditions are fulfilled at 12 GHz, but the output signal is the first harmonic at 24 GHz. Because we do not take out any power at 12 GHz the first harmonic is generated efficiently. Spurious spectral components, namely the fundamental and the higher harmonics are suppressed by use of a filter (Fig. 1). We measured an output power of about 3 dBm at 50 Ω behind the filter. The oscillator is designed for maximum power at 24 GHz and ideal oscillating conditions at 12 GHz.

The bias network in this design has to block both 12 GHz and 24 GHz. Therefore, two radial stubs were attached at each DC-line. The small stub is designed for 24 GHz and the big stub is for 12 GHz and for 24 GHz. The Design is realised in microstrip-technology on RT-Duroid.

To radiate the 24 GHz wave the oscillator is coupled to a common rectangular patch antenna, see Fig. 1. The patch is a $\lambda/2$ resonator. At the other side of the antenna a coupling structure is connected. Here the 24 GHz reference signal is injected into the oscillator via the antenna and the filter. The coupling line is realised as 100 Ω line to reduce the influence to the antenna. Furthermore a coupling slot is inserted, which allows for relatively weak coupling between the reference source and the resonant patch antenna in order to provide a sufficiently high loaded q-factor of the antenna. The simulated insertion loss of this coupler is 8.45 dB at 24 GHz.

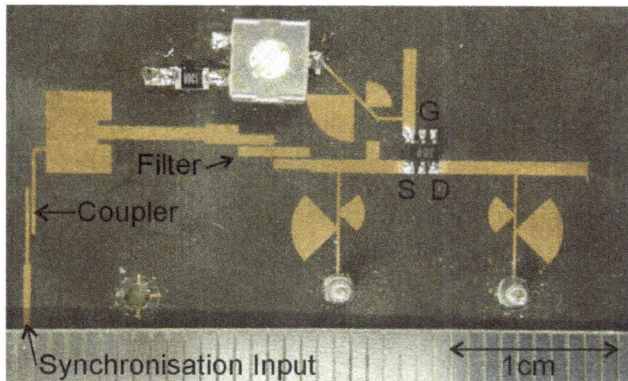

Fig. 1. Free running harmonic Oscillator with patch antenna, filter and coupler.

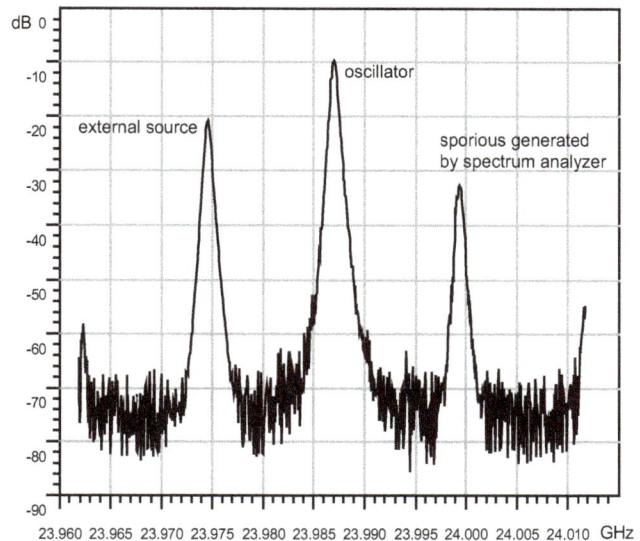

Fig. 2. Injected signal is not within the locking range of the oscillator.

3 Injection locking

Injection locking is the synchronisation in frequency and phase of a free running oscillator with a source. The mechanism of injection locking has been observed in a wide variety of oscillators (Adler, 1946; Kurokawa, 1973; York, 1993; Navarro and Chang, 1996). It can be shown, that the output phase of a simple oscillator with an injected signal is given by Adler's equation (Adler, 1946).

$$\frac{d\varphi}{dt} = \omega_0 - \omega_{inj} + \underbrace{\frac{\rho}{\alpha}\frac{\omega_0}{2Q}}_{\Delta\omega_m} \sin\underbrace{(\psi - \varphi)}_{\Delta\varphi} \qquad (1)$$

where ω_0 is the free-running frequency, ω_{inj} is the injected signal, a is the free-running oscillating amplitude, Q is the quality-factor of the oscillator's resonant circuit, ρ is the amplitude of the injected signal measured at the free-running oscillator. For phase synchronising the oscillator with the reference signal a steady state solution $d\varphi/dt = 0$ has to exist. Solving (3.1) for steady state gives

$$\Delta\varphi = \sin^{-1}\left(\frac{\omega_{inj} - \omega_0}{\Delta\omega_m}\right) \qquad (2)$$

which shows that an injection locked solution is possible only when the injected signal frequency lies within the "locking range" of the oscillator $\omega_0 \pm \Delta\omega_m$. Solving (3.2) will result in two possible solutions because of the inverse sine. Here a stability analysis is necessary to gain the right solution. The phase is distort from its free-running state (φ_0) by writing $\varphi = \varphi_0 + \delta\varphi$, which reduces (3.1) to

$$\frac{d\delta\varphi}{dt} = -\delta\varphi\Delta\omega_m\cos\Delta\varphi \qquad (3)$$

The distortion will decay in time provided that $\cos\Delta\varphi > 0$, which restricts the phase difference to the range $-\pi/2 \leq \Delta\varphi \leq \pi/2$. The locking range is proportional to the injected signal amplitude and inverse proportional to the Q-factor and the oscillator's amplitude. To get a large locking range in a practical system, low-Q oscillators, with large injected signals are required.

4 Injection locking at the first harmonic

In contrast to the described mechanism of injection locking above the injection locking in our case is different. The described mechanism requires that the injected signal is in the locking range of the free running oscillator's fundamental wave. Here, the oscillator is working as harmonic oscillator and the signal of interest is the first harmonic wave. The synchronisation signal, which is injected into the antenna, is also at the first harmonic of the oscillator, at a frequency of 24 GHz. So, what will happen by injecting the signal to this oscillator? The basic operation of a harmonic oscillator is that harmonics of the fundamental wave are generated by the nonlinearities of the active device. The only difference between a fundamental and a harmonic oscillator is the frequency where power is taken out of the oscillator. In our harmonic oscillator the first harmonic

$$2\omega_0 = \omega_h \qquad (4)$$

is delivered at the output, where ω_0 is the fundamental wave and ω_h is the first harmonic. Vice versa, the externally injected signal is handled.

The injected 24 GHz signal $\omega_{inj,h}$ is mixed down at the nonlinearities of the oscillator with its fundamental

$$\omega_{inj,h} - \omega_0 = \omega_{inj} \qquad (5)$$

where ω_{inj} is the half of the injected signal. Now, we apply the theory described above to our oscillator.

In Fig. 2 the external source is not within the locking range of the free running oscillator or the amplitude of the external signal is not large enough. For these conditions no interaction between oscillator and external signal is observed.

In Fig. 3 the injected signal is at the edge of the oscillator's locking range. Significant interaction between the signals results in harmonics of the beat frequency. Decreasing

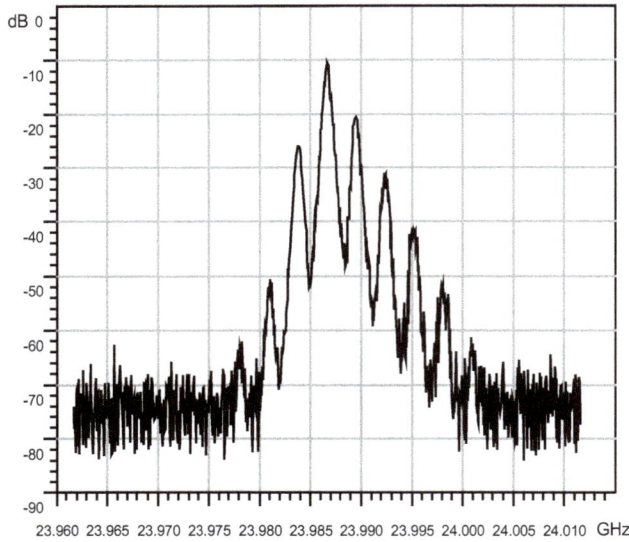

Fig. 3. Injected signal is very close to the locking range. Interference frequencies are produced.

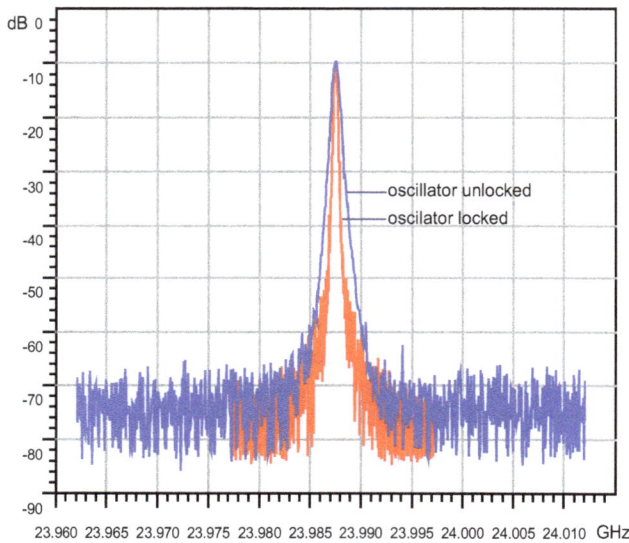

Fig. 4. Spectrum of free running and injection locked oscillator.

the frequency or increasing the power will lock the oscillator's frequency and phase to the injected signal (Fig. 4).

Another interesting point is the phase noise improvement due to the injection locking. The phase noise of the unlocked oscillator is -84 dBc @ 1 MHz, the phase noise of the locked oscillator is -88 dBc @ 1 MHz. The reduction of the phase noise can be analysed using the method given in Cao and York (1995). If we ignore AM noise, which is usually much less than the phase noise, the output voltage of an oscillator and the voltage of the injected signal can be written as

$$V_0 = A_0 \cos(\omega_0 t + \varphi_0(t)) \tag{6}$$

$$V_{inj} = A_{inj} \cos(\omega_{inj} t + \varphi_{inj}(t)) \tag{7}$$

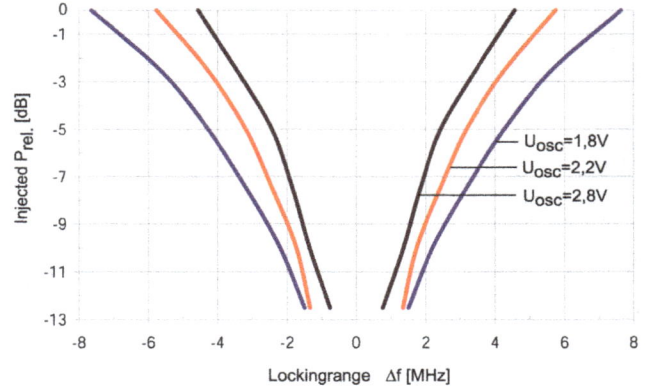

Fig. 5. Upper and lower edge of the locking range versus the power of the injected signal and DC bias of the oscillator's transistor.

With Adler's equation (3.1) and equation (4.3) and (4.4) it can be shown that the oscillator's phase variation follows the phase variations of the injected signal. As long as the oscillator is locked and the variations of the injected signal are within the locking range the phase variations follow

$$\frac{d\varphi_0(t)}{dt} = (\omega_{inj} - \omega_0) + \frac{d\varphi_{inj}(t)}{dt}, \tag{8}$$

i.e. $\varphi_0(t)$ and $\varphi_{inj}(t)$ are synchronous within the locking bandwidth (where $\omega_0 \omega_{inj} = 0$).

In Fig. 5 the dependencies of the injection locking parameters are plotted. The locking range δf depends on the injected power of the external signal and the amplitude of the free running oscillator (indicated by the DC drain-source-voltage U_{osc} applied to the transistor). The dependence of the locking range versus the injected power can easily be explained by analysing Adler's equation (3.1). Increasing the injected power results in a larger amplitude ρ. Increasing ρ will increase $\Delta\omega_m$. Because $\Delta\omega_m$ is the half of the locking range $\omega_0 \pm \Delta\omega_m$ the locking range becomes larger. On the other hand decreasing the oscillator voltage will decrease the output amplitude α and increase the locking range in the same way.

5   Summary

A harmonic free running oscillator at 24 GHz was demonstrated. The phase/frequency stability of the oscillator was enhanced by injection locking at the first harmonic, i.e. at 24 GHz. The oscillator can be used for applications in the ISM-band. The mechanism of injection locking at the first harmonic could be explained by an analogy to the theory for fundamental injection locking. Furthermore, the dependencies of the locking range versus power levels of the oscillator and the injected signal could be predicted and were proven by measurements.

References

Olbrich, M.: Development of a 24 GHz Doppler radar module, Diplomarbeit, Technische Universität München, April 2002.

Adler, R.: A study of Locking Phenomena in Oscillators, Proc. of the IRE, vol. 34, pp. 351–358, June 1946.

Kurokawa, K.: Injection-Locking of a solid-state microwave oscillators, Proc. IEEE, vol. 61, pp.1386–1409, Oct. 1973.

York, R. A.: Nonlinear Analysis of Phase Relationships in quasi-Optical Oscillator Arrays, Trans. IEEE MTT, vol. 41, no. 10, pp. 1799–1809, Oct. 1993.

Navarro, J. A. and Chang K.:, Integrated Active Antennas and Spartial Power Combining, K. Chang, Wiley, ch. 12, 1996.

Cao, X. and York, R. A.: Phase Noise Reduction in Scanning Oscillator Arrays, MTT-S, WE3F-F2, 1995.

34

A planar hybrid transceiving mixer at 76.5 GHz for automotive radar applications

M. O. Olbrich[1]**, A. Grübl**[2]**, R. H. Raßhofer**[3]**, and E.M. Biebl**[1]

[1]Technische Universität München, Fachgebiet Höchstfrequenztechnik, 80290 München, Germany
[2]now with Europäisches Patentamt
[2]now with TriQuint

Abstract. A growing number of applications for radar systems in automobiles demands for low-cost radar front-ends. A planar monostatic radar front-end is particularly suited for low cost applications as it uses only one antenna for transmission and reception and, thus, minimizes the needed chip area.

Generally, in a standard homodyne radar a radio-frequency (RF) signal generated by an oscillator is used for both, the transmitted signal and the local oscillator (LO). Well controlled distribution of the input power between antenna and mixer is crucial. A transceiving mixer at 76.5 GHz is presented, where this distribution is done by use of a rat-race coupler. In a conventional transceiver the oscillator signal is split into the transmitted and in the LO signal by a directional coupler. A second directional coupler is needed in order to merge the received and the LO signal at the mixer. In our design the purpose of splitting and merging the signals is realized with only one coupler. Elimination of the second coupler reduces losses significantly.

The received signal is down-converted to the intermediate frequency (IF) by use of a balanced mixer. For small relative speed in a CW-Doppler-radar or short distance in a FMCW-radar the IF is very small. Therefore $1/f$ noise is a significant value. In order to achieve good $1/f$ noise characteristics, Schottky diodes were used. The diodes were flip-chip bonded onto a microstrip circuit on a Al_2O_3 substrate.

The assembled transceiver was measured on-waver. An input power of 7 dBm was applied. The measured output power was 3 dBm and the conversion loss 9 dB. A noise figure of 15.3 dB was measured at 100 kHz.

1 Introduction

The demand for radar applications in the automotive sector is rapidly growing. Such applications are, e.g. distance measurement for the automotive cruise control, parking aid, side crash detectors or blind spot detection. Combining all the intended applications the car will be included in a radar bubble. The applications will support the driver and increase the road safety. First radar systems are already on the market. By now they are only implemented in upper class cars. However, in order to extend the applications and to open up the mass market for these systems, additional research work is still required and is already on it's way to be done (Dixit, 1997). For the mass market costs, reliability and easy fabrication are the most important factors. These needs can be satisfied best with hybrid or monolithically integrated circuits (Meinel, 1995).

By now no single chip implementation for the radar front-end exists or makes economic sense, and therefore the hybrid assembly offers some advantages. It is easy to combine different types of semiconductors like GaAs and silicon. Furthermore these active semiconductor devices can be mounted onto a cheaper substrate with the printed structures of the passive elements. Even more there is flexibility for the antenna design. The objective of the herein presented design was to reduce costs and losses by reducing the elements and the required substrate area. This is especially an advantage for MMIC circuits, but also for the less costly hybrid assembly reducing the size is a design goal.

2 Concept

In a traditional transceiver the oscillator signal is divided in the transmitted and in the LO signal by a directional coupler. In addition a second directional coupler is needed in order to merge the received and the LO signal at the non-linear devices of the mixer. In such a concept losses are considerable due to large structures. In this design the purpose of splitting and merging the signals is realized with only one directional coupler. A planar monostatic radar front-end is particularly suited for low-cost application as it uses only one antenna for transmission and reception and, thus, min-

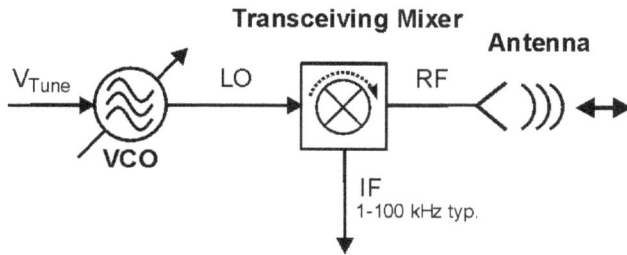

Fig. 1. Block diagram of a homodyne radar front-end.

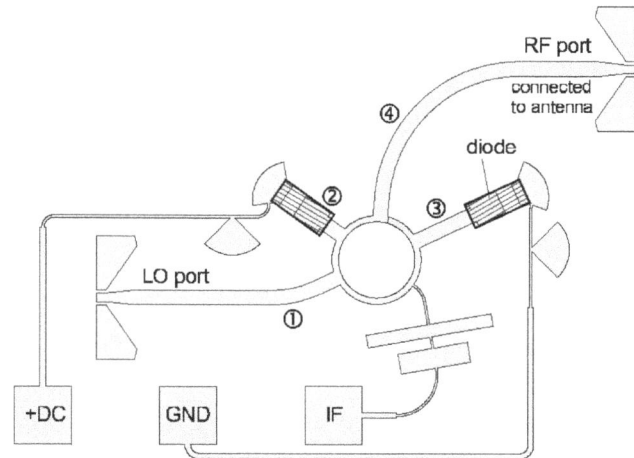

Fig. 2. Complete transceiving mixer layout.

imizes the needed chip area. For small relative speed in a CW-Doppler-radar or short distance in a FMCW-radar the IF is very small. Therefore 1/f-noise is a significant value. In order to achieve good 1/f-noise characteristics, silicon Schottky barrier diodes were used. Compared to GaAs based diodes they provide a lower 1/f-noise cut-off frequency. The diodes were flip-chip bonded onto a microstrip circuit on a Al_2O_3 substrate with microstrip waveguide structures. The flip-chip assembly using thermal compression bonding is suited for fully automated fabrication.

The largest noise contribution in a transceiving mixer is the LO AM noise. In order to gain good AM noise suppression at the IF port, a singly balanced mixer was used as the down-converting device. Thereby the signal is applied in phase to the diodes of the balanced mixer whereas the AM LO noise is applied out of phase to the diodes and therefore is rejected. The use of the singly balanced mixer principle is a good tradeoff between low losses and sufficiently high AM noise suppression. The singly balanced mixer can be realized by use of a 90° or a 180° hybrid. The designs basically differ in RF-LO isolation, but not in noise suppression. (Maas, 1993) The block-diagram of a homodyne radar front-end is illustrated in Fig. 1. This work deals with the transceiving mixer. The purposes of transmitting a part of the signal power to the antenna, driving the diodes and the mixing are all included in this element. The goal of the presented design was to realize these functioning with a minimum of components. Within this reduction the components usually allocated to a single function are merged into a single component. Therefore, in the resulting element, there is no more clear allocation between a special function and a discrete element.

Generally, in a standard homodyne radar a radio-frequency signal generated by an oscillator is used for both, the transmitted signal and the local oscillator. Well controlled distribution of the input power between antenna and mixer is crucial. There is a tradeoff between the output power, i.e. the fraction of power, which is transmitted, and the LO power level at the diodes determining the conversion loss of the mixer. Both parameters influence the achievable range of the radar.

The microstrip circuit was designed based on a full wave analysis. The whole front-end was simulated using a harmonic-balance method.

3 Realization

In a hybrid, the power from the input port (1) is equally delivered to two output ports (2,3). A fourth port (4) is isolated from the input port. At port (4) the antenna will be attached. The diodes are attached to ports (2,3). The concept of this transceiver is based on the bypass of the isolation between port (1) and port (4). This is achieved by the mismatch of the diodes at ports (2,3). Therefore the oscillator applied to port (1) feeds the diodes at ports (2,3) with LO power and in addition a fraction of it's power is transmitted to port (4) via the reflections on the mismatched diodes (Siweris, 1997).

The realization of a 90° hybrid is difficult due to the requirement of very low impedance lines. In contrast a 180° rat-race hybrid is simple to design and is transferred into a 90° hybrid by inserting a λ/4 long microstrip line between the rat-race and one of the diodes. Therewith the reflected signals are superimposed in phase at the antenna port (Grübl et al., 2002). The complete transceiving mixer layout with an indication of the diodes position is shown in Fig. 2.

The diodes are grounded by a virtual short realized with a radial stub. Since the properties of the diodes were not completely known, due to the absence of an appropriate model, the position of the virtual short was determined experimentally.

The IF is applied at the rat-race ring. The detachment of the IF from the RF matched part should not influence the symmetrical power distribution and, therefore, the AM noise suppression. A very low symmetry distortion was achieved by applying a standard RF block with a low-pass filter consisting of two stubs in a certain angle to the rat-race ring. The singly balanced mixer principle allows to serially bias the diodes. The DC-bias part is isolated from the RF by use of standard bandstop structures using radial stubs. For the measurement with a waver-prober, coplanar-to-microstrip transitions were included at the LO and the RF port.

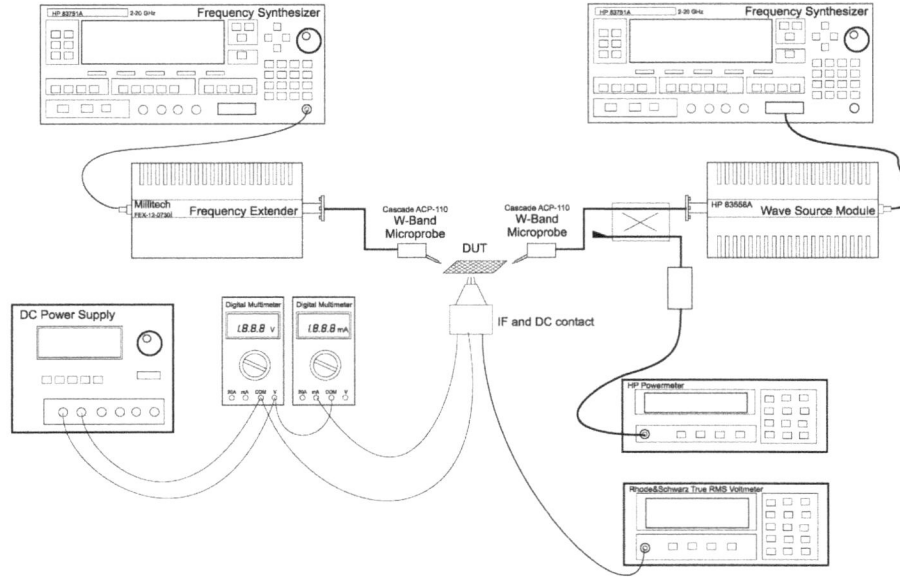

Fig. 3. Scheme of the measurement setup.

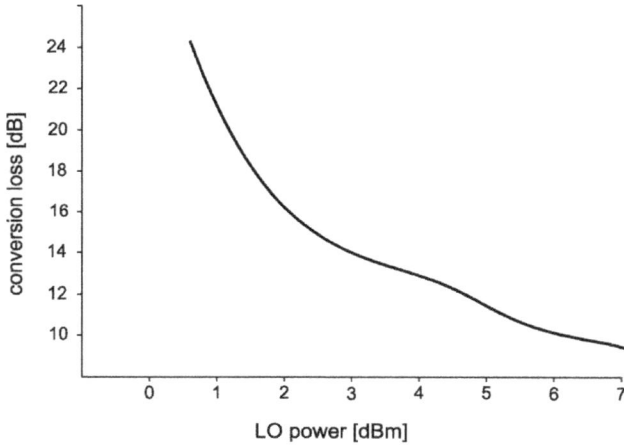

Fig. 4. Conversion loss vs. LO power.

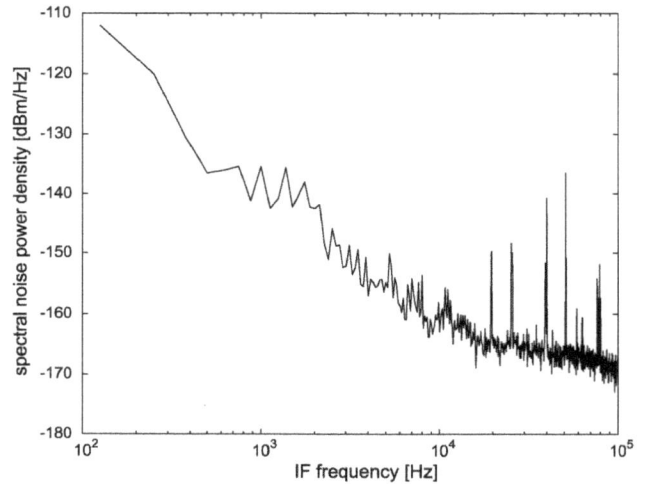

Fig. 5. Spectral noise power of the mixer.

4 Measurement

The assembled transceiver was measured on-waver with the setup illustrated in Fig. 3. To verify the simulation the passive circuit was first measured separately. In the following the complete mixer was measured with variations of the diode match and the virtual short. The measurement frequency was 76.5 GHz with an LO-power level of 7.1 dBm. The IF was measured with a True RMS Voltmeter. The optimum load impedance was determined to be 39 Ω.

The frequency shift of the applied RF signal compared to the LO frequency was 100 kHz with a RF power of -20 dBm. The bias voltage of the diodes was adjusted separately for each mixer under test. It depended on the match of the diodes and the resulting drive of the diodes by the LO. For both diodes in series the applied voltage was about 400 mV.

The conversion loss was calculated from the measured IF voltage and the impedance of the load. The best results were shown by a mixer without any matching network at the diodes. It's minimum conversion loss was 8.9 dB. Figure 4 shows the obtained results.

The noise figure of the mixer was characterized also. In order to measure the 1/f-noise the IF signal was amplified by use of a low noise amplifier and measured using an spectrum analyzer. Only LO signal and DC bias voltage were applied to the mixer. The RF port was terminated using a 50 Ω match. The measured noise power density is given in Fig. 5. The noise figure at 100 kHz was 15.3 dB. Herein the noise generated by the voltage source and the LO noise are included.

5 Conclusion

The presented transceiving mixer at 76.5 GHz takes advantage of the elimination of a second hybrid. Thus, losses and the required substrate area are reduced significantly compared to existing realizations. The LO-RF port isolation of the hybrid is bypassed via the reflections at the diodes of the singly balanced mixer. A matching of the diodes was found to decrease the performance, because of the additional losses. The measurements show low conversion loss of 8.9 dB at 3 dBm input power and excellent AM noise suppression.

References

Dixit, R.: Radar requirements and architecture trades for automotive applications, 1997 IEEE MTT-S Int. Micro-wave Symp. Dig., 3, 1253–1256, 1997.

Grübl, A., Herb, A. J., Raßhofer, R.H., Biebl, E.M.: A 76.5 GHz Transceiving Mixer Using Flip-Chip Mounted Silicon Schottky-Diodes for Automotive Radar Applications, EuMC Proceedings, 2002.

Maas, S. A.: Microwave Mixers, 2nd ed. Norwood, MA: Artech House, 1993.

Meinel, H. H.: Commercial Applications of Millimeterwaves History, Present Status, and Future Trends, IEEE Trans. Microwave Theory and Tech., MTT-43. 7, 1639–1653, 1995.

Siweris, H.-J.: Homodynes monostatisches Radarsystem, Patentschrift DE 196 10 850 C1, 1997.

35

An enhanced BSIM modeling framework for selfheating aware circuit design

M. Schleyer[1]**, S. Leuschner**[2]**, P. Baumgartner**[2]**, J.-E. Mueller**[2]**, and H. Klar**[1]

[1]Fachgebiet Mikroelektronik, Technische Universität Berlin, Germany
[2]Intel Mobile Communications GmbH, Munich, Germany

Correspondence to: M. Schleyer (martin.schleyer@tu-berlin.de)

Abstract. This work proposes a modeling framework to enhance the industry-standard BSIM4 MOSFET models with capabilities for coupled electro-thermal simulations. An automated simulation environment extracts thermal information from model data as provided by the semiconductor foundry. The standard BSIM4 model is enhanced with a Verilog-A based wrapper module, adding thermal nodes which can be connected to a thermal-equivalent RC network. The proposed framework allows a fully automated extraction process based on the netlist of the top-level design and the model library. A numerical analysis tool is used to control the extraction flow and to obtain all required parameters. The framework is used to model self-heating effects on a fully integrated class A/AB power amplifier (PA) designed in a standard 65 nm CMOS process. The PA is driven with +30 dBm output power, leading to an average temperature rise of approximately 40 °C over ambient temperature.

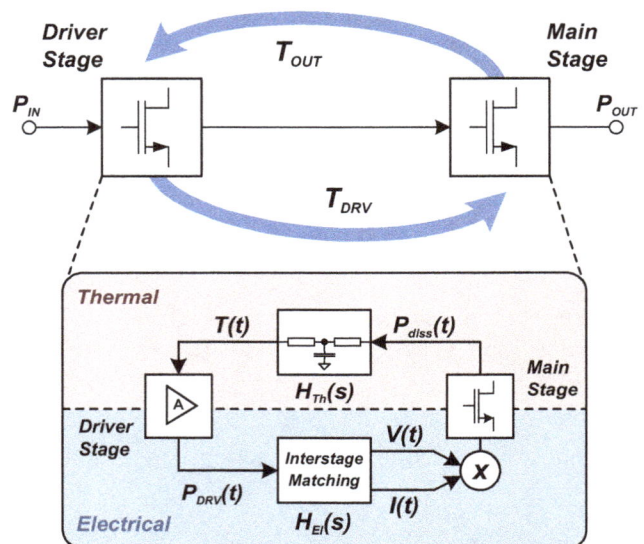

Figure 1. Memory effects due to electro-thermal resonances.

1 Introduction

In most analog and mixed signal radio frequency (RF) designs, static heat distribution is mainly a concern regarding the device matching and circuit performance. While self-heating is immanent in all RF integrated circuits (RFICs), its actual influence is negligible in most cases. As long as dynamic power dissipation is small compared to the dissipated DC power, no changes in the steady-state performance occur. However, the ongoing integration of system-on-chip environments, e.g. by integrating power amplifiers, adds notable dynamic heat sources on the same silicon die and affects the small signal properties other blocks, as heat is conducted through the chip. Therefore self-heating effects arise

as new challenge also in RFIC design. For this reason, tools are required to investigate and avoid thermal issues while designing such circuits.

Apart from output power and efficiency degradation due to self-heating, one important phenomenon is the occurrence of thermal memory effects in RF power amplifiers. The Joule effect translates electrical power dissipation into a heat flux Q. The die itself and the packaging have a thermal impedance, which determine the temperature increase ΔT due to Q. As the material has a certain mass and density, the overall thermal impedance is not purely real but has a capacitive component (Vuolevi et al., 2001). Hence, a large

Table 1. Modeling approaches – overview.

	Effort	Speed	Accuracy	Risk	Flexibility
(A) Adaption of BSIM4 C-Code	−	+	+	−	+
(B) Customized Verilog-A model	○	○	+	−	+
(C) Behavioral description	+	○	−	+	○
(D) Table Based Model	+	−	○	+	−
(E) Combined *BSIM2THERM* Model	+	○	○	+	○

thermal time constant is added to the system – typically in the order of a few kilohertz.

This temperature change due to electrical power dissipation directly modifies the properties of active devices: both electron mobility μ_e and threshold voltage V_{th} of a FET device decrease due to the rising temperature. Figure 1 illustrates the electro-thermal interaction in a typical two-stage PA design. Here, thermal time constants resonate with electrical memory effects in the baseband frequency domain and can cause severe memory to the power amplifier (Wolf, 2012, p. 78). These self-heating effects in power amplifiers have been studied extensively using behavioral models. Boumaiza et al. (2003) follow the basic concepts as presented by Vuolevi et al. (2001), and use a thermal network to control a simple model inheriting the gain reduction of an LDMOS amplifier. Boumaiza et al. (2003) verify their model also with measurements for pulsed signals. Mazeau et al. (2007) apply dynamic Volterra series and obtain a coupled behavioral electro-thermal model. These approaches allow an investigation of thermal memory effects based on measurements of an actual implementation and allow to model the influence of self-heating effects on non-linear distortions and spectral regrowth. Anyhow, due to their nature as behavioral models, their use for circuit designers is very limited. As they model the whole block, no actual interaction between individual devices is investigated.

To close this gap, customized device models or simulation tools have been developed. Heo et al. (1999) propose a MOSFET large signal model targeting at LDMOS device design. They extend the default equations by first order temperature dependencies for drain current and threshold voltage in a custom device model. Codecasa et al. (2002) perform a decent analysis on electro-thermal resonance effects. With their results, the SPICE level 3 MOSFET model is extended to incorporate the electro-thermal effects into the circuit design environment. Unfortunately, none of these and other published approaches (Jardel et al., 2006; Du et al., 2008) give a robust and generic CMOS device model as required for self-heating aware design in standard IC design flows: they all require non-standard device models or manual work to find the behavioral descriptions. Actually, the recent PSP Level 103.2 device model (Smit et al., 2013) indeed supports an external thermal equivalent network to anticipate self-heating effects. However, PSP Level 103.2 models are not available for

most standard CMOS technology nodes larger than 28nm. In contrast, the BSIM4 model (Xi et al., 2004) is still used in many wide-spread and cost-effective CMOS technologies – but does not allow dissipation-driven temperature changes.

This work proposes an extension to the widely used BSIM4 model. The BSIM4 model is enhanced using a Verilog-A wrapper module. It adds additional temperature and power nodes to convert the dynamically dissipated power of the particular device into a temperature change. This temperature change is used to determine variations of the device characteristics in addition to the original BSIM model equations.

2 Enhanced BSIM Modeling Flow

The BSIM4 model uses temperature-dependent equations to include thermal effects on various device properties. However, all those effects are modeled static, and the models cannot be used for dynamic electro-thermal simulations. The goal of the presented framework is to overcome this issue and to add support for external thermal equivalent networks.

2.1 Modeling Strategies

Several approaches have been already discussed within the introduction – each associated with its own advantages and drawbacks. A short summary is given in Table 1. The most straight forward implementation would be the adaption of the original BSIM4 source code (A). With this approach, a very generic and geometry independent self-heating aware model could be generated. The model is provided as source code using the SPICE API (Quarles, 1989). In total, the model contains approx. 25 000 lines of code. To support dynamic temperature changes, a major overhaul of this code would be required. Altering the model in such an intrusive way includes a severe risk in changing the numerical behavior and can lead to inconsistencies compared to the original models.

Next to the more complex C model, a Verilog-A compact model implementation of BSIM4 was investigated (B). This customized model is less complex, but is per se error-prone as the Verilog-A model cannot directly implement the same routines and calculations as the C code model. Another trade-off to consider is the reduced computational speed of the Verilog-A implementation. While still being compiled before

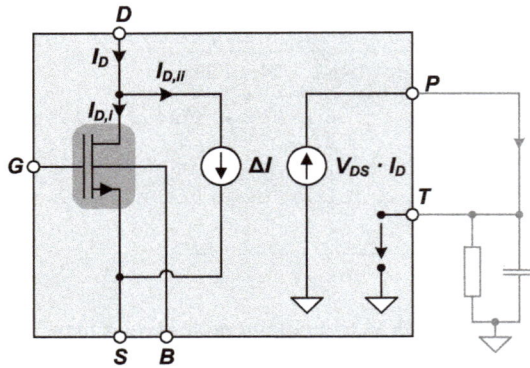

Figure 2. Equivalent circuit of the Verilog-A module.

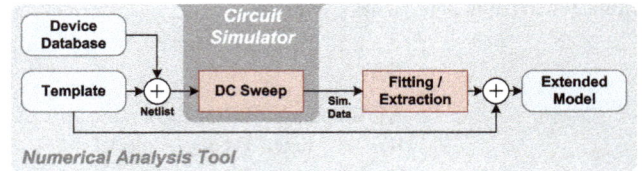

Figure 3. BSIM2THERM model generation flow.

run-time, it shows rather poor performance in comparison to the highly optimized binary implementation in C.

The danger of possibly deteriorating the model accuracy by altering its source code can be avoided by following a wrapper approach as e.g. proposed by Marbell and Hwang (2005). Behavioral sources are added to the underlying BSIM4 model (C). Although computationally efficient and with only slight implementation effort, a closed form description valid in all operation ranges is hard to find. The contradictory approach would be a table based model (D) which implements a look-up table based method for all operating conditions. The effort in terms of implementation and computations is very low, but a table-based model is generally less flexible and requires a huge amount of input data, if the complete operating range shall be covered. To allow implementation in both a reasonable time frame and with sufficient accuracy, a combined approach (E) is presented in this work, the so called *BSIM2THERM* framework. A behavioral source using a polynomial representation of drain current changes allows accurate modeling without modifying the BSIM4 source code. The coefficients for the polynomial representation are determined using fully automatized simulation and fitting routines.

2.2 *BSIM2THERM* Verilog-A module

In the recent years, the Verilog-A language superseded C and FORTRAN implementations for device models. Verilog-A based models do not require simulator- or vendor-specific coding when creating them (Troyanovsky et al., 2006). Thus, Verilog-A became the de facto language standard for compact device modeling – and has been used to implement e.g. the PSP or EKV device models. The *BSIM2THERM* modeling flow exploits the macro preprocessing capabilities of Verilog-A and splits the module in several parts. The core part of the wrapper contains descriptions of the device terminals and the branches required for current and voltage sensing and the controlled sources connected to the internal transistor device. Figure 2 shows the equivalent circuit

of the overall module, with an additional external thermal-equivalent RC network. The modeling equations, coefficients and intermediate variables are defined in an additional file and referenced with macro statements. Hence, structure and functionality are separated, which allows better maintenance of the model database. The controlled source is dependent of the temperature applied to the T node. The current source $\Delta I(T)$ incorporates current changes on the V/I characteristics of the device due to the dynamic temperature variations. In case of a short-channel CMOS device, the most notable mechanisms are electron mobility reduction and the velocity saturation.

3 Model generation

The frameworks aims to allow easy and fast characterization of the initial BSIM model. Figure 3 shows the basic control flow: A template is combined with device data stored in the *device database* to create the wrapper module with all connections and parameters as used by the original devices. The framework allows to parse foundry-provided model libraries and extract all BSIM4 based devices provided in the technology library. A fully characterized BSIM4 model uses 200+ parameters. The foundry-provided models normally calculate some parameters internally, others are left unaltered. The device database therefore contains a list of the parameters which need to be externally accessible. Furthermore, it holds default values and data types for these parameters, as Verilog-A does not allow undefined or empty values for instance parameters.

3.1 Device characterization

Based on the top-level netlist of a design, the framework determines all BSIM4 instances and their instance properties, such as geometry or device stress information. To complete the input netlist for the model characterization, the user needs to set limits and step sizes for the individual input variables. Typically, this would be a range from V_{GS} and $V_{DS} = 0\,V \ldots V_{DD}$. If the range is chosen too large, the fitting algorithm might not be able to properly fit sensitive areas – typically the transition between sub-threshold and linear region or linear and saturation region. The devices of interest are added to a Verilog-A based test bench. The simulator performs a nested DC sweep in the user-defined operating region. Investigations showed that a decent coverage and accuracy is reached with simulation times of 20 min, executed single-threaded on a 2.9 GHz Intel® Xeon E5-2690 machine.

3.2 Polynomial model representation

The simulation data obtained by DC characterization is processed within a numerical analysis tool to obtain a closed form expression of the V/I characteristics. A direct approach maps the current change due to $T \neq T_{\text{Nom}}$ into the source $\Delta I(T)$, such that

$$
\begin{aligned}
\Delta I(T) &= I_{\text{D}}(T) - I_{\text{D},i}(T = T_{\text{Nom}}) \\
&\triangleq f\left(T, V_{\text{DS}}, V_{\text{GS}}, V_{\text{BS}}, I_{\text{D},i}\right).
\end{aligned} \tag{1}
$$

Hence, a polynomial expression $P : X \to \Delta I$ with tupel $X = \{V_{\text{DS}}, V_{\text{GS}}, V_{\text{BS}}, I_{\text{D},i}, T\}$ can be used to approximate ΔI. If P is of degree N, and has $n = |X| = 5$ variables, it will require $k = \binom{N+n}{n}$ coefficients. For a higher-order degree calculation, this results in a high number of arithmetical calculations performed at each solver iteration step. The Verilog-A interpreter and compiler only does very basic code optimizations. It is therefore inevitable to reduce the arithmetic operations in forehand. A very simple technique is the *precalculation* of certain intermediate variables at runtime.

A second step is *iterative coefficient pruning*. Per default, the degree N is used for all input variables of X. If the degree N of one of the determinants is too high, the underlying QR decomposition delivers very small coefficients for high-order terms. As those coefficients c_k do not significantly contribute to the overall current ΔI, setting all $|c_k| < \varepsilon$ to zero directly reduces the computational effort while only marginally reducing the accuracy. The fitting algorithm removes those terms from the design matrix of P_i and repeats the QR decomposition delivering P_{i+1}, where i denotes the number of iterations starting from 0. If the approximation error $A_{i+1} = \left|1 - \frac{P(X)}{\Delta I(X)}\right|$ is not increased by more than an arbitrary chosen boundary E, a new acceptable representation $P_{i+1}(X)$ has been found. This iteration is repeated until E is finally crossed, and then P_i is kept as final representation for ΔI. With varying ε and E, the trade-off *accuracy vs. speed* can be set to an optimal point by empirical investigations. An additional weighting algorithm on the approximation error A further improves the overall model quality, as it penalizes errors in critical domains and adds relaxations in other regions. The representation P states that $I_{\text{D},i} \in X$. While not obvious in the regression model, $I_{\text{D},i}$ is actually dependent on all voltages as this is a boundary condition from the BSIM4 model. Anyhow, this redundancy has the advantage that the polynomial scales with $I_{\text{D},i}$. With $\widetilde{X} = X \setminus I_{\text{D},i}$, where $|\widetilde{X}| = \tilde{n} = n - 1$, the complexity can be reduced by costs of loosing the scaling property.

In Eq. (1), ΔI is a polynomial of the current change. It is obvious that $P(X)|_{T=T_{\text{Nom}}}$ results in $\Delta I = 0$. This property can be exploited to further reduce the computational effort. Instead of P, a new polynomial model $Q(\widetilde{X})$ is defined,

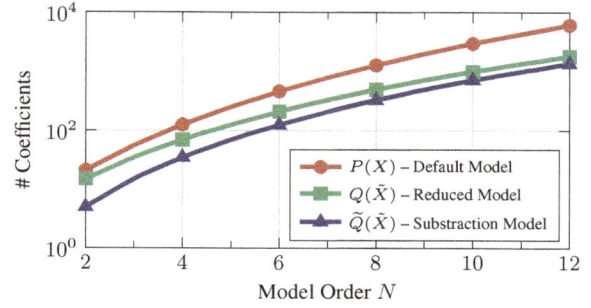

Figure 4. Comparison of different models.

where Q is a model of $I_{\text{D},i}(\widetilde{X})$, i.e. a polynomial representation of the underlying BISM4 model itself under bias and temperature conditions $\widetilde{X}$. Now, Eq. (1) can be expressed as

$$
\begin{aligned}
\Delta I(T) &= I_{\text{D}}(T) - I_{\text{D},i}(T = T_{\text{Nom}}) \\
&= Q(\widetilde{X}) - Q(\widetilde{X}, \quad T = T_{\text{Nom}}) = \widetilde{Q}(\widetilde{X}, T_{\text{Nom}})
\end{aligned} \tag{2}
$$

The polynomial $\widetilde{Q}$ is pre-calculated within the numerical analysis tool. A characteristic of $\widetilde{Q}$ lays in the subtraction: all terms of Q which are not related to T are equal in both sub-terms of $\widetilde{Q}$. It can be considered as

$$
\widetilde{Q}(\widetilde{X}, T_{\text{Nom}}) = \sum_{m=1}^{N} \left(T^m - T_{\text{Nom}}^m\right) \cdot \bar{Q}(\bar{X}), \tag{3}
$$

where $\bar{X} = \{V_{\text{DS}}, V_{\text{GS}}, V_{\text{BS}}\}$ is now independent of T. Thus $|\bar{X}|$ is reduced to $|\bar{X}| = \bar{n} = n - 2$. Hence, the polynomial $\widetilde{Q}$ has a reduced number of coefficients $\tilde{l}$ compared to l of the original $Q(\widetilde{X})$:

$$
\tilde{l} = l - \binom{\tilde{n}+N-1}{\tilde{n}-1} = \binom{\tilde{n}+N-1}{\tilde{n}} = \frac{N}{N+\tilde{n}} \cdot \binom{N+\tilde{n}}{\tilde{n}}. \tag{4}
$$

The advantage of this modeling method is directly stated in Eq. (4). Either the number of coefficients is reduced to $\tilde{l} = (l \cdot N)/(N + \tilde{n})$, or the model order can be increased to $\widetilde{N} = N + 1$ while $l = \tilde{l}$. If compared to k, the advantage is even bigger. It can be proven that $l = (k \cdot n)/(N + n)$ and finally

$$
\tilde{l} = k \cdot \frac{n \cdot N}{(N+n-1)(N+n)}. \tag{5}
$$

Figure 4 shows the number of coefficients N for the different modeling strategies. The model based on $P(X)$ is from hereon named *default model*, whereas the model based on $\widetilde{Q}(\widetilde{X})$ is named *subtraction model*. The plot also shows the $Q(\widetilde{X})$ model, here stated as *reduced model*. In the following, the two first mentioned will be investigated further by employing the modeling concept on an exemplary design.

4 Application example: CMOS RF Power Amplifier

To evaluate the capabilities of the framework, a fully integrated RF CMOS power amplifier for WCDMA operation

(a) Absolute error with $V_{GS} = 0.6$ V.

(b) Relative error with $V_{GS} = 0.6$ V.

(c) Absolute error with $V_{GS} = 3$ V.

(d) Relative error with $V_{GS} = 3$ V.

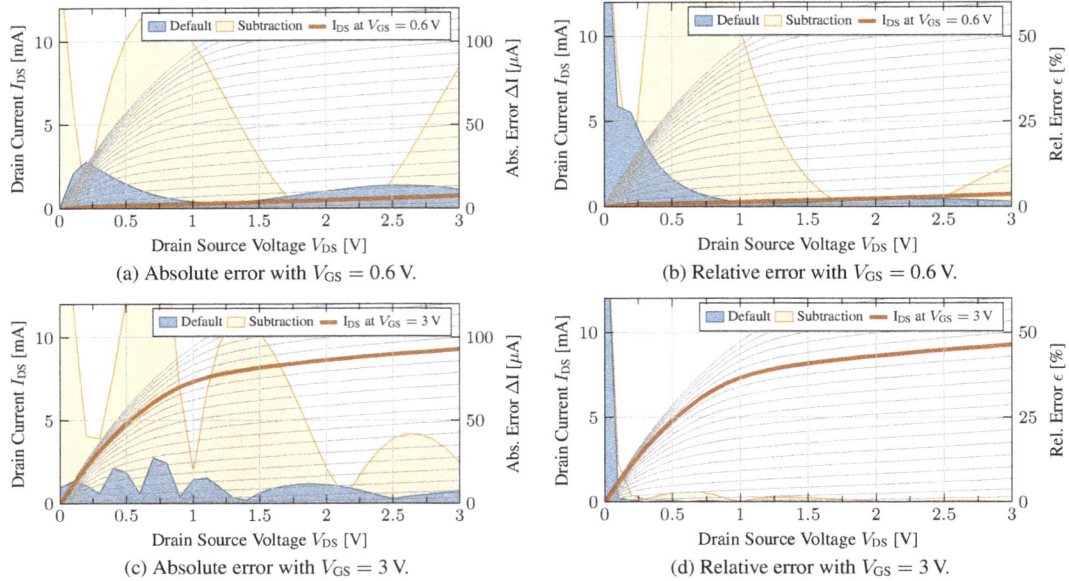

Figure 5. Output characteristics and modeling error at $V_{GS} = 0.6$ V and 3 V.

is analyzed regarding its electro-thermal properties. The PA is biased in class AB operation. It is built in a differential two-stage stacked-cascode structure with on-chip matching networks (Leuschner et al., 2011). The maximum linear output power $P_{Out,max}$ is $+27.9$ dBm with a PAE of ≈ 48 %. The dissipated DC power $P_{Diss,DC}$ is 340 mW typical and rises to $P_{Diss,Max} \approx 1.3$ W for large signal operation.

While not in the scope of this work, a decent modeling of the thermal properties is an important issue to achieve accurate simulation results. For the evaluation at hand, input and output stage have been connected to a single-stage RC network, based on estimations regarding thermal impedance of the package, giving a first approximation with acceptable modeling effort.

4.1 Model accuracy

The model accuracy is investigated by evaluating a thin-oxide I/O NMOS device with $W \approx 10\,\mu m$ and $L = 190$ nm. Both models use a polynomial of order $N = 5$. *Iterative coefficient pruning* was enabled with a boundary of 10^{-9}. For the *default model*, 125 additions and 237 multiplications are required to calculate the drain current change within the Verilog-A module. Due to the reduced number of coefficients and the iterative pruning, only 52 additions and 104 multiplications are required for the *subtraction model*.

The model was evaluated in the overall characterization range. Figure 5 shows absolute and relative errors at two different operating points – one closer to $V_{GS} = V_{th}$, the other with the device fully open. In both cases, $V_{BS} = 0$ V is applied. The shaded areas show the maximal errors over the complete temperature range from $0 \ldots 100\,°C$. In Fig. 5, the $I_{D,i}$ scaling effect of the polynomial is clearly visible, as the

default model shows significantly reduced absolute errors for low currents. Furthermore, the trade-off between no. of coefficients and accuracy is present: the subtraction model reduced the computational effort significantly less than ≈ 0.5 of the default model, but shows less accuracy especially in low current regions with small V_{th}.

4.2 Simulation results and performance

The obtained models where used to simulate the presented transistor stack and to investigate typical thermal issues. First, the operating point due to thermal runaway is simulated. Figure 6 shows that the bias current has an increasing offset with higher gate voltages. As expected, the additional self-heating reduces the overall current due to changes in electron mobility and velocity saturation. The gray line (right ordinate) implies that the die temperature increases by $\approx 17.8\,°C$ for a DC operating point of $I_{DC} \approx 200$ mA. These numbers illustrate that the framework is a valuable enhancement for bias design and temperature-independent biasing structures. A commonly seen issue in power amplifier design is the reduced saturation power due to self-heating of the power transistors. For continuous-wave operation, the power amplifier has a reduced gain compared to pulsed operation with small duty cycles. Figure 6 shows that this effect can be foreseen in simulations. Using the results of the DC simulation, the PA has been biased to a DC current of ≈ 100 mA in the output stage. The large signal behavior is evaluated in a single-tone *harmonic balance* simulation to obtain the AM/AM characteristics of the PA. The expected drop in output power is ≈ 0.9 dB at an output power level of $+20$ dBm. Here, the temperature increase is estimated to $38.8\,°C$.

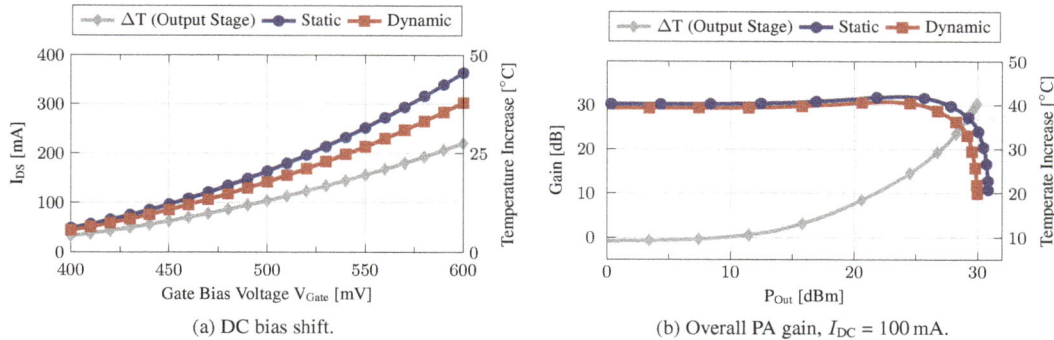

Figure 6. Simulated power amplifier characteristics with subtraction model.

Both simulations were executed multi-threaded on a 2.9 GHz Intel® Xeon E5-2690 machine. The time for the combined DC and HB simulations decreased slightly from 395.4 s to 429.4 s. Thus, it can be stated that the additional time constants due to the thermal RC network does not have a major impact on the overall simulation performance and the circuit showed good convergence.

5 Conclusions

The modeling framework presented in this work uses model data provided by the semiconductor foundries to obtain information about the thermal dependency of the drain current in the desired operation regions. An automated simulation procedure generates input data for a fitting process. The final result provides equations describing the dynamic temperature behavior of the device without altering the original model data. Hence, the approach is applicable for CMOS technologies where no foundry or vendor support for other models is available. An integrated CMOS power amplifier was evaluated and a good estimation of the self-heating behavior could be found with good simulation performance. Based on the framework, counter-measures to thermal-related issues can be taken before a first silicon is available for measurements.

Acknowledgements. The authors gratefully acknowledge research funding provided by the DFG (German Research Foundation), grant no. KL 918/8-1.

Edited by: D. Killat

References

Boumaiza, S., Gauthier, J., and Ghannouchi, F.: Dynamic electrothermal behavioral model for RF power amplifiers, in: Microwave Symposium Digest, 2003 IEEE MTT-S International, 1, 351–354, doi:10.1109/MWSYM.2003.1210950, 2003.

Codecasa, L., D'Amore, D., and Maffezzoni, P.: Modeling the thermal response of semiconductor devices through equivalent electrical networks, IEEE T. Circuits Syst., 49, 1187–1197, doi:10.1109/TCSI.2002.801279, 2002.

Du, B., Hudgins, J., Santi, E., Bryant, A., Palmer, P., and Mantooth, H.: Transient thermal analysis of power devices based on Fourier-series thermal model, in: Power Electronics Specialists Conference (PESC), 2008 IEEE, 3129–3135, doi:10.1109/PESC.2008.4592433, 2008.

Heo, D., Chen, E., Gebara, E., Yoo, S., Laskar, J., and Anderson, T.: Temperature dependent MOSFET RF large signal model incorporating self heating effects, in: Microwave Symposium Digest, 1999 IEEE MTT-S International, 2, 415–418, doi:10.1109/MWSYM.1999.779791, 1999.

Jardel, O., Quere, R., Heckmann, S., Bousbia, H., Barataud, D., Chartier, E., and Floriot, D.: An Electrothermal Model for GaInP/GaAs Power HBTs with Enhanced Convergence Capabilities, in: European Microwave Integrated Circuits Conference, The 1st, 296–299, doi:10.1109/EMICC.2006.282811, 2006.

Leuschner, S., Mueller, J.-E., and Klar, H.: A 1.8 GHz wideband stacked-cascode CMOS power amplifier for WCDMA applications in 65 nm standard CMOS, in: Radio Frequency Integrated Circuits Symposium (RFIC), 2011 IEEE, doi:10.1109/RFIC.2011.5940623, 2011.

Marbell, M. N. and Hwang, J.: A Verilog-based temperature-dependent BSIM4 model for RF power LDMOSFETs, in: Microwave Symposium Digest, 2005 IEEE MTT-S International, doi:10.1109/MWSYM.2005.1516882, 2005.

Mazeau, J., Sommet, R., Caban-Chastas, D., Gatard, E., Quere, R., and Mancuso, Y.: Behavioral Thermal Modeling for Microwave Power Amplifier Design, IEEE T. Microw. Theory, 55, 2290–2297, doi:10.1109/TMTT.2007.907715, 2007.

Quarles, T.: Adding Devices to SPICE3, Tech. Rep. UCB/ERL M89/45, EECS Department, Univ. of California, Berkeley, 1989.

Smit, G. D. J., Scholten, A. J., Klaassen, D. B. M., and van der Toorn, R.: PSP 103.2, Technical Note NXP-TN-2012-0080, 2013.

Troyanovsky, B., O'Halloran, P., and Mierzwinski, M.: Compact modeling in Verilog-A, in: Transistor Level Modeling For Analog/RF IC Design, edited by Grabinski, W., Nauwelaers, B., and Schreurs, D., 271–291, Springer Netherlands, doi:10.1007/1-4020-4556-5_10, 2006.

Vuolevi, J., Rahkonen, T., and Manninen, J.: Measurement technique for characterizing memory effects in RF power amplifiers, IEEE T. Microw. Theory, 49, 1383–1389, doi:10.1109/22.939917, 2001.

Wolf, N.: Charakterisierung von Leistungsverstärkern für die Entwicklung neuer und einfacher Vorverzerrungssysteme, Ph.D. thesis, Technische Universität Berlin, 2012.

Xi, X., Dunga, M., He, J., Liu, W., Cao, K. M., Jin, X., Ou, J. J., Chan, M., Niknejad, A. M., and Hu, C.: BSIM4.5.0 MOSFET Model, User Manual, 2004.

36

A software-radio front-end for microwave applications

M. Streifinger[1], T. Müller[2], J.-F. Luy[2], and E. M. Biebl[1]

[1]Technische Universität München, Fachgebiet Höchstfrequenztechnik, Arcisstr. 21, 80333 München, Germany
[2]DaimlerChrysler Forschungszentrum, Wilhelm-Runge-Str. 11, 89081 Ulm, Germany

Abstract. In modern communication, sensor and signal processing systems digitisation methods are gaining importance. They allow for building software configurable systems and provide better stability and reproducibility. Moreover digital front-ends cover a wider range of applications and have better performance compared with analog ones. The quest for new architectures in radio frequency front-ends is a clear consequence of the ever increasing number of different standards and the resulting task to provide a platform which covers as many standards as possible.

At microwave frequencies, in particular at frequencies beyond 10 GHz, no direct sampling receivers are available yet. A look at the roadmap of the development of commercial analog-to-digital-converters (ADC) shows clearly, that they can neither be expected in near future.

We present a novel architecture, which is capable of direct sampling of band-limited signals at frequencies beyond 10 GHz by means of an over-sampling technique. The well-known Nyquist criterion states that wide-band digitisation of an RF-signal with a maximum frequency f requires a minimum sampling rate of $2 \cdot f$. But for a band-limited signal of bandwidth B the demands for the minimum sampling rate of the ADC relax to the value $2 \cdot B$. Employing a noise-forming sigma-delta ADC architecture even with a 1-bit-ADC a signal-to-noise ratio sufficient for many applications can be achieved. The key component of this architecture is the sample-and-hold switch. The required bandwidth of this switch must be well above $2 \cdot f$.

We designed, fabricated and characterized a preliminary demonstrator for the ISM-band at 2.4 GHz employing silicon Schottky diodes as a switch and SiGe-based MMICs as impedance transformers and comparators. Simulated and measured results will be presented.

1 Introduction

Many applications of communication, sensor and signal processing systems would provide greater performance, if various standards could be handled with only one hardware device. Conceivable applications could be combined GSM / GPS devices (Müller et al., 2000) or a GPS receiver combined with a short-range radar. One first step towards this aim is the realization of a direct sampling receiver. The advantage is, that the input frequency and the channel bandwidth can be chosen in the receiver's digital part, fully independent of the used front-end.

In the following some analogue receiver architectures and the architecture of a direct sampling receiver will be introduced. The validity of the concept given and multiple simulations will be proved by measured results of a scaled demonstrator at 2.4 GHz.

2 Multiband capable receiver architectures

The requirements for multiband receivers lead to two main needs regarding the receiver topology: On the one hand the centre frequency has to be tuned over a large frequency range. On the other hand the bandwidths of the standards can differ by more than two decades.

The limit of heterodyne receivers regarding the tuneable bandwidth is given by the first IF frequency. Because of the non-linear character of the mixer, fourth order harmonics will arise between baseband and the double bandwidth of the input filter. This limits the maximum tuning bandwidth significantly or increases the number of IF stages up to an uneconomical number.

An attractive alternative is the direct conversion receiver. As shown in the block diagram (Fig. 1) the signal is mixed down to the baseband by two orthogonal carriers separately (Jentschel et al., 2000). This receiver can be built up very economically but several drawbacks prevent the usage for multimode receivers. Because of the always existing I-Q

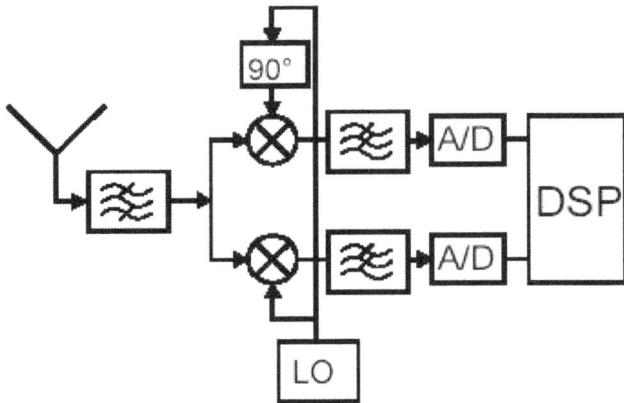

Fig. 1. Block diagram of the direct conversion receiver.

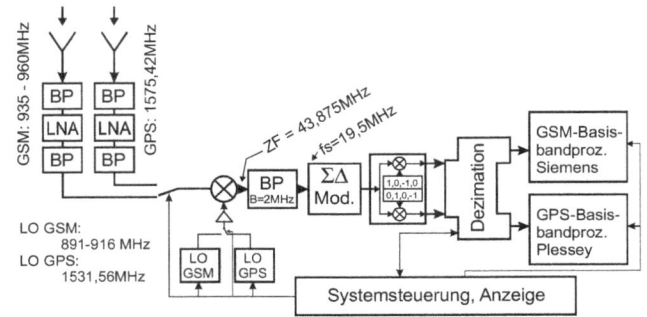

Fig. 2. Block diagram of a GSM – GPS receiver with common IF.

mismatch this topology is not suited for high order modulation schemes. A leakage of the LO leads to partly time variant DC bias after the down converting process which are costly to suppress. Because the bandwidth of the standard is defined by the lowpass filter, it has to be configured for multimode reception (Böhm et al., 1999).

The problem of LO leakage can be partly solved by using a very low IF for this type of receiver and implementing the shift into the baseband in the digital domain (Crols and Steyaert, 1998).

Looking at standards at mm-wave frequencies the Six-Port receiver becomes attractive (Tatu et al., 2001). The six-port can be considered as a black box with two inputs and four outputs. When the six-port itself is linear and the outputs are non-linear the relations between input signals and the output signals can be derived by a calibration procedure. The feasible tuning bandwidth is given by the characteristics of the six-port itself and the subsequent non-linear devices. The channel bandwidth is defined by the lowpass filters behind the non-linear outputs. A straight-forward approach to software configurable receivers is to sample the input signal at an IF or the input frequency itself. Bandpass-limited undersampling can be used to reduce the sampling rate and the effort in processing the digital signals. Unfortunately, the influence of sampling jitter and the requirements on the anti aliasing filter increases when doing this.

A disadvantage of sampling at IF is, that the IF filter defines the maximum channel bandwidth. When choosing a large IF bandwidth to receive any possible standard, the required dynamic range increases and leads to intermodulation specifications comparable to wide band sampling systems.

Figure 2 shows the block diagram of a combined GSM – GPS receiver, where both standards are received in time multiplex (Müller, 2002). The intermediate frequency was chosen in a way that the digital down-conversion can be realized by the multiplication with ones and zeros. The baseband processing is done by standard chipsets (Müller et al., 2000).

When sampling the RF signal directly, the analog-to-digital converter has to cope the high input bandwidth and all interferers after the antenna. The immense advantage is

that we obtain full flexibility in choosing the input frequency, channel bandwidth etc. by selecting the proper digital signal processing. The following chapter will present the design of a 24 GHz sampling head.

3   Concept of a direct sampling receiver

The concept of the direct sampling receiver is based on a band limited over sampling approach. The Nyquist-criterion is applied to the bandwidth of the signal of interest rather than to the complete incoming spectrum. Sampling of the whole spectrum of the input signal with a maximum frequency f, the minimum sample rate would be $2 \cdot f$. If the bandwidth B of the signal of interest is well known, the demands for the minimum sampling rate relax to the value of $2 \cdot B$ (Vaughan et al., 1991).

In case of oversampling the oversampling ratio (OSR) is defined as ratio of half the sampling rate related to the 3 dB-bandwidth of the filter. For an application in the ISM band ($f_0 = 24\,\text{GHz}$) the required bandwidth would be $B = 500\,\text{MHz}$. We assume a ratio of

$$\frac{f_0 \cdot 4}{f_s} = n = 9$$

and obtain a sampling frequency of $f_s = 10.66\,\text{GHz}$ and an OSR $= 10.66$ in this case. The expected resolution of this ADC is 3 bit (without noise shaping).

Using this concept, a noise shaping Sigma Delta Converter can be realized which will further improve the resolution (Norsworthy et al., 1997). The expected signal to noise ratio is shown in Fig. 3.

The aim is the realisation of the architecture shown in Fig. 4 for an input frequency of 24 GHz. As a first step, a scaled demonstrator was built up and analysed at 2.4 GHz. The following exposition mainly handles this demonstrator. If necessary and sensible, comments regarding the feasibility at 24 GHz will be made.

The key components of the receiver are the anti aliasing filter, the clocked sample & hold circuit (S/H) and the comparator or ADC itself as shown in Fig. 4 (Luy and Müller, 1999).

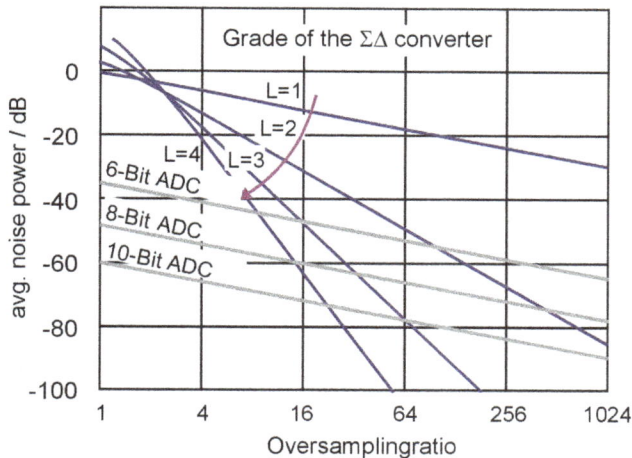

Fig. 3. Reachable signal to noise ratio of a 1-bit sigma delta converter and converters without noise shaping dependant on the OSR.

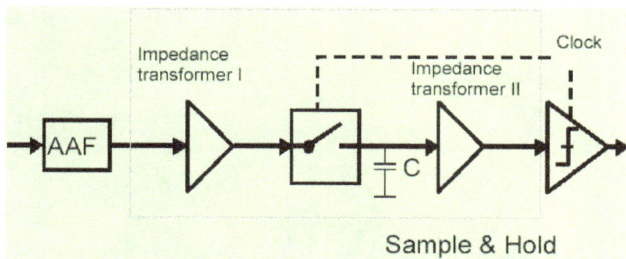

Fig. 4. Digital millimeter wave receiver architecture.

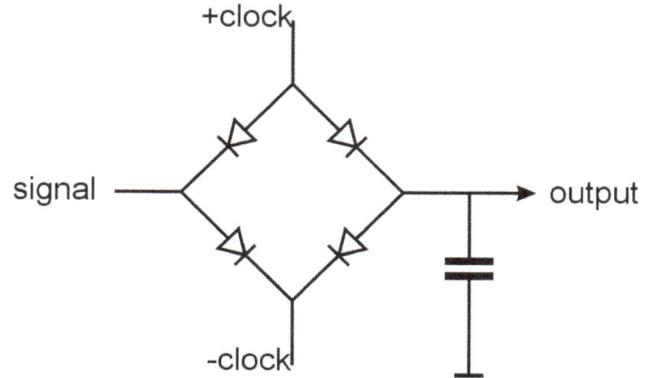

Fig. 5. schematic of the sample and hold circuit.

The anti-aliasing-filter (AAF) suppresses all out-of-band interferers in order to fulfil the Nyquist criterion at least regarding the bandwidth of the signal of interest.

The output voltage of the S/H switch must follow the input signal as long as the switch is closed (Fig. 6). Thus, the output impedance of the first amplifier driving the switch bridge has to be rather small. Good results are received by values of less then $10\,\Omega$. For this reason an impedance transformer is needed, as we usually have systems with a $50\,\Omega$ wave impedance. On the other hand, to hold the output voltage of the S/H switch constant as long as the switch is open, the input impedance of the second amplifier should be more than $200\,\Omega$. So we need a second impedance transformer, to transform the $200\,\Omega$ impedance back to a value dependant on the subsequent stage. Moreover both impedance transformers have to amplify the signals according to the dynamic range of the respective stages.

The clocked comparator and further signal processing parts are beyond the scope of this paper.

The key component of the receiver is the S/H circuit as shown in Fig. 5. As sampling switch a SIMMWIC Schottky diode quadruple bridge is used, followed by a shunt lumped holding capacitor (Jensen and Larson, 2001). The diodes are well suited because of their high reverse and their low forward resistance. As the diode switch contains four strong

nonlinearities generating many harmonics and intermodulation products an precise simulation of the circuit is rather difficult but nonetheless absolutely essential. Generally spoken the diode switch is the most broad-banded component of the whole front-end.

To prevent cross talk between the incoming signal and the clock signal, the diode quadruple has to be driven symmetrically. This is achieved by implementing a balun circuit in the clock signal path. Further very critical components are the length of the wave guide between the diode switch and the holding capacitor as well as the capacitor's physical dimensions. If one of them two is too large, oscillations emerge because of the parasitic inductivities and the holding capacitance.

4 Simulation of the sample and hold circuit

Having an exact diode model is crucial for the correct simulation of the sampling circuit. The following parameters where extracted from DC and RF measurements: Saturated reverse current: $I_S = 4,3\,\text{nA}$, serial resistance $R_S = 8,5\,\Omega$, ideality factor $N = 1.09$ and zero bias junction capacity $C_{j0} = 12\,\text{fF}...14\,\text{fF}$.

Simulations have shown that small changes of I_S, R_S and N do not affect the behaviour of the device much, whereas even small changes in the value of C_{j0} can deteriorate the functionality of the switch: During the hold stage, output voltage shows sinusoidal oscillations. Obviously, the isolation between signal input and the output is not sufficient due to the capacitive coupling by the junction capacity. As a countermeasure one can either apply a reverse DC bias or increase the clock power. Adding reverse DC bias increases the complexity of the circuit significantly, increasing the clock power is limited by the maximum forward current of the diodes. Simulations showed that we could expect sufficient isolation for reasonable clock power without reverse bias for working in the ISM Band at 24 GHz.

Figure 6 shows the voltage measured at the holding capacitor. As input signal an unmodulated carrier of 2.4 GHz was used. The SMA connectors and bonding wires were

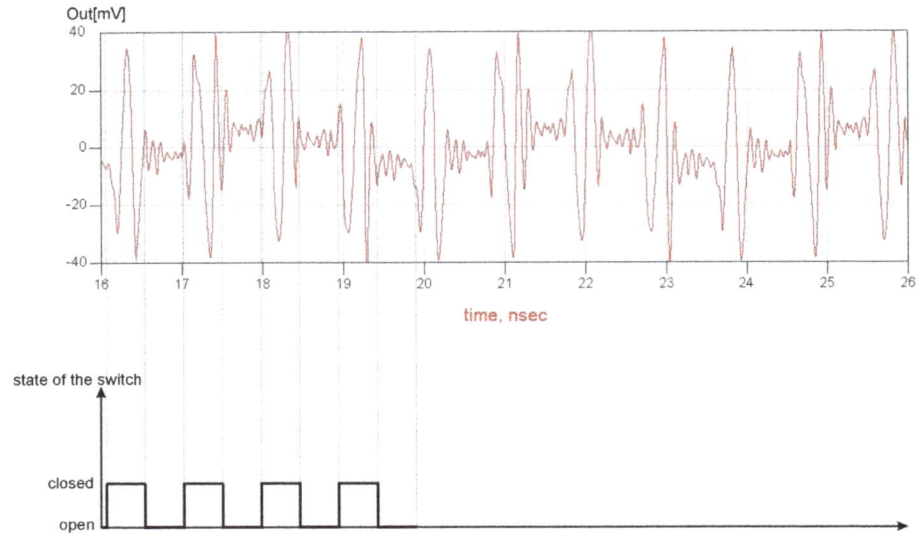

Fig. 6. Simulated voltage time curve at the holding capacitance and state of the diode switch.

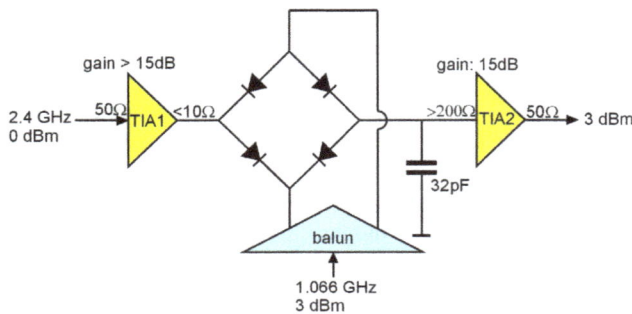

Fig. 7. Specification of the 2.4 GHz demonstrator.

Fig. 8. The chip module with both impedanc tranformers and the diode bridge.

not considered in this simulation. As can be seen, the output voltage follows the input signal as long as the switch is "closed". When the switch is open, the output voltage is not exactly constant. In fact, decaying high-frequency oscillations are observed. A comprehensive examination of this effect shows, that parasitic inductivities of the bonding wires and the via holes are responsible for that behaviour. As shown later (Fig. 9), a simulation including these parasitics predicts very well the measured results. This indicates that proper models are used for the hardware components and layout structure.

5 Demonstrator at 2.4 GHz

For being able to verify the simulated results a scaled demonstrator at 2.4 GHz was built. It was realized without AAF and comparator. Based on the results of the simulations showed, the following specifications (Fig. 7) were chosen.

The demonstrator was built according to the specifications given in Fig. 7 and characterized. Both impedance amplifiers were realized by ATMEL. To work around the prob-

lems of the bonding wires and their parasitic influences, both impedance transformers were assembled on one chip where the diodes were also flip chip bonded on. In future, also the holding capacity will be integrated in this chip module.

The chip module was mounted on TMM10i substrate and connected to the balun and the bias circuits. The connections to the sources and to a sampling oscilloscope were done via SMA connectors mounted on the demonstrator's backside.

Figure 9 shows the output voltage, measured at the holding capacitance. This signal corresponds well to the simulated output voltage shown in Fig. 6. A problem; which has still to be solved, are the oscillations created by the parasitics. They are mainly caused by the externally mounted chip capacitor. In future, it will be integrated on the chip module containing the diode bridge and the impedance transformers.

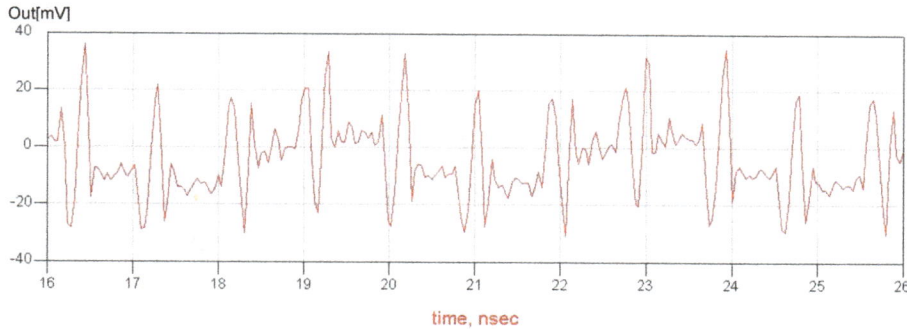

Fig. 9. Measured output voltage.

6 Conclusion

An overview on existing mm-wave receivers was given. The direct sampling receiver based on band-limited oversampling seems to be the best and most flexible solution, but as it is the most ambitious approach, it still needs extensive development. The critical components have been analysed and specified. It appears that a realisation for applications working in the ISM band at 24 GHz seems possible using present technologies.

References

Böhm, K., Pfau, J., Schnepp, H.: A tunable low power BiCMOS continuous-Time Gm-C lowpass filter, IEEE Radio and Wireless Conference, T5.6, Denver, Colorado, USA, August 1999.

Crols, J. and Steyaert, M. S.: Low-IF Topologies for High-Performance Analog Front Ends of Fully Integrated Receivers, IEEE Tr. on Circuits and Systems – II: Analog and Digital Signal Processing, 45, 3, 1998.

Jensen, J. C. and Larson, L. E.: A Broadband 10-GHz Track-and-Hold in Si/SiGe HBT Technology, IEEE Journal of Solid State Circuits, 36, 3, 2001.

Jentschel, H.-J., Berndt, H., and Pursche, U.: Direct Conversion Receivers – Expectations and Experiences, in RF Front End Architectures, IEEE MTT-S 2000, Boston, Workshop, 2000.

Luy, J. F. and Müller, Th.: Front-End Architectures in Millimeter-Wave Systems: From Analogue to Digital Concepts, EuMC 1999, Munich, pp. 309–312, 1999.

Müller, T., Biebl, E. M., Böhm, K., and Luy, J.-F.: Zwischenfrequenz – digitalisierender GPS / GSM Empfänger, Kleinheubacher Berichte 2000, Band 43, pp. 301–307, 2000.

Müller, T.: Direktabtastende Architekturen fr Hochfrequenzempfänger, Dissertation, Fachgebiet Höchstfrequenztechnik, TU München, 2002.

Norsworthy, S., Schreier, R., and Temes, G.: Delta-Sigma Data Converters, Theory, Design and Simulation, IEEE Press 1997, Piscataway, NJ, 1997.

Tatu, S. O., Moldovan, E., Wu, K., and Bosisio, R. G.: A New Direct Millimeter-Wave Six-Port Receiver, IEEE Tr. on Microwave Theory and Techniques, 49, 12, 2001.

Vaughan, R. G., Scott, N. L., and White, D. R.: The Theory of Bandpass Sampling, IEEE Transactions on Signal Processing, 39, 9, 1991.

37

Determination of the input impedance of RFID transponder antennas with novel measurement procedure using a modified on-wafer-prober

M. Camp[1,2]**, R. Herschmann**[1,2]**, T. Zelder**[1]**, and H. Eul**[1]

[1]Leibniz University of Hannover, Institute of Radiofrequency and Microwave Engineering, Appelstraße 9a, 30167 Hannover, Germany
[2]Smart Devices GmbH & Co. KG, Schönebecker Allee 2, 30823 Garbsen, Germany

Abstract. This paper shows a new method to determine the input impedance of RFID transponder antennas with a combination of on-wafer-prober and network analyzer. It is shown that the results are in a good agreement with FEM simulations (HFSS) for a large part of the examined antenna structures.

1 Introduction

Today RFID technology (Finkenzeller, 2003) allows identifying arbitrary objects by implementation of low cost transponders. Use of RFID transponders to the identification of different products with different electromagnetic properties requires the development of new antenna systems which are able to compensate the detuning effects caused by material in the antenna near-field region. For example broadband antenna systems or antenna structures with variable transponder chip positioning on the antenna are promising approaches (Camp et al., 2006). In the second case the chip is implemented according to the desired application for an optimal matching between the antenna and the transponder chip. Numerical simulation methods are suitable for the development of the antenna systems (Herschmann et al., 2005). The verification of the realized antenna impedances must be carried out by measurement because of a variety of production methods and their specific properties used at present for the realization of the antenna structures (Fahlbusch et al., 2006).

Goal of the investigation is the development of a new procedure to measure the input impedance of planar RFID transponder antennas. A modified on-wafer-prober is used to perform the measurement.

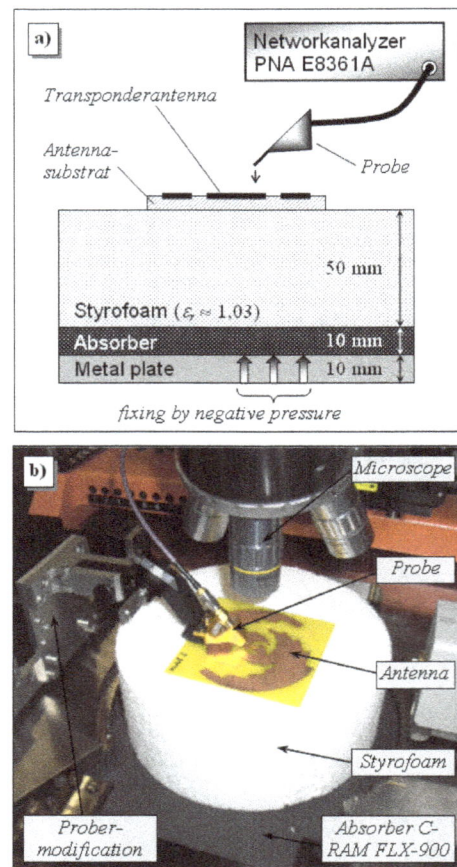

Fig. 1. Measurement setup for the determination of RFID transponder antenna impedances with on wafer prober PA 200 HS (Süss) and network analyzer PNA E8361A (Agilent).

Correspondence to: M. Camp
(camp@ieee.org)

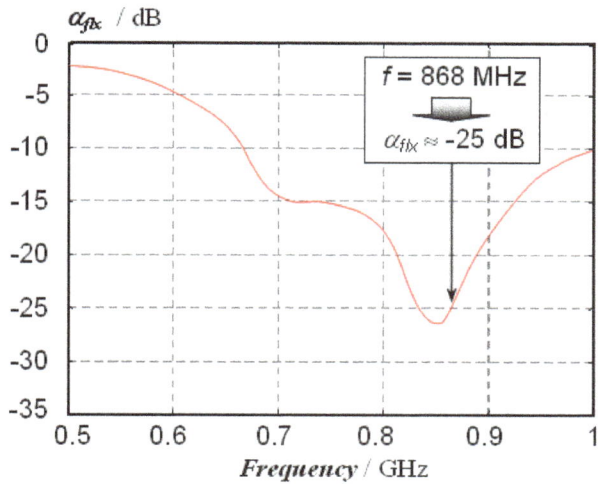

Fig. 2. Reflection loss α_{flx} of the absorber C-RAM FLX-900.

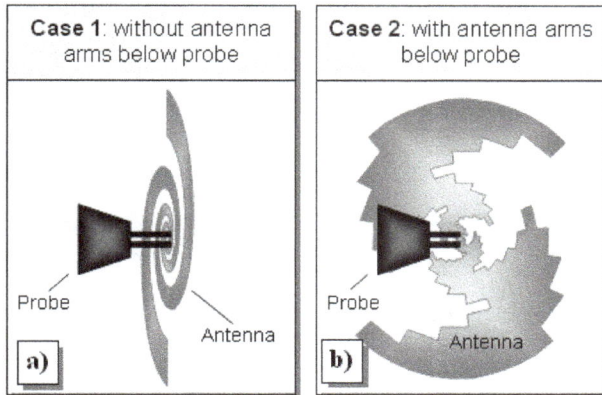

Fig. 3. Distinction of cases: measurement with antenna arms below probe (**a**) and without antenna arms below probe (**b**).

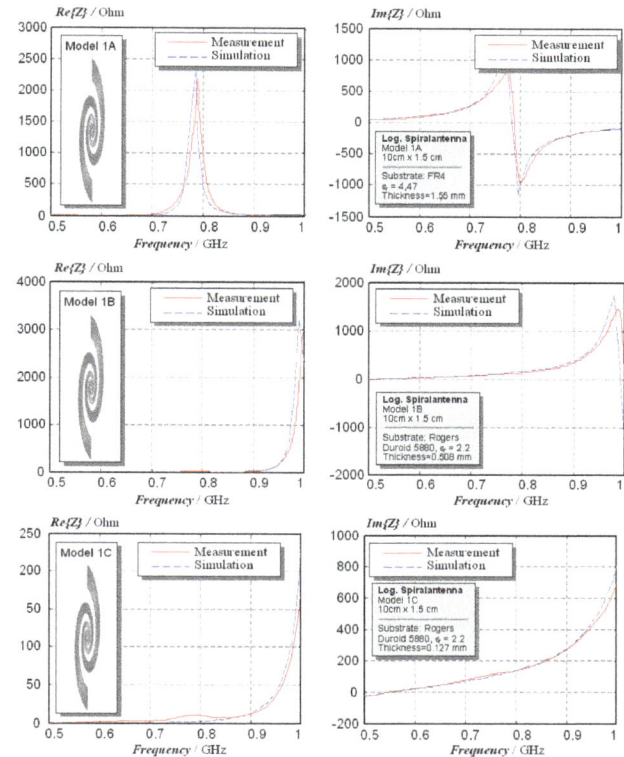

Fig. 4. Comparison of measured and simulated (FEM) antenna impedances of logarithmic spiral antennas on different substrates without antenna arms below probe.

Table 1. Absorber.

Name	Thickness	Specification
C-RAM MT	3.2 mm 0.125"	Lossy, Carbon Filled, Flexible Foam Sheet Stock
C-RAM FF-2	3.2 mm 0.125"	High Loss Silicone Rubber Sheet Absorber für UHF and Microwave Frequencies
C-RAM FDSS	3.2 mm 0.125"	High Loss Silicone Rubber Sheet Absorber for Supression fo UHF Surface Waves
C-RAM GDSS	3.2 mm 0.125"	High Loss Silicone Rubber Sheet Absorber for Supression fo UHF Surface Waves
C-RAM FLX	1.6 mm 0.062"	Flexible Magnetic Sheet Absorber

2 Measurement setup

For the determination of the antenna impedance of different RFID transponder antennas a new measurement procedure was implemented. A modified on-wafer-prober (PA 200 HS) was combined with a network analysator (PNA E8361A) to perform the measurement. First the probe level of the on wafer prober is elevated by means of mechanical spacers over the Chuck. The Chuck is then covered with absorber material of the type C RAM FLX 900 and is fixed by negative pressure. On the absorber material a styrofoam layer and finally the antenna substrate is placed with the transponder antenna to be measured. That way, the antenna impedance can be measured approximately under free space conditions. Figure 1 shows the principle measurement setup.

The styrofoam layer can alternatively be exchanged with the desired underground material of the antenna. In this case it has to be taken care that the thickness of the test material is at least 3 cm or in addition styrofoam is inserted between absorber and test material to avoid changes of the antenna impedance by the absorber material. With the absorber material used in this investigation (C RAM FLX 900) a reflection loss <-20 dB can be achieved in the frequency range from 865 MHz to 870 MHz. For other frequency domains absorber

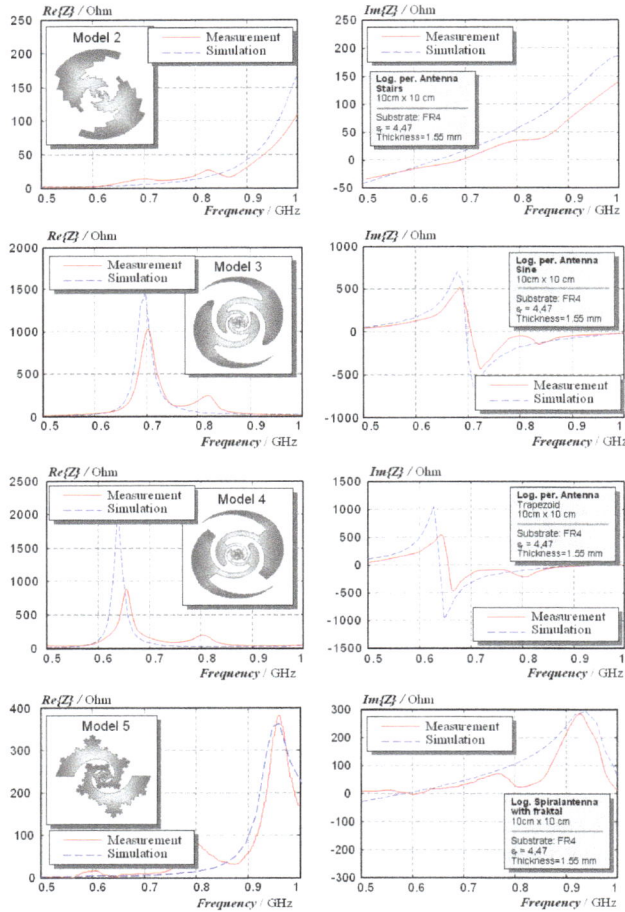

Fig. 5. Comparison of measured and simulated (FEM) antenna impedances of logarithmic periodic antennas with antenna arms below probe.

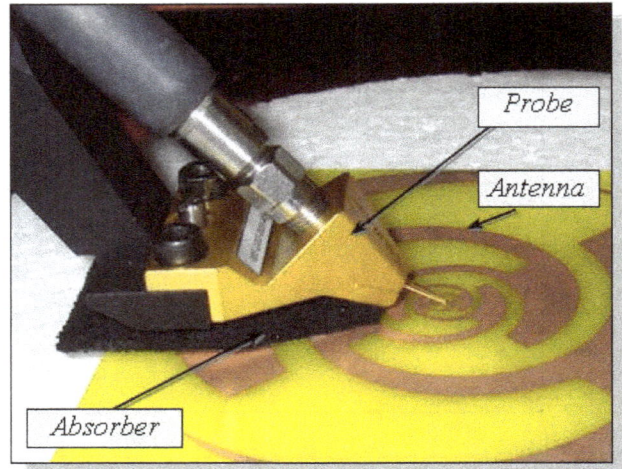

Fig. 6. Measurement setup with absorber between probe and antenna.

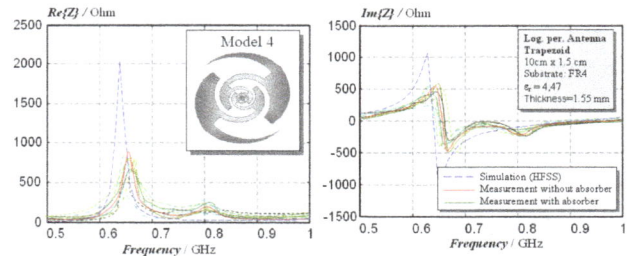

Fig. 7. Measurement of the antenna impedance with different absorber materials between probe and antenna.

materials are also available with high reflection losses. The used styrofoam material has a relative dielectric constant of about $\varepsilon_r \approx 1.03$ in the frequency range 10 MHz–10 GHz. Figure 2 shows the reflection loss of the absorber material in the frequency range 500 MHz–1 GHz.

Antennas with parts of the antenna body directly under the test probes due to their physical dimensions (Fig. 3) represent an essential problem at the measurement of the antenna impedance with the introduced method. The main metal body of the currently available probes is only a few millimetres above the test level. This results in unwanted coupling effects between the probe and the antenna with deviations of the measured antenna impedance.

3 Measurement results

Figure 4 show the measurement results of logarithmic spiral antennas on different base substrates in comparison to sim-

ulation results obtained with the programme HFSS (FEM). The simulation and measurement results are generally in an good agreement. Figure 5 show the results of different logarithmic periodic antennas with large areas of the antenna arms below the probes. Due to unwanted coupling effects deviations between measurement and simulation are visible. Further a resonance appears by the measurement equipment at all examined antennas in the area of 810 MHz.

To reduce the influence of the probe, different absorber materials were placed between probe and antenna. Figure 5 shows the measurement setup with the absorber C-RAM-MT26/PSA between probe and a logarithmical periodical antenna with trapezoid base function. Table 1 shows the absorber materials used.

The problem using absorber material is the low distance between probe and antenna of about 4 mm. For this reason only thin absorber sheets can be placed between probe and antenna. Further the antenna impedance is affected by the absorber material due to the close distance. Figure 7 shows the change of the impedance of a logarithmical periodical antenna with trapezoid base function e.g. The absorber materials represented in Table 1 were used. With none of the materials used, a clear improvement of the measurement result of the antenna impedance could be reached.

4 Conclusions

Goal of this investigation was the development of a new procedure to measure the input impedance of planar RFID transponder antennas. A combination of an modified on-wafer-prober and a network analyzer was used to perform the measurements. The comparison of the measurement results with results achieved by numerical field simulation methods (FEM) has shown, that for a large part of the examined antenna structures the results are in a very good agreement. Deviations between simulation and measurement results are mainly due to unwanted electromagnetic coupling between antenna and measurement probe at antenna structures with extensive metallic area below the probe.

References

Finkenzeller, K.: RFID Handbook, John Wiley and. Sons, New York, New York, 2003.

Camp, M., Herschmann, R., Fahlbusch, T., Eul, H., and Overmeyer, L.: Optimierung der Anpassung zwischen integrierter Schaltung und Antennensystem bei UHF-RFID-Transpondern durch variable Chippositionierung auf der Antenne, ITG-Fachbericht 195, 2. Workshop RFID, VDE-Verlag GmbH, Erlangen 2006, 4–5 Juli, ISBN: 978-3-8007-2976-0, 2006.

Herschmann, R., Camp, M., and Eul, H.: Design und Analyse elektrisch kleiner Antennen für den Einsatz in UHF RFID Transpondern, Kleinheubacher Tagung (URSI), 26-30 September 2005, Miltenberg, Germany, Sitzung B.3: Sensorik durch elektromagnetische Felder und Wellen, 2005.

Fahlbusch, T., Herschmann, R., Camp, M., Meier, D., and Overmeyer, L.: Verfahren zur Produktion von UHF-Labeln auf laserstrukturierten Substraten, ITG-Fachbericht 195, 2. Workshop RFID, VDE-Verlag GmbH, Erlangen 2006, 4–5 Juli, ISBN: 978-3-8007-2976-0, 2006.

38

Design and Analysis of Fully Integrated Differential VCOs

M. Prochaska, A. Belski, and W. Mathis

University of Hannover, Institute of Electromagnetic Theory and Microwave Technology, Appelstraße 9A, 30167 Hannover, Germany

Abstract. Oscillators play a decisive role for electronic equipment in many fields – like communication, navigation or data processing. Especially oscillators are key building blocks in integrated transceivers for wired and wireless communication systems. In this context the study of fully integrated differential VCOs has received attention. In this paper we present an analytic analysis of the steady state oscillation of integrated differential VCOs which is based on a nonlinear model of the oscillator. The outcomes of this are design formulas for the amplitude as well as the stability of the oscillator which take the nonlinearity of the circuit into account.

1 Introduction

The demand for ever-higher frequencies and higher levels of integration poses a challenge for the design and implementation of high frequency oscillators. This holds especially in communication area, for example for UMTS receivers, or data acquisition devices like flash ADCs. Since at gigahertz frequencies most off-chip solutions are rendered impractical, on-die solutions are necessary. Recent developments in the field of semiconductor technology lead to the realizability of integrated LC-Tank differential VCOs (Fig. 1) (van den Tang et al., 2003). Due to the easy implementation of the differential operation and relatively good phase noise, differential VCOs play a decisive role for application (Hajimiri, 1999) – for example in high speed PLL circuits. In this paper a bifurcation analysis of integrated LC oscillators is presented. By using symbolic algorithms, which we have implemented by computer algebra, we analyze the stability as well as we calculate an approximate amplitude. By using analytic methods we can provide a functional dependence of the results on circuit parameters. Our analysis leads to design formulas which can directly be used for the implementation of LC-Tank oscillators.

2 Nonlinear Analysis

For the analysis and the design of electrical oscillators mostly linear models are used. It is one of the most important restrictions of the linearization that the amplitude is not computable (Tietze and Schenk, 2004). On this account for the calculation of an approximate amplitude are several "ad hoc" methods – e.g. harmonic balance – are common practice (Odyniec, 2002). But the difference between the nonlinear oscillatory system and its linearization is more fundamental. The nonlinearity of an oscillator is an integral part of its functionality (Mathis, 2000). So we use a nonlinear approach in order to take the complete behavior of nonlinear components into account.

The mathematical justification for the significance of the nonlinearity is given by the Hartman Grobman Theorem (Guckenheimer and Holmes, 1983): It is well-known that a nonlinear dynamic system is stable, if its linear part has only eigenvalues with negative real parts. If at least one eigenvalue has a positive real part, an oscillatory circuit is unstable. In those cases a system is called hyperbolic. It is an also well-establish condition for a steady state oscillation that a system has a pair of conjugate complex eigenvalues with vanishing real parts. Related criteria were presented by Barkhausen and Nyquist (Tietze and Schenk, 2004; Odyniec, 2002; Parzen, 1983). In this case a circuit model is called a non-hyperbolic system. The Hartman Grobman Theorem tells us that, if a system is non hyperbolic, we cannot neglect the nonlinearity. The neglect of the nolinearity leads to a loss of accuracy of the underlying circuit model. Thus the necessity of a nonlinear analysis of oscillators is originated in the Hartman Grobman Theorem. On this account we present a nonlinear design methodology for integrated LC-Tank oscillators.

So it is suitable to model electrical oscillators by using a nonlinear dynamic system

$$\dot{\mathbf{x}} = \mathbf{f}(\mathbf{x}), \tag{1}$$

where $\mathbf{f}: \mathbb{R}^n \to \mathbb{R}^n$ and $\mathbf{x} \in \mathbb{R}^n$. The vector x corresponds to time depending currents or voltages of the circuit while f is a nonlinear vector field containing the influence of the gain element.

3 Andronov Hopf Theorem

In order to apply the Andronov Hopf Theorem it is useful to get an overview of the possible solutions. A planar nonlinear

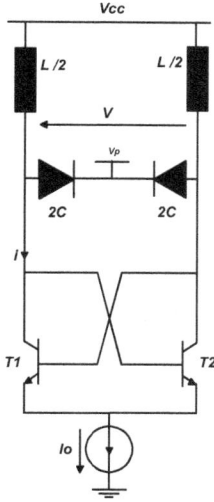

Fig. 1. Fully integrated LC-Tank VCO.

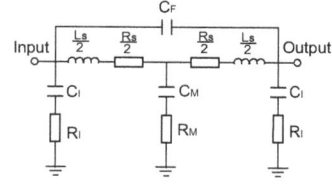

Fig. 2. Model of an Octo-Coil.

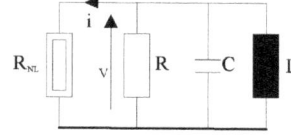

Fig. 3. Fully integrated LC-Tank VCO.

dynamical system has primarily two different types of solutions: equilibrium points and periodic solutions. Equilibrium points are the roots of the nonlinear field $\mathbf{f}(\mathbf{x})$ while periodic solutions are close trajectories in the state space. If a periodic solution is isolated, i.e. isolated in the sense that there are no other close trajectories in its neighborhood, this curve will be called a limit cycle. The relationship between an equilibrium point and a limit cycle is given by the Andronov Hopf Theorem. Originally this theorem was proved by Andronov in 1935 for the analysis of tube oscillators (Mathis, 1998). In this context we consider a dynamic system which additionally depends on the parameter μ:

$$\dot{\mathbf{x}} = \mathbf{f}(\mathbf{x}, \mu) \tag{2}$$

For example, μ represents a component of the circuit like a load resistor. Comprising, the Andronov Hopf Bifurcation describes the birth of a limit cycle depending on μ in the neighborhood of zero and under the following conditions:

- $\left.\frac{\partial \mathbf{f}}{\partial \mathbf{x}}\right|_{\substack{\bar{\mathbf{x}}=0, \\ \mu=0}}$ has a pair of conjugate complex eigenvalues

- all other eigenvalues possess negative real parts

- $\left.\frac{d}{d\varepsilon} \Re \lambda_1(\mu)\right|_{\mu=0} > 0$

- the equilibrium point is asymptotic stable

If these conditions are satisfied, there is a stable equilibrium for $(-\mu_1, 0)$ and a stable limit cycle for $(0, \mu_2)$. Thus, the Andronov-Hopf theorem is the basis for the operating mode of oscillators (Mees and Chua, 1979).

4 Circuit Model

For a nonlinear analysis of LC-Tank oscillators modeling of the tuning diodes, the bipolar transistors and the monolitic

inductors is necessary. In order to get an adequate approximation of our model we shape the transistors by the Ebers-Moll model. We assume that both transistors are the same and that their parameters are sufficiently known. Furthermore, we neglect the nonlinearity of tuning diodes which we approximate as ideal capacitors.

The on-chip inductors perform a critical role in integrated RF circuits, since their Quality factors are much lower than that of off-chip components. Since the model of integrated inductors depends primarily on the IC process and the implementation, it is not easy to specify a generally applicable model. Moreover, a suitable inductor model leads to heavy calculations of high-dimensional dynamical systems. For example, an adequate equivalent inductor model of an Octo-Coil, which can be used for the implementation of a fully integrated LC tuned VCO using a $0.12\,\mu\text{m}$ CMOS process, consist of several dynamic components in a double-π-circuit (Fig. 2) (Konstanznig et al.,2002). That is why we represent the on-chip inductors by ideal electrical devices.

Strategies for the modelization of integrated inductors are given in Arcioni et al. (1999), Brunch et al. (2002) and Danesh and Long (2002). In Prochaska and Mathis (2004) we have presented some ideas for the reduced order modeling of high dimensional oscillators. Figure 3 shows a useful nonlinear model for differential VCOs (Buonomo and Schiavo, 2003).

In order to gain such equivalent circuit of cross-coupled LC-Tank oscillators, first we calculate the bias-emitter voltage of the transistors. After that, we are able to find an equation for the current

$$i(v) =$$
$$\frac{\alpha_F I_0}{1+e^{v/V_T}} \left(1+\frac{1}{\beta_F}e^{v/V_T}\right) + \frac{I_s}{\alpha_R}\left(e^{v/V_T}-e^{-v/V_T}\right) + \frac{I_s}{1+e^{v/V_T}}$$
$$\left[\alpha_F\left(e^{-v/V_T}-e^{2v/V_T}\right) + \left(4-3\alpha_F-\frac{1}{\alpha_F}\right)\left(1-e^{v/V_T}\right)\right], \tag{3}$$

which enables us to define a simple equivalent circuit. R_{NL}

represents the nonlinearity of the bipolar transistors. So we can calculate the model of the oscillator

$$\begin{pmatrix} \frac{dv}{dt} \\ \frac{di_L}{dt} \end{pmatrix} = \begin{pmatrix} -\frac{1}{RC} & -\frac{1}{C} \\ \frac{1}{L} & 0 \end{pmatrix} \begin{pmatrix} v \\ i_L \end{pmatrix} + \begin{pmatrix} \frac{1}{C} \\ 0 \end{pmatrix} i(v). \tag{4}$$

The model also shows that the popular maximally loaded condition for the design of oscillatory circuits fails (Mathis and Weghorst, 1995). In fact, Odyniec (2002) desribes the general oscillator – i.e. the case of a non-maximum loading.

5 Analysis of the Stability

The Andronov Hopf Theorem tells us, that the linear part of an oscillatory system, which is given by the Jacobian $D_x\mathbf{f}(\mathbf{0})$ evaluated at the equilibrium point, has to have a pair of conjugate complex eigenvalues with vanishing real parts. The equilibrium point of Eq. (4) is given by $i=i(I_0)$ and $v=0$. In order to guarantee the existence of a pair of pure imaginary eigenvalues, we have to calculate a multidimensional power series of the nonlinear vector field of Eq. (4). We get

$$\begin{pmatrix} \frac{dv}{dt} \\ \frac{di_L}{dt} \end{pmatrix} = \begin{pmatrix} -\frac{1}{RC} + \frac{i'(I_0)}{C} & -\frac{1}{C} \\ \frac{1}{L} & 0 \end{pmatrix} \begin{pmatrix} v \\ i_L \end{pmatrix} + \begin{pmatrix} \frac{1}{C} i(I_0) \\ 0 \end{pmatrix}, \tag{5}$$

where only the linear part of the expansion is treated. If the condition

$$\frac{i'(I_0)}{C} - \frac{1}{RC} = 0 \tag{6}$$

holds, Eq. (5) has eigenvalues

$$\lambda_{1,2} = \pm j\omega = \pm j \frac{1}{\sqrt{LC}}. \tag{7}$$

The first condition given by the Andronov Hopf Theorem is fulfilled which is equivalent to the Barkhausen condition. In this case we find

$$I = I_0 = -\frac{2\left((2\alpha_R V_T - 4I_s R + 4I_s R\alpha_R)\alpha_F - I_s R\alpha_R\right)}{\alpha_R\alpha_F\left(-1 + 2\alpha_F\right)R}. \tag{8}$$

If we choose $I_0 = I$ the equilibrium point of Eq. (5) is in the origin and so a necessary condition for the applicability of the Andronov Hopf Theorem is also satisfied. The next point, we have to proof that the equilibrium point is asymptotic stable. For the further calculation it is useful to simplify the linear and nonlinear part of our given system. The initial point for the simplification is Eq. (5). The idea is to choose a coordinate transformation so as to simplify the terms of the vector field. In order to simplify the linear part of $\mathbf{f}$, we diagonalize the linear part by

$$\dot{\mathbf{y}} = \mathbf{J}\mathbf{y} + \mathbf{T}^{-1}\tilde{\mathbf{f}}(\mathbf{T}\mathbf{y}), \tag{9}$$

with the diagonal matrix $\mathbf{J} = \mathbf{T}^{-1}\mathbf{A}\mathbf{T}$, $\mathbf{x} = \mathbf{T}\mathbf{y}$ and $\tilde{\mathbf{f}}$ the nonlinear part of $\mathbf{f}$. The initial point of the analysis of stability of our oscillator is Eq. (5), which can be rewritten as

$$\begin{pmatrix} \frac{dx_1}{dt} \\ \frac{dx_2}{dt} \end{pmatrix} = \begin{pmatrix} \mu(I_0) & -\frac{1}{C} \\ \frac{1}{L} & 0 \end{pmatrix} \begin{pmatrix} x_1 \\ x_2 \end{pmatrix} + \begin{pmatrix} \frac{1}{C}\gamma x_1^3 \\ 0 \end{pmatrix}, \tag{10}$$

where $\mathbf{x} = [v\, i_L]^T$ and the bifurcation parameter μ is given by

$$\mu(I_0) = \frac{i'(I_0)}{C} - \frac{1}{RC} \tag{11}$$

and the constant

$$\gamma(I_0) = -\frac{1}{48}\frac{I_0}{V_T^3} + \frac{1}{24}\frac{I_0}{V_T^3}\alpha_F + \frac{1}{3}\frac{I_S}{\alpha_R V_T^3}$$
$$-\frac{1}{2}\frac{I_S}{V_T^3}\alpha_F + \frac{1}{6}\frac{I_S}{V_T^3} - \frac{1}{24}\frac{I_S}{\alpha_F V_T^3} \tag{12}$$

With $\mathbf{T} = [0\ 1;\ -1/C/\omega_0\ 0]$ we find:

$$\dot{\mathbf{y}} = \begin{pmatrix} 0 & j\omega_0 \\ -j\omega_0 & 0 \end{pmatrix} \mathbf{y} + \begin{pmatrix} 0 \\ -\frac{1}{C}\gamma y_1^3 \end{pmatrix}. \tag{13}$$

After that, in order to simplify the nonlinear terms we try to find a sequence of coordinate transformations which remove terms of increasing degree from the Taylor series (Mathis, 1995). So a dynamic system which has eigenvalues $\alpha(\mu) \pm j\omega(\mu)$ can be expressed in the so called Poincare normal form

$$\dot{\mathbf{y}} = \begin{bmatrix} \alpha(\mu) & \omega(\mu) \\ -\omega(\mu) & \alpha(\mu) \end{bmatrix} \mathbf{y} + \sum_{i=1}^{\infty} \left(y_1^2 + y_2^2\right)^i \begin{bmatrix} a_i & b_i \\ -b_i & a_i \end{bmatrix} \begin{bmatrix} y_1 \\ y_2 \end{bmatrix} \tag{14}$$

The reader should note that the equilibrium point is asymptotical stable, if the so-called Poincare coefficient a_1 is negative ($a_1 < 0$). The calculation of the Poincare normal form of Eq. (13) for $\mu = 0$ gives:

$$\dot{\mathbf{y}} = \begin{bmatrix} 0 & \omega \\ -\omega & 0 \end{bmatrix} \mathbf{y} + \left(y_1^2 + y_2^2\right) \begin{bmatrix} -\frac{3}{8}\gamma & 0 \\ 0 & -\frac{3}{8}\gamma \end{bmatrix} \begin{bmatrix} y_1 \\ y_2 \end{bmatrix} + O\left(\|\mathbf{y}\|^5\right). \tag{15}$$

We get the Poincare coefficient $a_1 = -3/8\gamma$. For $\gamma > 0$ the equilibrium is asymptotic stable. So we can guarantee the asymptotic stability of the equilibrium point. In order to demonstrate the birth of a limit cycle vividly, we analyze the product of the voltage and current of the nonlinear resistor $R_N L\ iv = f(v, I_0)$ which is shown in Figs. 4 and 5. If we choose $I_0 = I$, the term iv has locally a negative slope. When iv is interpreted as power the negative slope indicates a negative differential resistor in the neighborhood of the zero solution. This leads to an oscillation since the loss of the circuit will be compensated. It must be pointed out that Eq. (8) also satisfies the condition $d(\Re\lambda_1(\mu)|_{\mu=0})/d\varepsilon > 0$ given by the Andronov Hopf Theorem, which can be tested by an easy computation.

6 Predicting of the Amplitude

In order to calculate the amplitude of a sinusoidal oscillator it is suitable to transform the reduced system to polar coordinates. Starting from Eq. (9) we obtain the following system for $\mu = 0$ ($\mathbf{y} \Leftrightarrow (r, \Theta)$):

$$\begin{bmatrix} \dot{\Theta} \\ \dot{r} \end{bmatrix} = \begin{bmatrix} \omega \\ 0 \end{bmatrix} + \mathbf{f}_{PC}(\Theta, r). \tag{16}$$

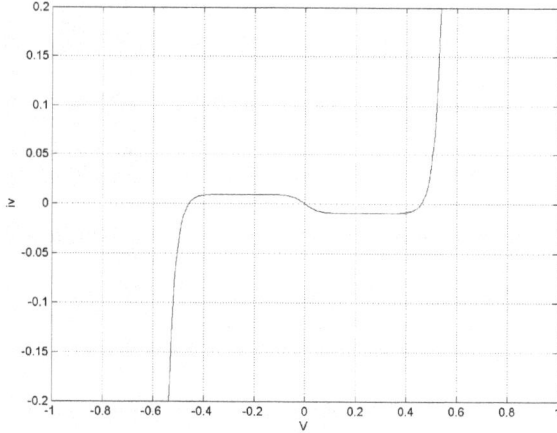

Fig. 4. Analysis of the power iv for $I_0=I$.

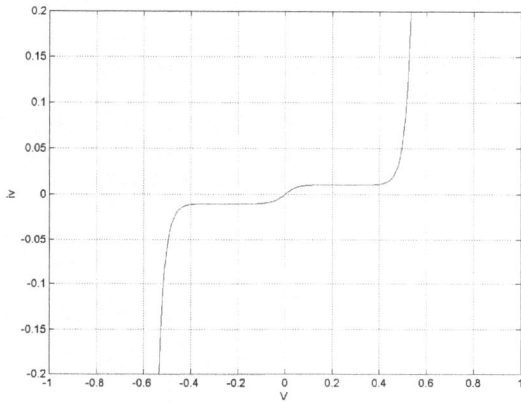

Fig. 5. The power iv for $I_0=I$.

Since the equation $\dot{r}=0$ is mostly a function of Θ, i.e. both equations of Eq. (16) are coupled; we cannot calculate the amplitude directly. To produce a relief we use an average technique – a perturbation method (Guckenheimer and Holmes, 1983). Through this method the trajectory of the limit cycle of one period is averaged. Instead of popular "ad hoc" methods averaging is an analytic method; the error of the approximation can be calculated. By means of this technique it is our goal to eliminate the action of Θ in the second equation of Eq. (16). The method hinges on the identification of a small parameter ε which marks the perturbation. We assume that a LC circuit is perturbed by a small nonlinearity. So we get in Cartesian coordinates for a planar system with $\mu=0$

$$\dot{\mathbf{y}}=\begin{bmatrix} 0 & \omega \\ -\omega & 0 \end{bmatrix}\mathbf{y}+\varepsilon\mathbf{f}_S(\mathbf{y})=\mathbf{J}\mathbf{y}+\varepsilon\mathbf{f}_S(\mathbf{y}). \tag{17}$$

In polar coordinates we can write the following system, where mostly both equations are coupled

$$\begin{bmatrix} \dot{\Theta} \\ \dot{r} \end{bmatrix}=\begin{bmatrix} \omega \\ 0 \end{bmatrix}+\varepsilon\begin{bmatrix} R(\Theta,r) \\ T(\Theta,r) \end{bmatrix}. \tag{18}$$

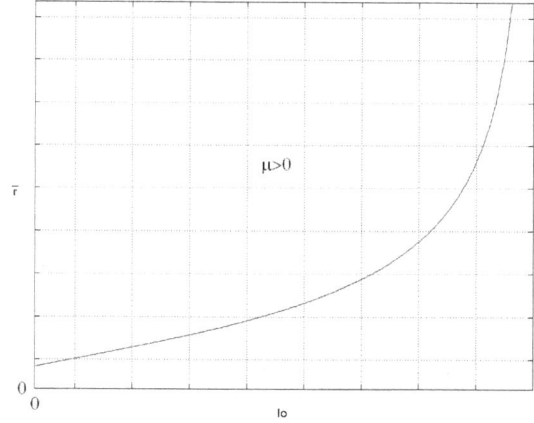

Fig. 6. Qualitative behavior of the amplitude $\bar{r}_1=f(I_0)$ in the neighborhood of the bifurcation point for $\mu>0$.

By using a suitable transformation we find the averaged system

$$\begin{bmatrix} \dot{\bar{\Theta}} \\ \dot{\bar{r}} \end{bmatrix}=\begin{bmatrix} \omega \\ 0 \end{bmatrix}+\varepsilon\begin{bmatrix} \bar{R}(\bar{r}) \\ \bar{T}(\bar{r}) \end{bmatrix}. \tag{19}$$

Here we have assumed that already one averaging gives a usable result. But sometimes a sequence of coordinate transformations is necessary. Then our approach, which is based upon Lie series, is advantageous (Mathis, 1987). So we can transform Eq. (13) to polar coordinates:

$$\begin{bmatrix} \dot{\Theta} \\ \dot{r} \end{bmatrix}=\begin{bmatrix} \omega_0 \\ 0 \end{bmatrix}+\varepsilon\begin{bmatrix} r^2\sin\Theta\cos\Theta \\ r\mu+\mu\cos^2\Theta+\dots \end{bmatrix} \tag{20}$$

Since both equations of Eq. (20) are coupled averaging is necessary. We calculate

$$\begin{bmatrix} \dot{\bar{\Theta}} \\ \dot{\bar{r}} \end{bmatrix}=\begin{bmatrix} 1 \\ \varepsilon\bar{r}\mu-\varepsilon\frac{3}{8}\gamma\bar{r}^3 \end{bmatrix}, \tag{21}$$

where already one averaging removes the terms depending on Θ. The roots for $\dot{\bar{r}}=0$ are

$$\bar{r}_1=0,\quad \bar{r}_2=\pm\sqrt{\frac{8}{3}\frac{\mu(I_0)}{\gamma(I_0)}}. \tag{22}$$

The value $\bar{r}_1=0$ defines the averaged amplitude for an unstable equilibrium. If the Andronov Hopf Theorem is fulfilled, the oscillator has the amplitude $|\bar{r}_2|$, which depends on $\mu(I_0)$ and the parameters of the transistors. Thus, by means of current I_0 the amplitude can be tuned in the neighborhood of the bifurcation point (Fig. 6). Furthermore, Eq. (22) shows the dependence of the amplitude on the nonlinearity of the transistor.

7 Conclusion

In this work we have presented a complete bifurcation analysis of integrated differential LC oscillators. We have shown the analysis of the stability of the circuit as well as the calculation of an approximate amplitude. We have presented the results in an analytic form – designers are able to implement differential VCO by means of the given results even if other models of the components of the circuit are required. In this case possibly the calculations are more complicated. However, our methodology can be used in the same way. So it represents a guide for the analysis and the design of electrical oscillators whereas the nonlinearities of circuit components are an integral part of the design process.

It turns also out that geometric methods are powerful tools for investigations of nonlinear oscillators. Since they provide a survey of the solution set we get a deeper insight in the behavior of the network. In contrast to numerical computation the shown analytical methods have the advantage that the results are interpretable by the parameters of the network. Furthermore, we have implemented the geometric methods by means of computer algebra. Because of these routines the calculations proceed in an automated way.

References

Arcioni P., Castello, R., Perregrine, L., Sacchi, E., and Svelto, F.: An innovative modelization of loss mechanism in silicon integrated inductors, IEEE Trans. On Cir. And Sys., Vol. 46, No. 12, 1453–1460, Dec. 1999.

Brunch R. L., Sanderson, D. I., and Ramon, S.: Quality factor and inductance in differential IC implementations, IEEE Mircrowave Mag., Vol. 3, No. 2, 82–92, June 2002.

Buonomo A. and Schiavo, A. Lo: Determining the oscillation of differential VCOS, Proc. of the 2003 Intern. Sym. on Cir. and Sys. 2003 (ISCAS2003), Vol. 3, 25–28, May 2003.

Danesh, M. and Long, J. R.: Differential driven symmetric microstrip inductors, IEEE Trans. On Microwave Theo. And Tech., Vol. 50, No. 1, 332–341, Jan. 2002.

Guckenheimer, J. and Holmes, P.: Dynamical Systems and Bifurkation of Vector Fields, Springer Verlag New York, 1983.

Hajimiri, A and Lee, T. H.: Design Issues in CMOS Differential LC Oscillators, IEEE Journal. of Solid State Circuits, Vol. 34, No 5, May 1999.

Konstanznig, G., Pappenreiter, T., Maurer, L., Springer, A., and Weigel, R.: Design of a 1.5 V, 1.1 mA Fully Integrated LC-tuned Voltage Controlled Oscillator in a 4 Ghz-Band using a 0.12 μm CMOS-Process, Asian-Pacific Microwave Conference, 1471–1474, Nov. 2002.

Mathis, W.: Transformation and Equivalence, in: The Circuits and Filters Handbook, edited by Chen, W. K., CRC Press Boca Raton, 1995.

Mathis, W.: Historical Remarks of the History of Electrical Oscillators, Proc. MTNS-98 Symposium, July 1998, 309–312, Padova 1998.

Mathis, W.: Nonlinear electronic circuits – An overview, Proc. 7th., Gdynia, Poland, MIXDES 2000.

Mathis, W. and Weghorst, I.: A Nonlinear Theory for Maximum-Loaded-Oscillators with the Andronov Hopf Bifurcation Theorem, SIAM Fall Meeting, 28-30 Oct. 1985, Arizona State University, 1985.

Mathis, W. and Voigt, I.: Applications of Lie Series Averaging in Nonlinear Oscillations, Proc. IEEE Intern. Symp. Of Cir. And Sys. (ISCAS1987), Philadelphia, 1987.

Mees, A. I. and Chua, L. O.: The Hopf Bifurcation and its Application to nonlinear Oscillations in Circuits and Systems, IEEE Trans. Circuits and Systems, vol. 26, 235–254, 1979.

Odyniec, M.: RF and Microwave Oscillator Design, Artech House, Boston, 2002.

Parzen, B.: Design of Crystal and Other Harmonic Osillators, Wiley & Sons, 1983.

Prochaska M. and Mathis, W.: On limit cycles in singularly perturbed electrical circuits, Proc. of the 16th Int. Sym. on Math. Theory of Netw. and Sys. (MTNS), Leuven 2004.

van der Tang, J., Kasperkovitz, D., and van Roermund, A.: High-Frequency Oscillator Design for Integrated Transceivers, Kluwer Academic Publishers, Boston, 2003.

Tietze, U. and Schenk, C.: Electronic Circuits: Handbook for Design and Application, Springer Verlag, New York, 2004.

39

MOS capacitances used in mixed-signal circuits and their operative range

W. Kraus, B. Stelzig, T. Tille, and D. Schmitt-Landsiedel

Institute for Technical Electronics, TU-Munich, Germany

Abstract. To avoid additional layers for high linearity capacitances in modern CMOS process families, compensated depletion mode MOS capacitances can be used. As shown in previous publications, these MOS capacitances are suitable for low voltage applications.

But there exist limitations concerning the linearity of these capacitances. In this work, the impact of the nonlinearity of the capacitances on different kinds of circuits is investigated. Several examples will be discussed to show how to choose the right capacitance topology.

1 Introduction

In state of the art highly integrated CMOS process families strong design constraints exist concerning the area consumption of a circuit. Often capacitances are the most area intensive components, if you look at mixed-signal circuits, like the sample and hold, the frequency compensation of opamps, or more complex circuits like filters or $\Sigma\Delta$-converters. So it's obvious that it is necessary to search for a method to reduce the area for these capacitances. As shown in Tille et al. (2000) you can use compensated MOS capacitances, because they have a very thin gate oxide, which leads to high area efficiency. Another positive aspect is the knowledge about matching of MOS transistors, because it can easily be transfered to the MOS capacitances. In addition to that no extra process steps are required in contrast to MIM capacitances.

In previous work we have presented $\Sigma\Delta$-modulators using compensated MOS capacitances. The purpose of this work is to investigate the basic analog building blocks of mixed-signal circuits with respect to their sensitivity to the nonlinearity occuring in compensated MOS capacitances. This gives guidelines to the designer for the decision, which type of capacitance is appropriate for a given application.

2 Device characteristics of MOS capacitances

2.1 C-V characteristic

The usable voltage range of MOS capacitances can be divided in three parts: Accumulation, inversion and depletion. A physical C-V curve of an p-channel MOSFET in an n-well is given in Fig. 1. It shows, that the first two parts behave more linear and have a higher absolute capacitance value than the depletion mode. However, these two voltage ranges are too high for low voltage applications. Concerning the operating point voltages the depletion mode capacitances must be used, but there the value strongly depends on the voltage. The depletion range can be broadened by applying a source bulk voltage (see Fig. 2), but the voltage dependency is still there. In order to reduce this dependency you have to combine two MOS transistors in an antiparallel or antiserial arrangement.

2.2 Compensated MOS capacitances

Most of the nonlinearities of depletion mode MOS capacitances can be eliminated by an antiserial or antiparallel connection of two capacitors (Tille et al., 2000). The exact circuit diagram and the remaining nonlinearity are shown in Fig. 3. Comparing the two methods, you can see that the more area efficient parallel compensation has less linearity than the serial compensation. In both cases you can broaden the usable voltage range by applying an additional source bulk voltage. Just 0.5 V are enough for the serial compensation to get a working range of ± 1.5 V which is enough for low voltage applications.

The floating source bulk voltages are hard to generate. As the source and drain contacts have to be negative referring to the bulk potential, we can simply connect them to the constant voltage V_{SS}. By doing that we ensure that the drain-bulk and the source-bulk diodes are always reverse-biased. With source and drain on a constant potential and the contacts A and B being variable (see Fig. 4), the depletion broad-

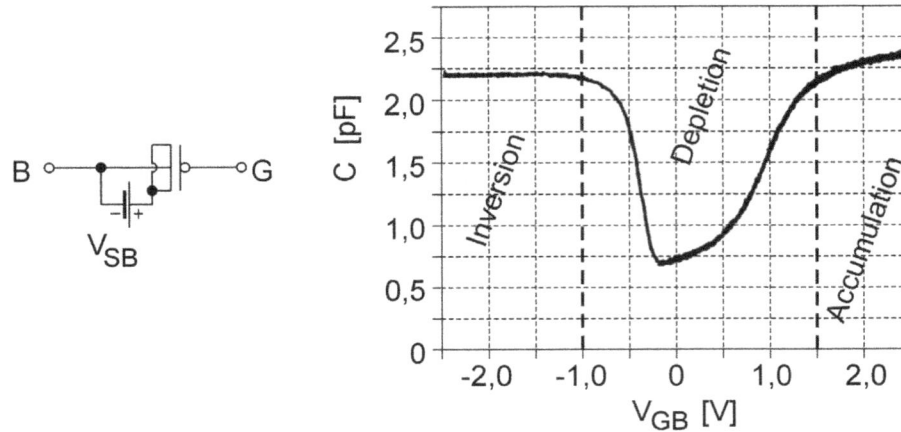

Fig. 1. Measured C-V characteristic of a MOS capacitance and its different voltage ranges.

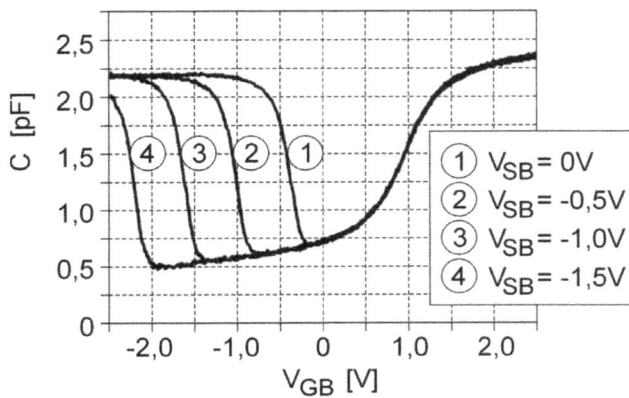

Fig. 2. Measured C-V characteristics of a MOS capacitance with depletion broadening caused by several source bulk voltages.

Table 1. Comparison of the area consumption of different types of capacitances in different technology generations

| Process: | 0.35 μm | 0.25 μm | 0.18 μm |
Type	$[fF/\mu m^2]$	$[fF/\mu m^2]$	$[fF/\mu m^2]$
Poly-Poly	1.30	-	-
Poly-Metal	0.049	0.10	0.10
Metal-Metal	0.041	0.05	0.09
MIM cap.		0.70	0.70
MOS cap. (serial)		0.43	0.49
MOS cap. (parallel)		1.60	1.97

ening is modulated. This does not cause a problem, because in the voltage range $V_{DD} - V_{SS} = 1V$ depletion can be maintained in all operating points, see Fig. 4. With antiserial compensation, the remaining nonlinearity in the voltage range of $[-0.5\,V; 0.5\,V]$ is below 0.5%. For antiparallel compensation we obtain a nonlinearity of nearly 10%, when the full voltage range is used. However, this can be reduced by limiting the voltage amplitude on the capacitance. Within

$[-0.3\,V; 0.3\,V]$, the nonlinearity is only around 3%.

To complete the introduction of compensated depletion mode MOS capacitances, Table 1 gives a comparison of the area consumption of different types of capacitances in different technology generations. Here it has to be kept in mind, that MIM capacitances have a big drawback. They need additional process steps, which leads to higher process complexity.

3 Analog circuit blocks with compensated depletion mode MOS capacitances

To demonstrate the usability of MOS capacitances, four different analog circuit blocks have been simulated, representing the basic components of mixed-signal circuitry. To determine which circuits are sensitive to nonlinearity, the parallel compensation was assumed as worst case. This shows the impact of area efficiency on the linearity of mixed-signal circuits.

3.1 Sample and hold

In the sampling mode, the capacitance is charged up to the input voltage. With a nonlinear capacitance, the amount of charge is nonlinearly dependent on the input voltage. In the hold mode, the sampled voltage is evaluated by connecting it to a high-ohmic node, most often to the input of an operational amplifier. Therefore the nonlinearity of the $C(V)$ and $Q(V)$ characteristics do not influence the circuit function. To check this in a complete system, we exchanged the sampling capacitance in a sample and hold circuit between an anti-aliasing filter and an A/D-converter and replaced it with an antiparallel compensated MOS capacitance. The simulations showed no decrease of SNDR, confirming that there is no degradation of linearity.

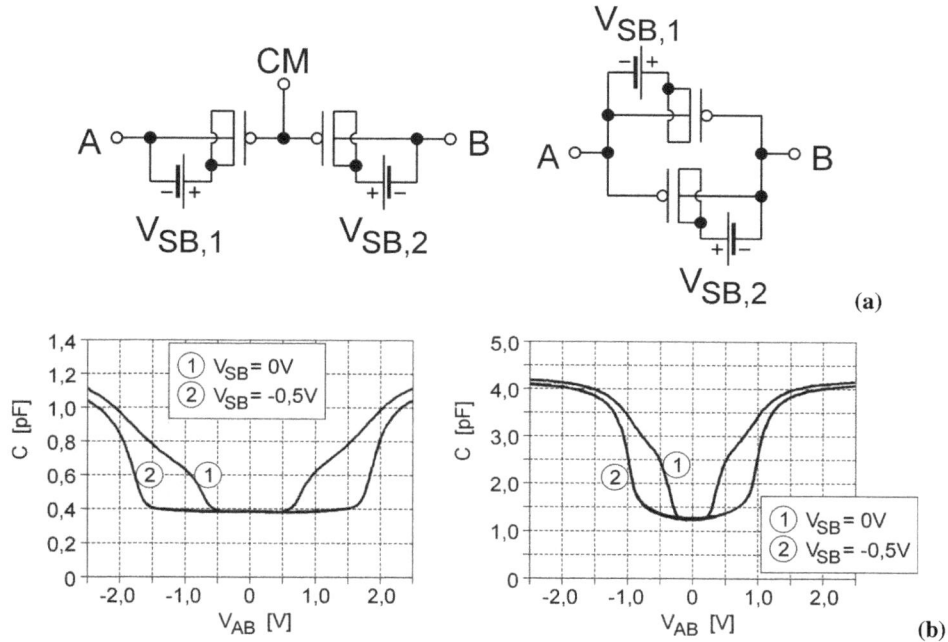

Fig. 3. Circuit diagram and measured C-V characteristics of an antiserial (left) and an antiparallel (right) compensated depletion mode MOS capacitance.

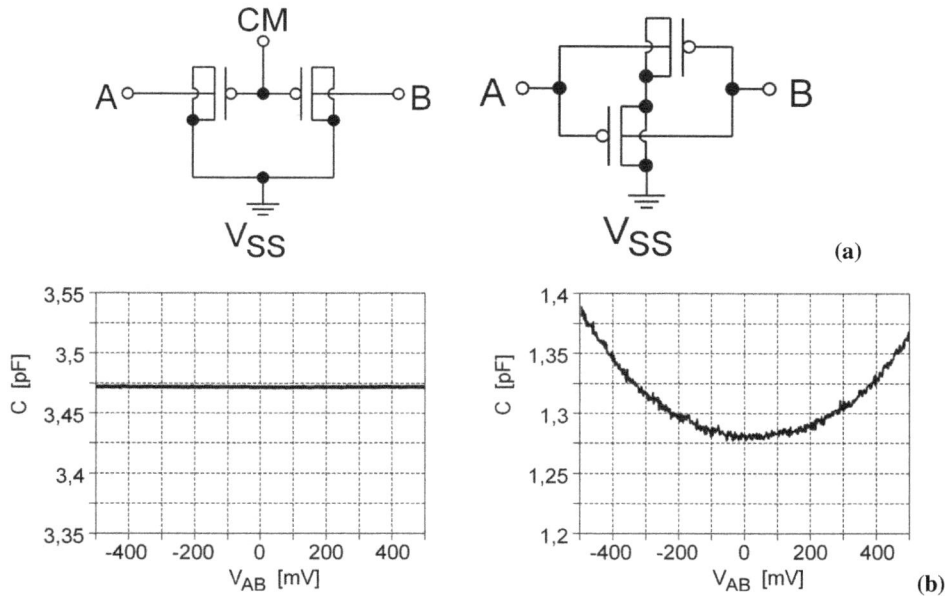

Fig. 4. Circuit diagram and measured C-V characteristics of an antiserial (left) and an antiparallel (right) compensated depletion mode MOS capacitance without floating sources.

3.2 Frequency compensation of operational amplifiers

The design constraint for the frequency compensation of opamps is stability. As an example we discuss the lead lag compensation method using parallel compensated MOS capacitances as shown in Fig. 5.

The lead lag network adds a pole and a zero to the transfer function of the opamp. In this example the pole is at about 10 kHz and the zero at about 100 kHz. Both of them depend on the compensated capacitance. A problem arises, if the pole/zero-shift, caused by the nonlinearity of the capacitance, influenced the phase margin. But Fig. 5 also shows that a small shift would have only little effect on the 0 dB limit and the phase will only change in part 1 of the frequency range. The 0 dB frequency is placed in part 2, so the pole/zero-shift will have no effect on the stability. That's

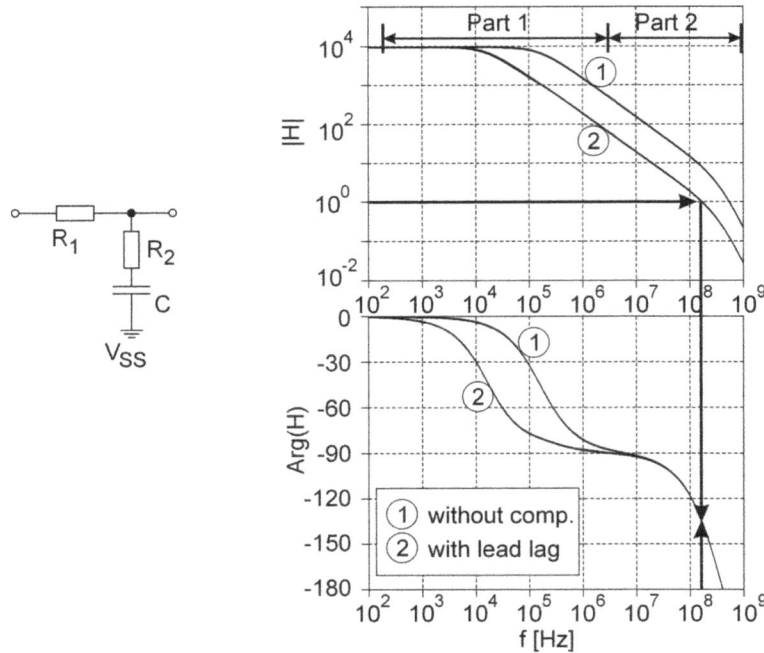

Fig. 5. RC-network and bode diagram of the lead lag frequency compensation method for operational amplifiers.

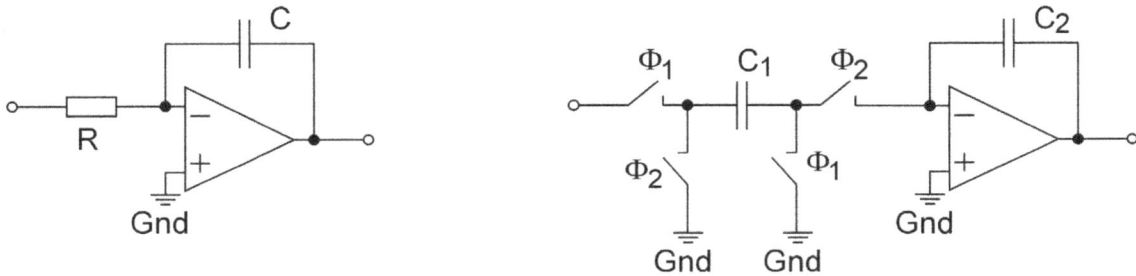

Fig. 6. A simple RC integrator on the left and a SC integrator on the right side.

why we can use the most area efficient method, the parallel compensation (Sauerbrey et al., 2002).

3.3 Switched capacitor low pass filter

For this investigation a second order switched capacitor filter with a bandwidth of $f_g = 100 \, \text{kHz}$ and a sampling frequency of $f_S = 5 \, \text{MHz}$ has been used, where all linear capacitances were replaced by parallel compensated MOS capacitances. The usability has again been evaluated with the help of the signal to noise and distortion ratio.

In Table 2, the signal to noise and distortion ratios for different input amplitudes are given. The SNDR decreases for increasing input amplitude. This can be explained regarding the operation mode of the filter in the two clock phases of the SC system. In the first phase, the input capacitance C_1 is charged with $Q_{C1}(V_{in})$, where a certain nonlinearity is presented in the function $Q(V)$. This charge is transferred to the output capacitance C_2, giving rise to an output voltage determined by $Q_{C2}(V_{out}) = Q_{C1}(V_{in})$. As C_1 and C_2 usu-

ally are not equal, their nonlinearities will not cancel. This leads to harmonics in the output signal and therefore reduce the SNDR. This effect increases with the amplitude, as can be seen from Fig. 4. The example of Table 2 represents the worst case, as only parallel compensated capacitances were used. This is sufficient for a SNDR of 60 dB or a resolution of about 10 bit. A much higher accuracy can be obtained by using serial instead of parallel compensated MOS capacitances. In this way the area consumption can be optimized concerning the needed accuracy.

3.4 Integrators

The analysis of the RC integrator (Fig. 6) is relatively easy. You just have to realize that the inverting input of the opamp is at virtual ground. So the C-V dependency of the integration capacitance appears directly at the output of the integrator.

Concerning the SC integrators $C_2(V_{out})$, the same dependency exists as in the RC integrator. In addition to that, there

Table 2. Simulated SNDR of a second order SC low pass filter with a bandwidth of $f_g = 100\,\text{kHz}$ and a sampling frequency of $f_S = 5\,\text{MHz}$, realized with parallel compensated depletion mode capacitances

Input Amplitude	Signal to Noise and Distortion Ratio [dB]
50 mV	84.6
100 mV	71.4
200 mV	60.9

is a voltage dependent deviation caused by the input capacitance $C_1(V_{in})$. So with very small input amplitudes the SC-Integrator will be as good as the RC integrator, but it gets worse with increasing input amplitudes.

For the integrators we can also summarize, that the area efficiency strongly depends on the needed accuracy for the individual application.

4   Conclusion

The mixed-signal circuit designer has not only to consider the functionality of a circuit, but also the amount of resources needed. In some circuit blocks there is no problem using the highly area efficient parallel compensated depletion mode MOS capacitances, like in the sample and hold or the frequency compensation of opamps. In other blocks the appropriate capacitance topology dependent on the specifications of the application has to be chosen. A combination can be suitable in complex mixed-signal circuits like $\Sigma\Delta$-modulators. They are quite insensitive to parameter variations due to their special feedback structure. So you just have to spend serial compensated MOS capacitances at a few key instances, where particularly high linearity is needed. All other capacitances can be realized by area efficient parallel compensated MOS capacitances.

This work shows, that compensated depletion mode MOS capacitances are suitable for low voltage applications in future technology generations.

References

Sauerbrey, J., Tille, T., Schmitt-Landsiedel, D., and Thewes, R.: A 0.7 v mosfet-only switched-opamp sigma-delta-modulator, Proceedings of International Solid-State Circuits Conference, 2002.

Tille, T., Sauerbrey, J., Mauthe, M., Kraus, W., and Schmitt-Landsiedel, D.: A 1.8 v mosfet-only sigma-delta modulator using compensated mos-capacitors in depletion with substrate biasing, Proceedings of 26th European Solid-State Circuits Conference (ESSCIRC), 76–79, 2000.

40

A new BIST scheme for low-power and high-resolution DAC testing

H. Li[1]**, J. Eckmueller**[1]**, S. Sattler**[2]**, H. Eichfeld**[1]**, and R. Weigel**[3]

[1]SMS TI MT MS, Infineon AG, Postfach 80 09 49, D-81609 München, Germany

[2]CTT TS ADT, Infineon AG, Postfach 80 09 49, D-81609 München, Germany

[3]Lehrstuhl Technische Elektronik, Friedrich-Alexander-Universität Erlangen-Nürnberg, Cauerstr. 9, 91058 Erlangen, Germany

Abstract. A BIST scheme for testing on chip DAC is presented in this paper. We discuss the generation of on chip testing stimuli and the measurement of digital signals with a narrow-band digital filter. We validate the scheme with software simulation and point out the possibility of ADC BIST with verified DAC

1 Introduction

Due to the increasing complexity of electronic systems and the capabilities of very deep sub-micron technologies, in recent years more and more system functionality has been integrated onto a single chip (SoC). Consequently an increasing number of chips that combine digital and analogue functions is designed. High-performance applications of mixed-signal (MS) integrated circuits (IC's) in many areas such as wireless telecommunications, data exchange systems and satellite communications have been realized. But this development makes big challenge on the testing of Mixed-signal circuits for such applications is more and more difficult, because it becomes both expensive and time consuming. Then an attractive alternative for the Mixed-signal testing is the Built-in Self-Test (BIST) approach in which testing (test generation and test application) is accomplished through built-in hardware features. Recently, several BIST concepts for DA and AD converters were proposed. In Toner and Roberts (1995), the authors proposed a BIST approach for testing the SNR of a delta-sigma ADC. This method employs a sine wave generator based on delta- sigma technique and digital signal processing techniques for data analysis. But the technique needs both on chip ADC and DAC, and powerful computation ability on chip, which is not always possible. The BIST approach in Sunter and Nagi (1997) refers to approximating the transcharacteristic of an ADC with the best fitting a 3rd order polynomial. Then the offset, gain, and harmonic distortion

will be computed out from this polynomial. Its drawback is that it is very sensitive to the noise level and needs also a powerful DSP-core on chip. By the oscillation test method proposed in Arabi and Kaminska (1996), test vector generation problem is eliminated and the test time is very small. Nevertheless, the impact of control logic delay and the imperfect analog BIST circuitry on the test accuracy are not clear. In Azaies et al. (2000), the authors present a histogram BIST method for ADC, which shows how many times each different digital code word appears on the outputs. Because ADC errors will modify the output code count and so impact the histogram shape, offset, gain, DNL and INL can be extracted from the results. However, the number of input patterns is so huge that the testing time is much longer than that in other methods. The work in Ohletz (1991) proposes a BIST approach for testing DAC-ADC pair with a Pseudo-Random Bit Sequence (PRBS) generated by a Linear-Feedback Shift Register (LFSR) as stimulus and a second LFSR, which compacts the digital output of the ADC to generate a signature compared to a value stored in memory. The drawback is again the need of a powerful DSP core on chip. And all the above methods are devoting either to the testing of ADC or to the testing of ADC-DAC pair. In Hajjar and Roberts (1998), the authors show a BIST method for testing a single DAC without ADC on chip based on IEEE 1149.4. It takes advantage of the period character of testing stimuli to trade the analog nature of the DAC output. Nevertheless, it is not suitable to the testing of high-speed and low voltage DAC because the testing time is too long and the requirement on the comparator is too fine to implement on chip.

Our work is concerned with an improved BIST scheme for the testing of a high-speed and low voltage DAC. By this work, we use a full digital way to produce a sine wave with a good Signal-to-Noise Ratio (SNR). Since we do not relay on the on chip ADC and DAC to produce our testing stimuli, our BIST strategy does not need the presence of on chip ADC and DAC at the same time, which means our approach is more feasible and flexible. For DAC testing, we employ the one-points-multi-comparison algorithm

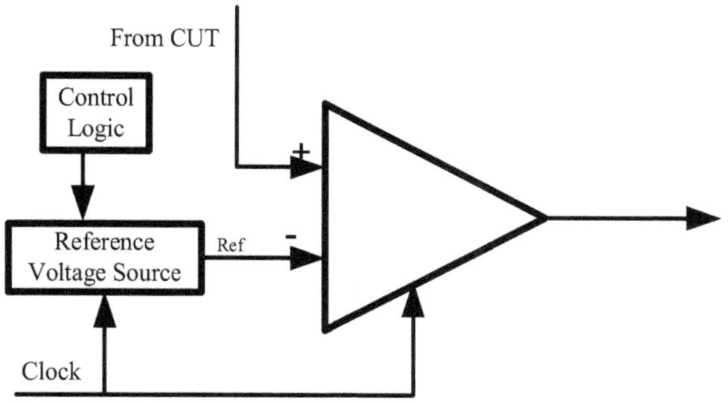

Fig. 1. On-Chip Digitizer.

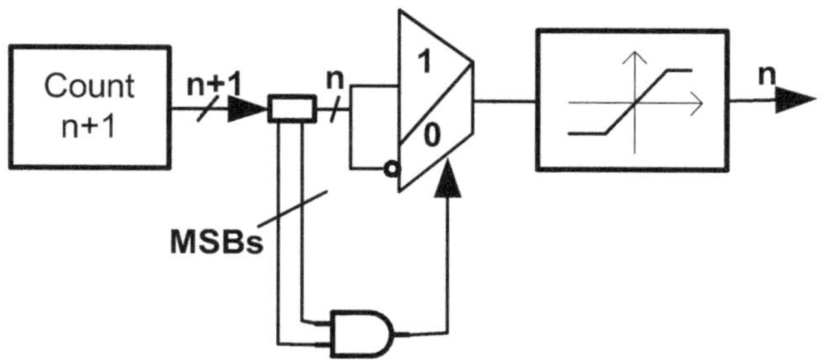

Fig. 2. Trapezoid Wave Generator.

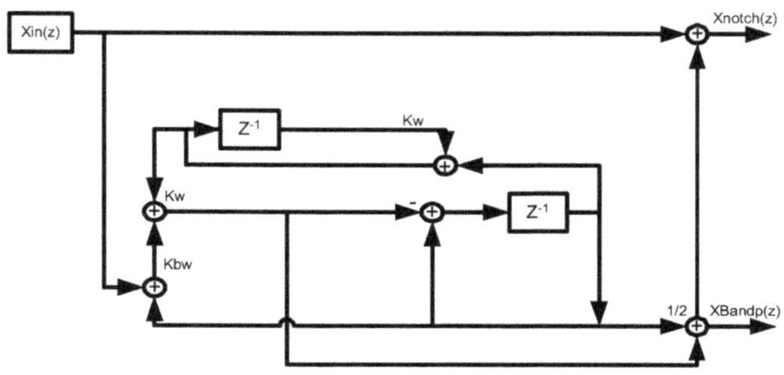

Fig. 3. Digital Notch Filter.

instead of multi-points-one-comparison algorithm in Hajjar and Roberts (1998). Moreover, we apply two-step comparison to decrease the requirement of the comparator resolution.

For the evaluation of the digitized signal, we use a digital notch filter to separate the signal from noise, which is based on an approach for on chip time recursive implementation of an arbitrary transform described in Padmanabhan and Martin (1993). Thus we can avoid the application of a DSP on chip, which makes our proposal more comm.

This paper is organized as follows. The method in Hajjar and Roberts (1998) and some discussion about it are presented briefly in Sect. 2. Then, we introduce some preliminaries of our work. The BIST approach we use for on chip DAC testing is present in Sect. 4. Simulation results are shown in Sect. 5 to demonstrate our ideas. We conclude the work in Sect. 6.

2 Present DAC BIST method

In Hajjar and Roberts (1998), an on chip signals extractor is presented. The extractor consists of a single comparator and a robust on chip reference voltage generator. Subsampling technique is used to enable the capture of analog waveforms using a single comparator clocked appropriately (Fig. 1). Arbitrary amplitude resolutions are achieved by varying the reference comparison level input to the comparator, Vref, which is held constant for a duration equal to the time it takes the comparator to compare all samples of the UTP (Unit Testing Period) to this reference level. Once all comparisons are made, Vref is incremented to the next quantization level, and the process is repeated.

However, this technique can not be used for the high-resolution and low voltage DAC. For instance, for a 16 bits DAC with the full scale of 3.7 V (+1.85 V $\sim$ −1.85 V), whose least significant bit (LSB) is 56 uV, if the testing signal is a 1 kHz sine waveform (a typical situation for the testing of voice-band chip), the testing time will be at least 1 ms × 2^{16} = 65.536 s, and that does not include the time for post-processing. Besides this, the amplitude resolution of the comparator will be half the LSB, namely 28 uV. Both of these terms are no good specification for our testing and design.

We must make some improvement in order that we can employ such similar structure to the applications with high resolution and low supply voltage on chip.

3 Preliminaries

3.1 Testing stimuli generation

For the generation of testing stimuli, we use the following method. Every periodic waveform $f(x)$ can be expressed by a Fourier series, which is an infinite sum of sine and cosine waveforms (Oppenheim and Schafer, 1989). When a filter is used to attenuate higher harmonics, a sinusoidal waveform with the same period wave will be obtained at the output. Because the trapezoid wave has very good transition band character, if its discrete period is a multiple of 6, we will employ such a trapezoid wave by saturating a triangular wave to generate our testing stimuli (Fig. 2). After downsampling the frequency to decorrecate the sampling and generated frequency, the signal is fed into a low pass filter (LPF) to filter out the harmonic frequencies (Crochiere and Rabiner, 1983) so that we get a sinewave, which is lastly interpolated to the frequency of the DAC under test.

3.2 Delta-Sigma modulation based reference voltage generator

For reference voltage generator in Fig. 1, we employ the delta sigma modulation based approach proposed in Hawrysh and Robert (1996). A software delta-sigma modulator converts the desired signal into a 1-bit stream. Then this bit stream

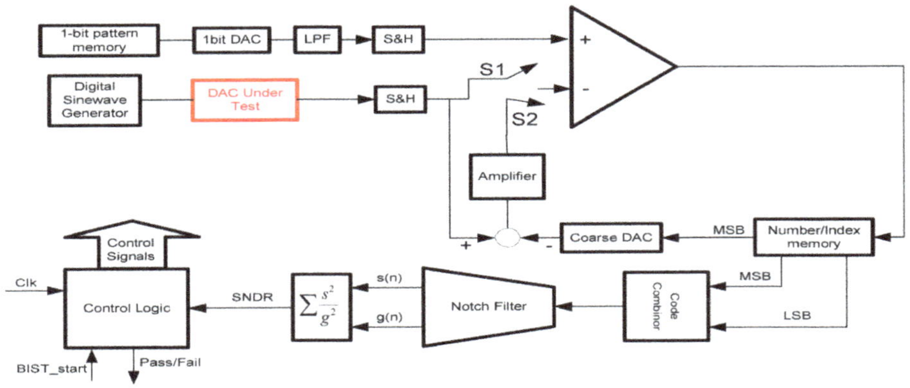

Fig. 4. The BIST scheme.

is transmit into a 1-bit DAC which is followed by an analog low pass filter. The LPF attenuates the out-of-band high-frequency modulation noise and therefore restores the original waveform. In our implementation, we extract a set of points from the bit stream which contains an integer number of signal periods under the terms of some criteria, since the bit stream of a periodic signal is even not a periodic one. Then we store it in the on chip memory, which is applied to 1-bit DAC and low-pass filter periodically to generate the desired signal.

3.3 SNDR estimation via digital notch filter

The standard industry method of SNR measurement involves taking an FFT of the digitized signal. But it needs a DSP core on chip, which is not always possible by BIST for mixed-signal circuit. We will give another approach – digital narrow band method – proposed in Padmanabhan and Martin (1993). This filter showed in Fig. 3 has two outputs: one is band-pass output tuned to the desired signal frequency; another one is band-stop output including all the harmonic frequency and noise. $X_{in}(z)$ is the filter input, and $X_{bandp}(z)$ is the band-pass output, and $X_{notch}(z)$ is the band-stop output. If we note $A_2(z)$ as:

$$A_2(z) =$$
$$= \frac{k_{bw} + k_w + (1 + k_{bw})z^{-1} + z^{-2}}{1 + k_w(1 + k_{bw})z^{-1} + k_{bw}z^{-2}} \quad (1)$$

Then we can get the following equation:

$$H_{notch}(z) = \frac{X_{notch}(z)}{X_{in}(z)} = 1 + A_2(z)$$

$$H_{bandp}(z) = \frac{X_{bandp}(z)}{X_{in}(z)} = 1 - A_2(z) \quad (2)$$

where k_w and k_{bw} are determined by the desired frequency f_0 and 3 dB bandwidth $\omega_{3\,dB}$:

$$K_w = -\cos(2\pi f_0 T)$$
$$K_{bw} = \frac{1 - \tan(\frac{w_{3dB}}{2})}{1 + \tan(\frac{w_{3dB}}{2})} \quad (3)$$

The narrow band filter is connected to the output of on chip digitizer. If $s(n)$ is the digital sequence of band-pass output and $g(n)$ is the sequence of band-stop output, then the SNDR of input is given by:

$$SNDR = 10\log_{10}\left(\frac{\sigma_s^2}{\sigma_g^2}\right) = 10\log_{10}$$

$$\left(\frac{\frac{1}{M-1}\sum_{n=1}^{M}(s(n))^2}{\frac{1}{M-1}\sum_{n=1}^{M}(g(n))^2 - \frac{1}{M(M-1)}\sum_{n=1}^{M}(g(n))^2} \right) \quad (4)$$

Mathematically, it has been demonstrated that this SNDR can be used as a preliminary estimate of the desired SNDR.

3.4 Two steps AD conversion technique

By AD conversion, the two steps AD conversion technique is usually used when the resolution requirement is too small. This two-steps architecture consists of a sampling-and-hold circuit, a coarse flash ADC, an analog substractor, a fine ADC, a DAC stage and a digital bits-combinator. By two-step testing, first the S&H tracks the analog input and holds it for coarse conversion and subtraction operation. Then the coarse flash ADC makes a coarse digital estimate of the analog input (discrete-time signal) to yield a small voltage range around the input level. The DAC stage converts this digital estimate into an analog signal, which is deducted from the original analog signal through the substractor. The fine ADC subsequently digitizes the residue signal. Lastly the digital outputs of the coarse ADC and fine ADC are combined to the final output.

4 The New BIST Scheme

The overall BIST topology for the proposed DAC BIST scheme is depicted in Fig. 4. The required functional blocks and control signals are as following (assuming that the full-scale voltage of Mbits-DAC is 2Vref):

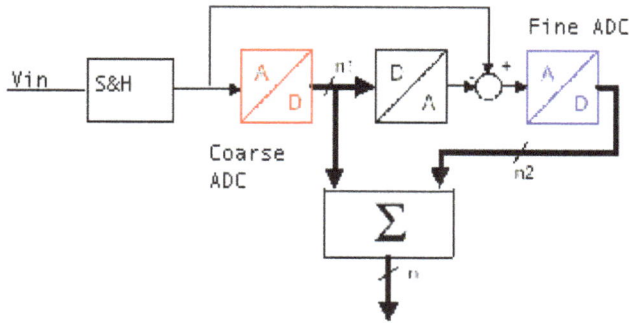

Fig. 5. Two-step conversion.

- 1-bit pattern memory, 1-bit DAC and LPF together produce the reference voltage, which rise from $-$Vref to $+$Vref with 2^M small stages and each stage lasts UTP/2^M.

- Digital Sine wave generator generates a digital sinus stimulus using the method we have introduced in Sect. 3.

- The two S&H before the comparator resample the signal with coherent sampling technique (Mahoney, 1988) in order that the DAC output will hold constant while the reference voltage rises through the full scale, which is on the contrast to Hajjar and Roberts (1998), because the testing signal in our application is 1 kHz.

- Notch filter and SNDR meter are the signal evaluation systems on chip.

- Control logic controls the BIST processing and gives the testing result out.

- Comparator, Code/Index memory, coarse DAC, amplifier and substractor build up a two step conversion unit. In fact, we employ a switched-capacitor ADC reported in Song and Lee (1990) to eliminate coarse DAC, substractor, S&H and amplifier.

- Switch S1 and S2 control the comparison step. By coarser comparison, S1 close and S2 open: while by fine comparison, S2 close and S1 open.

5 Simulation results

To validate the proposed technique, we perform numerical simulation by applying the sine wave signal to DAC under test. Our DAC is a 16 bit DAC (13 effective bits) with the full scale voltage of 3.7 V. The testing frequency is 1 kHz with a peak to peak voltage of 3.7 V, too. Thus the SNR of our testing stimuli must be not less the SNR of our DAC, which is $6.02 \times (13+2) + 1.76$ dB $= 92.06$ dB according to Norsworthy et al. (1997). The testing points must be not less than $2^{14} = 16384$. If we test so many points within

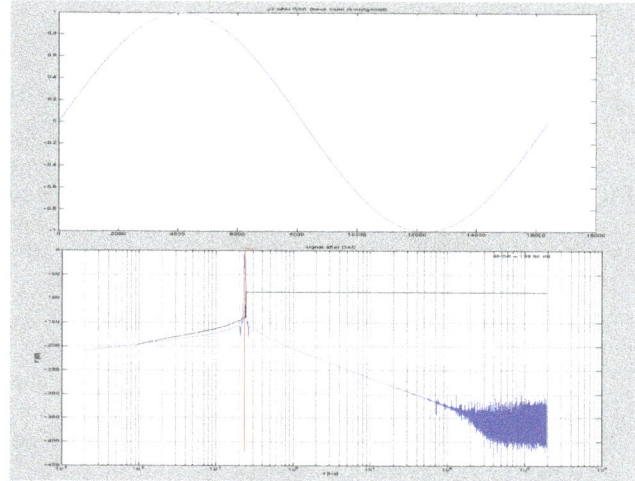

Fig. 6. Simulation on Sinewave Generator.

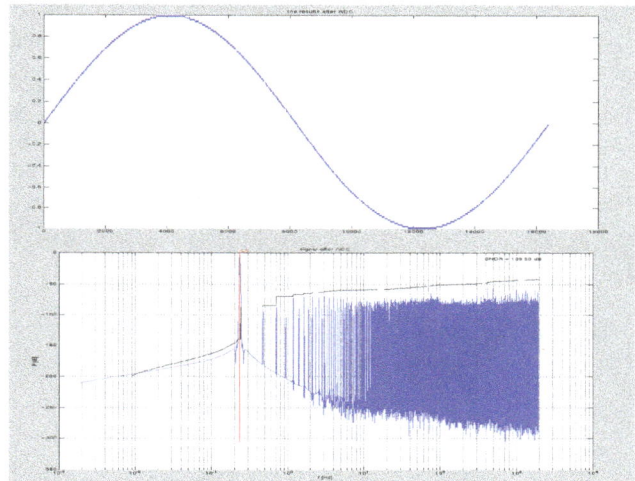

Fig. 7. Simulation on BIST scheme.

1001 testing periods. Therefore, the total testing time is 1 ms $\times$ 1001 $=$ 1001 ms $=$ 1.001 s. Without two steps multilevel comparison, we should complete each comparison in $1.001/(2^{14} \times 2^{14})$ s $= 3.73$ ns, and the min. voltage resolution of comparator is $3.7/2^{14} \times 2^{14}$ V $= 0.55$ mV. That is a big challenge for design and layout. For two steps, the coarse comparison takes place firstly (get the MSB of the "ADC") and then LSB will be worked out through the fine comparison. The coarse and fine multi-level comparison times are both 2^7, thus the comparison time for each level is $1.001/(2^{14} \times 2 \times 2^7)$ s $= 238.67$ ns, and the voltage resolution of comparator is $3.7/2^7$ V $= 29$ mV. This is obviously not a difficulty for design engineers.

First, we make a simulation on the digital sinus generator, which can achieve a SNDR over 110 dB (Fig 6.). Then we simulate the whole system in Matlab/Simulink (Fig. 7), and the simulation results demonstrate that our BIST method works very well.

6 Conclusion and discussion

We present a BIST scheme for testing on chip DAC. The main advantage are (1) it does not need the presence of a powerful DSP core on chip, (2) does not need both ADC and DAC on chip, and (3) can test low power and high resolution DAC. We also show how to get an accurate testing stimulus and how to evaluate digital signal without DSP on chip. And the simulation results verify our BIST scheme.

If there is ADC on chip and the resolution of DAC is better than that of ADC, we can test it with our verified DAC by connecting the their analog part together, then send digital signal to DAC and collect/analyze the output digital signal from the ADC.

Considering the difference between the ideal model under Matlab and the real situation, we are devoting to design a test chip for verifying our ideas further. Owing to the nonlinearly and mismatch of the capacitor by CMOS technology, we will first implement a 10 bits DAC BIST design. Moreover, future work involves also the development error model of BIST components and the research of their effect on final testing results.

Acknowledgements. This work has been funded within the AZTEKE project under label 01M3063A by the German Ministry for Education and Research (BMBF = Bundesministerium für Bildung und Forschung). Their support is gratefully acknowledged.

References

Arabi, K. and Kaminska, B.: Oscillation Built-in Self-test (OBIST) Scheme for Functional and Structural Testing of Analog and Mixed-Signal Integrated Circuits, IEEE Design & Test of Computers, Fall 1996.

Azaies, F., Bernard, S., Bertrand, Y., and Renovell, M.: Towards an ADC BIST Scheme using the Histogram Test Technique, IEEE European Test Workshop, Cascais, Portugal, May, 23–26, 2000.

Crochiere, R. E. and Rabiner, L. R.: Multi-rate Digital Signal Processing, Prentice-Hall Signal Processing Series, 1983.

Hajjar, A. and Roberts, W.: A High Speed and Area Efficient on-chip Analog Waveform Extractor", International Test Conference 1998, pp. 688–697, 1998.

Hawrysh, E. M. and Robert, G. W.: An integrated memory-based analog signal generation into current DFT architecture, Proc. Int. Test Conf., 1996, pp. 528–537, 1996.

Mahoney, M.: DSP based Testing of Analog and Mixed-signal Circuit testing, IEEE Computer Society Press, 1988.

Norsworthy, S. R., Schreier, R., and Temes, G. C. (Eds.): Delta-Sigma DataConverters: Theory, Design, and Simulation, New York IEEE Press, 1997.

Ohletz, M. J.: Hybrid BIST for Mixed Analogue/Digital Integrated Circuits, Proc. 1991 European Test Conference, pp. 307–316, April 1991.

Oppenheim, A. V. and Schafer, R. W.: Discrete-Time Signal Processing, Prentice-Hall Signal Processing Series, 1989.

Padmanabhan, M. and Martin, K.: Filter Banks for Time-Recursive Implementations of Transforms, IEEE Trans. Circuits and systems, 40, 1, 41–50, 1993.

Song, B. and Lee, S.: A 10b 15 mHz COMS recycling Two step ADC, IEEE J. Solid-State Circuit, 25, 6, 1328–1338, 1990.

Sunter, K. and Nagi, N.: A Simplified Polynomial-Fitting Algorithm for DAC and ADC BIST, International Testing Conference 1997, 389–395, 1997.

Toner, F. and Roberts, W.: A BIST SNR, gain tracking and frequency response test of a sigma-delta ADC, IEEE Trans. Circuits Syst. II, 42, 1–15, 1995.

41

Noise Considerations of Integrators for Current Readout Circuits

B. Bechen, A. Kemna, M. Gnade, T. v. d. Boom, and B. Hosticka

Fraunhofer Institute of Microelectronic Circuits and Systems, Duisburg, Germany

Abstract. In this paper the noise analysis of a current integrator is carried out and measures to reduce the overall noise are presented. The effects of various noise sources are investigated and their dependence on the input capacitance and on the gate area of the input transistors of the OTA used for the readout is shown. Both, input capacitance and gate area, should be kept as small as possible. Moreover, the linearity of the integrator is examined. In addition to that, the available application of such sensor readout circuit, which is a CMOS photodetector readout, is introduced. It uses an automatic gain switching, so that the dynamic range is extended.

1 Introduction

For various special applications in industrial measuring techniques sensor readout circuits are required. In the available consideration the current readout, particularly of a CMOS photodetector, is examined. An advantage of CMOS technology is the possibility to integrate sensor readout and photodetector together on a single chip. The pixels, consisting of photodiode and readout electronics, are arranged in an array, which enables an optional random access to each pixel. Before investigating such pixels, a general integrator is investigated in Sects. 2 and 3. As shown in Fig. 1 the integrator is realized with an operational transconductance amplifier (OTA), which has a capacitive feedback. This integrator converts the current of the current source, particularly of the photodiode, into an output voltage. In Sect. 4 a method to reduce the requirements on the OTA is presented.

2 Noise analysis

In the following section the noise behaviour of a feedback integrator as a current-to-voltage converter is investigated (see Fig. 1). The noise equivalent circuit diagram is shown in Fig. 2. Noise sources are grey-shaded. Here $\sqrt{i_{n1}^2}$ represents

the shot noise of the current source, $\sqrt{u_{n2}^2}$ in each case the thermal noise of the switches and $\sqrt{u_{n3}^2}$ is the input-referred OTA noise.

C_E is the total input capacitance, which is the sum of all capacitances at the OTA input node. It is mainly determined by the gate area capacitance of the OTA input transistors and, of course, by the current source capacitance. The single-ended OTA is modelled by the transconductance g_m, the output conductance g_L, and the load capacitance C_L.

As can be seen, the operation is separated in the two switched capacitor operation phases, i.e. the reset phase and the integration phase.

2.1 Reset phase

During the reset phase the integrator output is short-circuited via the reset switch with its series resistance $R_{S,R}$ to its inverting input. By using transfer functions $H_{e,i}(f)$ from the noise sources to the input node, the noise power can be calculated as

$$P_{e,res,i} = \int_0^\infty |H_{e,i}(f)|^2 \cdot W_i(f) df \tag{1}$$

with i as an index for the different types of noise sources. W_i are the different spectral noise densities, which can be defined as

$$W_{iph} = 2 \cdot e \cdot I_{in} \tag{2}$$

for the shot noise of the current source,

$$W_{s,th} = 4 \cdot k \cdot T \cdot R_{S,R} \tag{3}$$

for the thermal noise of the reset switch,

$$W_{O,th} = \frac{16}{3} \cdot \frac{k \cdot T}{g_m} \tag{4}$$

for the thermal noise of the OTA and

$$W_{O,1/f} = \frac{2 \cdot K_f}{C_{ox} \cdot f} \tag{5}$$

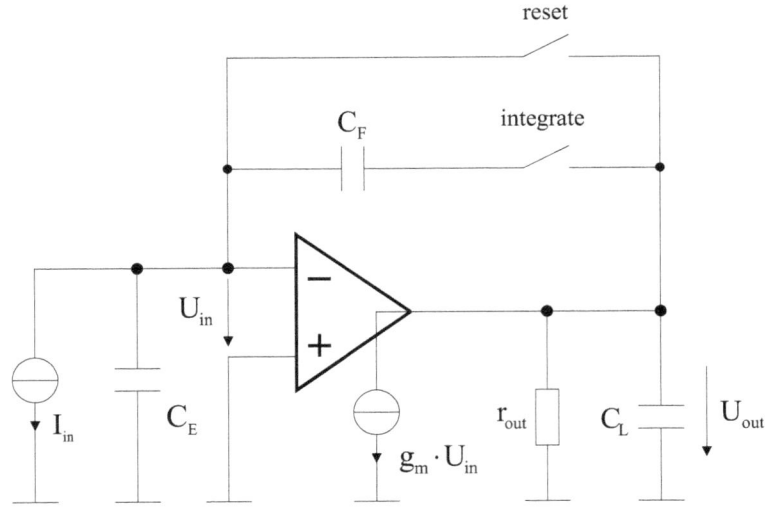

Fig. 1. Principle of the current integrator.

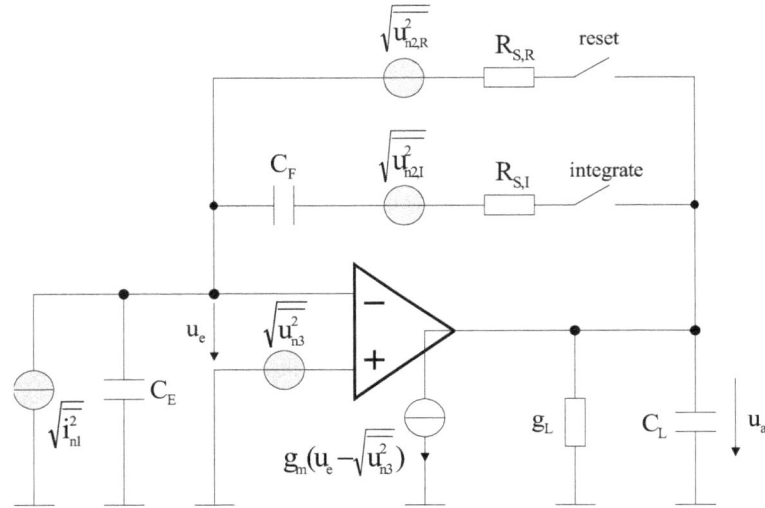

Fig. 2. Noise equivalent circuit diagram of the integrator.

for the 1/f-noise of the OTA. Together with Kirchhoff's rules the noise transfer functions of the single noise contributions can be described as follows:

$$H_{e1}(\omega) = \frac{u_e}{\sqrt{i_{nl}^2}}$$

$$\frac{1 + R_{S,R}g_L + R_{S,R}j\omega C_L}{g_m + g_L + j\omega(C_L + C_E) + R_{S,R}j\omega C_E(g_L + j\omega C_L)} \quad (6)$$

for the transfer function of the noise current source to the input node,

$$H_{e2}(\omega) = \frac{u_e}{\sqrt{u_{n2,R}^2}}$$

$$= \frac{-g_L - j\omega C_L}{g_m + g_L + R_{S,R}g_L j\omega C_E + j\omega(C_L + C_E) - \omega^2 C_L C_E R_{S,R}} \quad (7)$$

for the transfer function of the reset switch noise source to the input node and

$$H_{e3}(\omega) = \frac{u_e}{\sqrt{u_{n3}^2}}$$

$$= \frac{g_m}{g_m + g_L + R_{S,R}g_L j\omega C_E + j\omega(C_L + C_E) - \omega^2 C_L C_E R_{S,R}} \quad (8)$$

for the input-referred OTA noise to the input node.

In the following integration phase the input referred noise is transferred to the output by the following equation:

$$P_{a,res,i} = \left(\frac{C_E}{C_F}\right)^2 \cdot P_{e,res,i}. \quad (9)$$

2.2 Integration phase

During the integration phase the integration capacitor C_F is switched into the feedback path by the integration switch with its series resistance $R_{S,I}$. The noise equivalent circuit is still shown in Fig. 2. In difference to the reset phase, the noise is calculated in the integration phase as referred to the output. Similarly, the transfer functions have to be determined:

$$H_{a1}(\omega) = \frac{u_a}{\sqrt{\overline{i_{nl}^2}}} = \frac{(g_m + j\omega C_E)(1 + j\omega C_F R_{S,I}) - [j\omega C_F + j\omega C_E(1 + j\omega C_F R_{S,I})]}{(g_L + j\omega C_L)[j\omega C_F + j\omega C_E(1 + j\omega C_F R_{S,I})] + j\omega C_F(g_m + j\omega C_E)} \tag{10}$$

for the transfer function of the noise current source to the output node,

$$H_{a2}(\omega) = \frac{u_a}{\sqrt{\overline{u_{n2,I}^2}}} = \frac{g_m C_F + j\omega C_F C_E}{g_L C_E + g_m C_F + g_L C_F + j\omega(C_F C_E R_{S,I} g_L + C_L C_E + C_E C_F + C_L C_F) - \omega^2 C_L C_F C_E R_{S,I}} \tag{11}$$

for the transfer function of the integration switch noise source to the output node and

$$H_{a3}(\omega) = \frac{u_a}{\sqrt{\overline{u_{n3}^2}}} = \frac{g_m C_F + g_m C_E + j\omega C_F C_E R_{S,I} g_m}{g_L C_E + g_m C_F + g_L C_F + j\omega(C_F C_E R_{S,I} g_L + C_L C_E + C_E C_F + C_L C_F) - \omega^2 C_L C_F C_E R_{S,I}} \tag{12}$$

for the input-referred OTA noise to the output node. Thus the noise of the integration phase $P_{a,int}$ and the total noise power can be calculated:

$$P_{tot} = \sum P_i = \sum \left(P_{a,int,i} + \left(\frac{C_E}{C_F}\right)^2 \cdot P_{e,res,i} \right) \tag{13}$$

2.3 Results of noise analysis

Although the circuit under investigation is a sampled-data circuit we can still use a continuous-time noise analysis if we take into account that all undersampled noise is aliased. Thus we find all noise in the baseband.

A noise analysis that involves a detailed evaluation of Eqs. (1–13) is quite tedious. In order to simplify the analysis we assume that following approximations hold: $g_m \gg g_L$, $R_{S,R} \cdot g_L \ll 1$, $R_{S,I} \cdot g_L \ll 1$, $\omega \cdot R_{S,R} \cdot C_L \ll 1$ and $\omega \cdot R_{S,I} \cdot C_L \ll 1$. These are valid for high open-loop gain of the OTA and low ON-resistance of the switches, respectively. It is interesting to note that then the signal bandwidth (and, hence, the noise bandwidth) is dictated mainly by the product $g_m/(C_L + C_E)$ in the reset phase and by $g_m/[C_E + C_L \cdot (1 + C_E/C_F)]$ in the integration phase. Increasing C_E to lower the bandwidth is not recommended, since it introduces a zero in two of the noise transfer functions (see Eqs. (11) and (12)) and, also, it raises the DC gain of the OTA noise transfer function to the output (see Eq. (12)). Hence the most important measure to reduce the readout noise is to reduce the integrator bandwidth

$$BW = \frac{g_L}{2 \cdot \pi \cdot C_L} \tag{14}$$

by keeping C_L as large as possible. Unfortunately the bandwidth limitation due to C_L affects the slew rate and the available output current of the OTA. This analysis shows that the input capacitance C_E has the most deleterious effect on the noise performance of the integrator. This has been corroborated by simulations, as shown in Fig. 3.

It can be seen that the input capacitance C_E has to be realized as small as possible to minimize the noise. For the available application (photodetector readout), a reduced capacitance of the photodiode is presented in Kemna (2003) by using a dot diode. It must be noted that the dark current also has to be reduced to ensure maximum signal resolution.

In single stage amplifier designs, the OTA performance is mainly determined by the MOS transistors of the input stage (Uhlemann, 1994). According to that, noise powers are plotted in Fig. 4 in dependence of the gate area of the input transistors. An increase in the gate area would reduce the 1/f noise, but the 1/f noise of the OTA contributes only very little to the overall noise. Hence the gate area should be chosen small.

3 Linearity of the integrator

A Laplace transformation of the transfer function of a simple current integrator, yields

$$\frac{U_{out}(s)}{I_{in}(s)} = \frac{A_0 \left(1 - \frac{sC_F}{g_m}\right)}{s\{[C_F(A_0 + 1) + C_E] + sr_{out}(C_L C_E + C_L C_F + C_F C_E)\}} = \frac{A_0 \left(1 - \frac{s}{z}\right)}{\frac{s}{p_1}\left(1 + \frac{s}{p_2}\right)} \tag{15}$$

with the open loop gain $A_0 = g_m \cdot r_{out}$. The abbreviations z, p_1 and p_2 are defined as

$$z = \frac{g_m}{C_F}, \qquad p_1 = \frac{1}{(A_0 + 1) \cdot C_F + C_E}, \qquad p_2 = \frac{(A_0 + 1) \cdot C_F + C_E}{r_{out} \cdot (C_L \cdot C_E + C_L \cdot C_F + C_F \cdot C_E)} \tag{16}$$

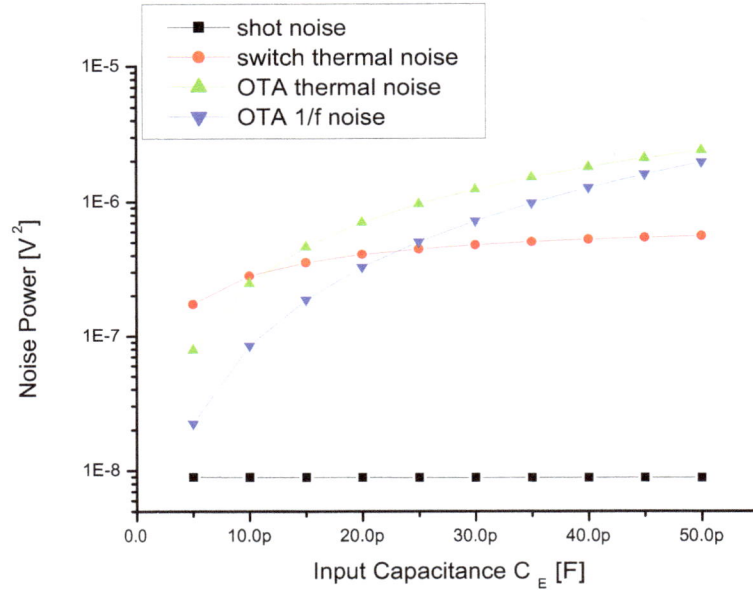

Fig. 3. Noise power of the integrator stage versus the input capacitance.

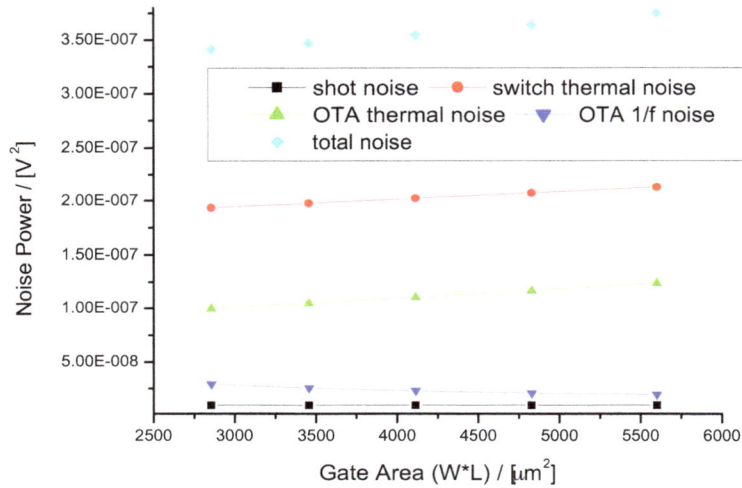

Fig. 4. Noise power of the integrator stage versus gate area of the input transistors.

Transforming it back and assuming the current is a step function, results in an output voltage as follows:

$$u_{out}(t) = I_{in0} \cdot \left\{ A_0 \cdot p_1 \cdot \left[t - \frac{\left(1 + \frac{p_2}{z}\right)}{p_2} \cdot \left(1 - e^{-p_2 \cdot t}\right) \right] \right\}. \quad (17)$$

After one integration period T, the integral nonlinearity can be calculated as

$$INL = \frac{\left(1 + \frac{p_2}{z}\right) \cdot \left(1 - e^{-p_2 \cdot T}\right)}{p_2 \cdot T}. \quad (18)$$

In Fig. 5 the nonlinearity INL is plotted over the integration time T. The limit for a zero integration time results in an INL of 1.

4 An application example

In the available application a photocurrent integrator is investigated, which operates with two automatically selectable gains, in order to relax the requirements on the OTA. To implement an automatic gain control (AGC) an auto zero comparator is used, which controls the two feedback capacitors. The principle is shown in Fig. 6. If the output voltage exceeds the threshold voltage of the comparator, a second capacitor with a 31 times larger capacitance is switched in parallel. Then the integrator has turned from high into low gain configuration, because signal charge is shared between

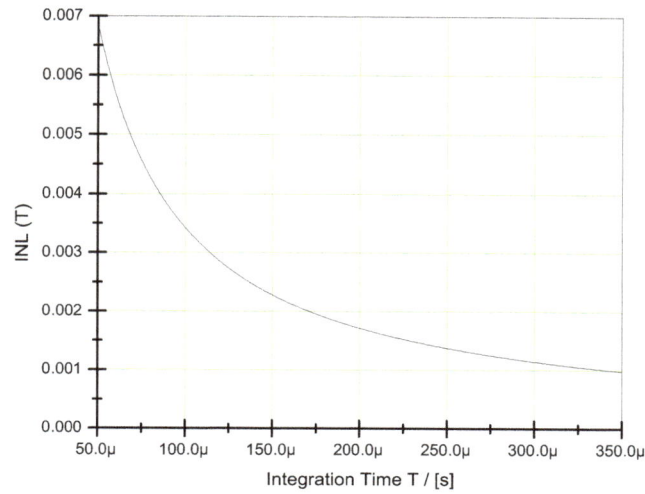

Fig. 5. INL over integration time T.

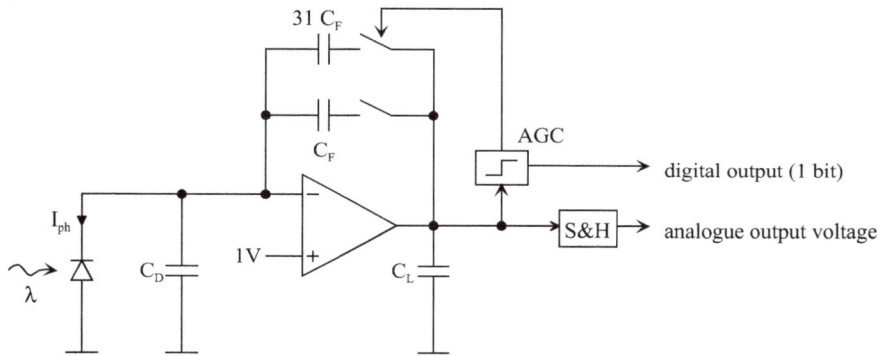

Fig. 6. Pixel readout electronics.

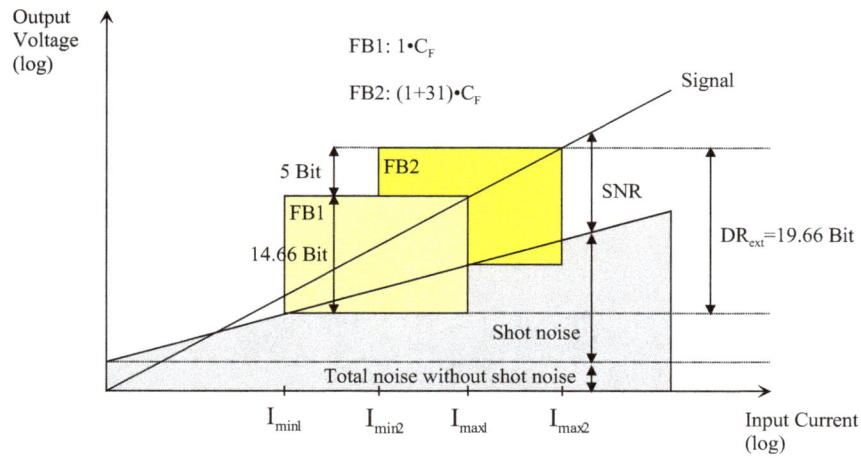

Fig. 7. Dynamic range extension by automatic gain control.

Table 1. Specifications of the photocurrent integrator.

Minimum Signal (theoretical)	99.33 fA	Dynamic Range (theoretical)	19.66 Bit
Minimum Signal (noise determined)	491.83 fA	Dynamic Range (noise determined)	17.35 Bit
Maximum Signal	82.29 nA	Linearity (FB1)	10.33 Bit
Integration Time	350 μs	Linearity (FB2)	13.99 Bit

the capacitors correspondingly to their ratio. After a frame period, the signal is held by a sample and hold stage (S&H) and can be read out during the next frame period using a row and column multiplexer circuitry, whereas the used gain is determined by the digital output.

In photocurrent integration, the minimum input signal that can be processed is defined by the input-referred noise, and the maximum output voltage swing defines the maximum output signal. This leads to the conclusion, that noise determines the dynamic range (DR) and the signal-to-noise ratio (SNR). Due to the shot noise, the maximum SNR is signal dependent, and the maximum SNR is not equal to the dynamic range.

A dynamic range extension is possible by using the automatic gain control. This is visualised for some exemplary values in Fig. 7, in which the feedback FB2 has a feedback capacitance 32 times larger than feedback capacitance FB1. The signal and the noise at the output are plotted over the input current. The noise consists of the signal-dependent shot noise and the remaining constant noise. The extension of the dynamic range is determined by the ratio of the feedback capacitors.

The specifications used in this application example are listed in Table 1. Here the specifications concerned to the minimum input photocurrent are distinguished between the theoretical capability of the integrator and the value limited by the noise and the dark current of the photodiode.

5 Summary and conclusion

The principle of a noise analysis of a current integrator has been demonstrated. This noise analysis led to the conclusion, that the input capacitance of the photodiode has to be kept at minimum. Also decreasing the bandwidth is an important measure to reduce noise. These conclusions can be extended to all current readout circuits.

References

Kemna, A.: Low Noise, Large Area CMOS X-Ray Image Sensor for C. T. Application, Proceedings of IEEE Sensors, Vol. 2, 1260–1265, 2003.

Uhlemann, V.: Rauscharme Auslesesysteme zur Verarbeitung elektrischer Ladung in integrierter CMOS-Technik, Fortschr.-Ber. VDI-Reihe 9, Nr. 199, VDI-Verlag, 1994.

42

Increasing the time resolution of a pulse width modulator in a class D power amplifier by using delay lines

M. Weber, T. Vennemann, and W. Mathis

Institute for Theoretical Electrical Engineering, Leibniz Universität Hannover, Hannover, Germany

Correspondence to: T. Vennemann (vennemann@tet.uni-hannover.de)

Abstract. In this paper, we present a method to increase the time resolution of a pulse width modulator by using delay lines. The modulator is part of an open loop class D power amplifier, which uses the ZePoC algorithm to code the audio signal which is amplified in the class D power stage. If the time resolution of the pulse width modulator is high enough, ZePoC could also be used to build an high accuracy AC power standard, because of its open loop property. With the presented method the time resolution theoretically could be increased by a factor of 16, which means here the time resolution will be enhanced from 5 ns to 312.5 ps.

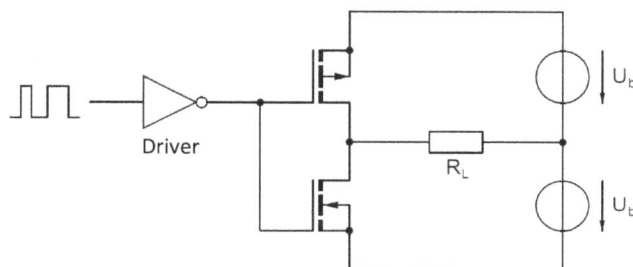

Figure 1. Class D power stage.

1 Introduction

The mean advantages of a complete digital class D power amplifier based on the zero positon coding (ZePoC) are the low switching rate and the separated baseband. ZePoC was initially used and implemented for audio coding by the Institute for Theoretical Electrical Engineering in Hannover. Much research was done over years resulting in a number of publications. A good overview could found in the white paper from Texas Instruments (2005). The binary signal computed by the ZePoC algorithm implemented on a digital signal processor (DSP) is generated by a pulse width modulator (PWM) unit inside the DSP.

The DSP is an ADSP-21369 from Analog Devices, which PWM unit is clocked at 200 MHz (Analog Devices ADSP-21369 (2009)). At this time there are no DSPs on the market, which go far beyond this clock rate. To use ZePoC in an high accuracy AC power standard, the open loop structure of the ZePoC system must be preserved. The only way to increase the accuracy is to increase the time resolution of the PWM signal (Wellmann et al., 2010). In this paper we present a

way to increase the time resolution without increasing the clock rate of the PWM unit.

2 Basics

Figure 1 shows a class D power stage, which consists of an inverting driver, two complementary MOSFETs and a symmetric power supply. The driver switches the MOSFETs one at a time on and the other one off according to the input signal. This means that the theoretical efficiency of the power stage is 100 %, but in reality there are power losses because of the internal resistance of the MOSFETs and switching losses. Because of the low switching rate generated by the ZePoC algorithm, the efficiency of the power stage is > 90 %. Basically a class D power stage can only amplify binary signals. As a result, the power stage has an absolute nonlinear transmission behaviour. Therefore, a modulation and demodulation as shown in Fig. 2 of the signal is necessary to get an overall linear transmission behaviour.

In a class D amplifier, a passive low pass filter (LPF) is used for demodulation. The correctly reconstruction of the audio signal is only possible, if there is a gap in the spectrum

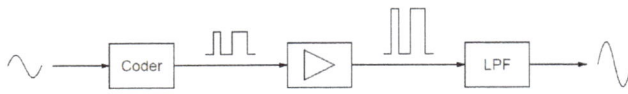

Figure 2. Signal transmission in class D amplifier.

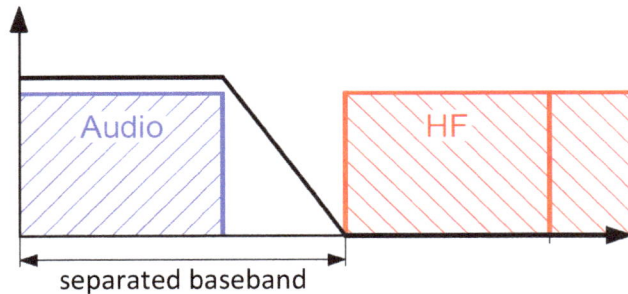

Figure 3. Spectrum of the ZePoC binary signal.

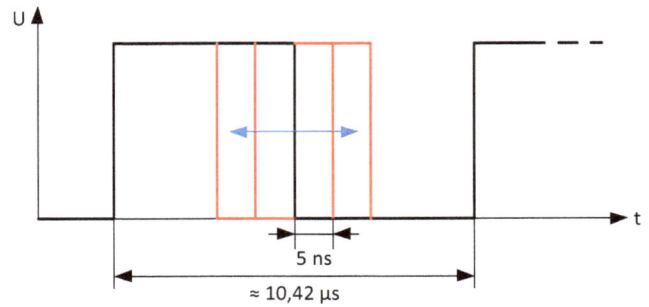

Figure 4. Pulse width modulated signal.

of the binary signal between the audio band and the high frequency distortions caused by the modulation process. ZePoC is a modulating algorithm with a separated baseband. This means that the spectrum of the generated binary signal is free of distortions below a specified frequency. The maximum of this frequency is half the switsching frequency.

Figure 3 shows the ZePoC spectrum with the audio band inside the sparated baseband. The black line illustrates the transfer function of the LPF used for signal reconstruction. Only the audio signal could pass the LPF, the high frequency (HF) distortions are blocked completely. The gap between the audio band and the HF part of the spectrum must be as wide as the transition band of the LPF. For that reason a switching frequency of 96 kHz is used.

The PWM signal shown in Fig. 4 demonstrates the need of increasing the time resolution of the PWM modulator. The PWM unit inside the DSP on which the ZePoC algorithm is implemented works with a clock speed of 200 MHz, so the value of the binary signal could only change every 5 ns. This means that the edges of the PWM signal are in a 5 ns time grid. With a switching rate of 96 kHz (T = 10.42 μs) the number of possible duty cycles of the PWM signal is limited to 2084. This equals to a resolution of approx. 11 bits. Compared to the resolution of an audio compact disc (CD) of 16 bits, the resolution of this class D amplifier must be increased by 5 bits to get the same resolution as a CD. 5 bits means a factor of $2^5 = 32$.

3 Approach

Figure 5 shows the entire signal chain of the class D amplifier. Out of a pulse code modulated (PCM) audio signal a PWM signal is generated by the DSP using the ZePoC algorithm. This binary signal is amplified by the class D power stage. A loadspeaker is connected to the output of the LPF

which reconstructs the audio signal. The amplitude of the signal (volume) could be controlled by the supply voltage of the class D power stage.

To increase the time resolution of the PWM modulator, an additional module will be inserted into the signal chain at the position of the red arrow. From the DSP board to the class D power stage the PWM signal is transferred via a low voltage differential signaling (LVDS) interface. For that reason the additional module must have a LVDS input and output.

The concept of the time resolution enhancement is displayed in Fig. 6. The PWM signal with the low time resolution at LVDS IN is received simultaneously by four LVDS receivers. Each output of the receivers drives a 75 Ω transmission line of an individual length. Every transmission line will delay the PWM signal by a specified time, this is why we call them delay lines. At the end the delay lines are terminated by 75 Ω resistors. The four trimmer capacitors are used to calibrate the delay times of the particular signal pathes.

Every delay line is connected to an input of an analogue multiplexer (Analog Devices ADG704, 1999) controlled by the DSP. Exactly one delay line will be routed through the multiplexer to the LVDS transmitter. The input of the class D power stage will be connected to the PWM signal with the higher time resolution at LVDS OUT. Thus, the modulated edge of the PWM-signal can be delayed for a defined time controlled by the DSP.

4 Implementation

The time resolution enhancement module (TREM) consists of two delay stages as shown in Fig. 6 connected in series. Figure 7 demonstrates how the time slot of 5 ns will be divided into 16 sections by the two stages. The blue marks show the delay times of the first stage and the red marks show the delay times of the second stage in combination with the first stage. So the first delay stage generates a 1.25 ns time grid and both together a grid of 0.3125 ns.

For the delay lines a coaxial cable of type RG179 is used. The velocity factor for this type of cable is 0.7. With this information, the length of the delay lines can be calculated. The coax cables must be a little bit shorter than calculated

Figure 5. Signal chain of the class D amplifier.

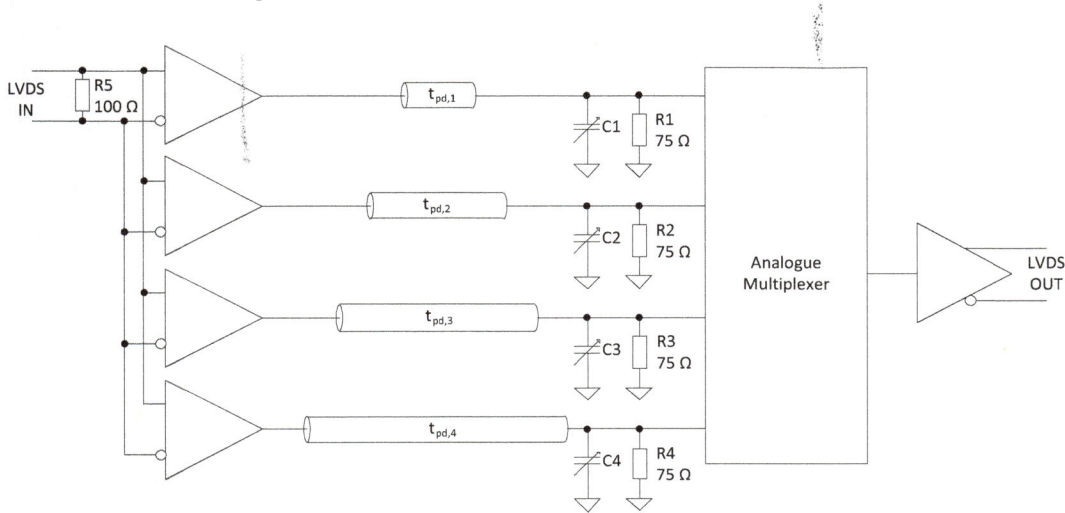

Figure 6. Concept with delay lines.

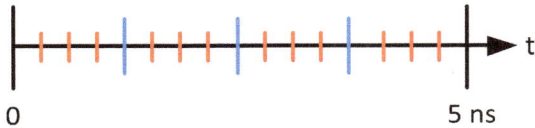

Figure 7. Time sections.

Table 1. Delay times and cable lengths of stage 1.

line	delay time	length
1	0 ns	0
2	1.25 ns	263 mm
3	2.5 ns	525 mm
4	3.75 ns	788 mm

Table 2. Delay times and cable lengths of stage 2.

line	delay time	length
1	0 ns	0
2	0.3125 ns	66 mm
3	0.625 ns	131 mm
4	0.9375 ns	197 mm

5 Measurement results

For each of the 16 possible signal paths the delay times were measured 1000 times and than averages were calculated. The blue line in Fig. 9 shows the measurement results. The red line shows the calculated delay times listed in Tables 1 and 2. There are some differences between the calculated and the measured delay times. The deviation between these two values are depicted in Fig. 10 for all 16 signal paths.

The highest deviation was measured for path 8. Its size is about 200 ps. In comparison to the shortest delay time of 312.5 ps, the time resolution could not reach the maximum value of 4 bits. It is not possible to reduce the deviation of path 8 without changing other deviations. The analysis of the standard deviation of the jitter for each path shows that all values are uniformly distributed and less than 50 ps. That means that the jitter has no problematic effect on the delay times.

because of the trimming capacitors which add an capacitive load to each delay line. The delay times and the resulting lengths for each delay line of the two different stages are shown in Tables 1 and 2.

Figure 8 shows the prototype which includes the two delay stages. On the top side of the printed circuit board (PCB) are all surface mounted devices (SMD) and trimming capacitors. The connectors to the DSP board and the class D power stage are on the bottom side of the PCB. All delay lines are coaxial cables of type RG179 and are also mounted on the bottom side.

Figure 8. Top and bottom side of the prototype board.

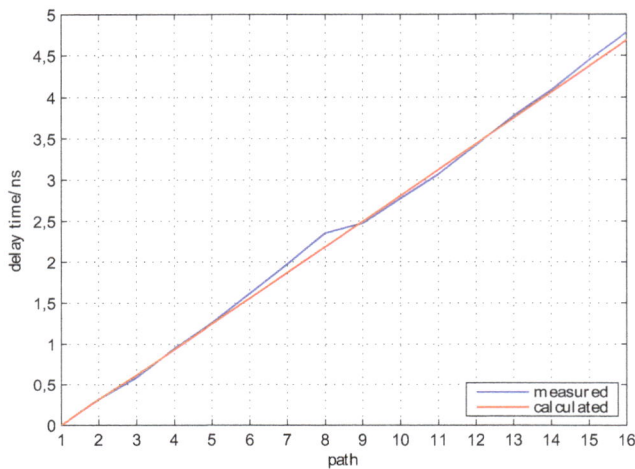

Figure 9. Averages of the delay times (over 1000 values).

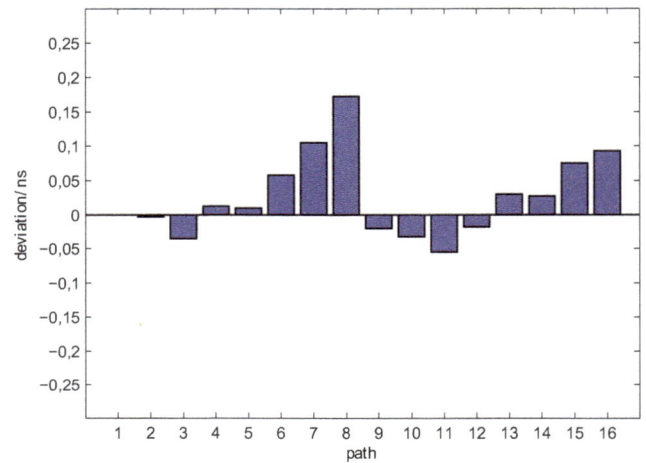

Figure 10. Averages of the deviations (over 1000 values).

6   Conclusions

In this paper, a method to increase the time resolution of a pulse width modulator that is used in a class D power amplifier was presented. The measurement results of the prototype shown that it is possible to increase the time resolution. However, the theoretical factor of 16 (4 bits) was not reached because the delay lines could not be tuned exactly. The exact time resolution improvement factor of the presented module can be determined by measurement of the signal to noise ratio (SNR) of the audio amplifier.

References

Analog Devices: Data Sheet Low Voltage 4 Ω, 4-Channel Multiplexer ADG704, Rev. A, 1999.

Analog Devices: Data Sheet SHARC Processors ADSP-21367/ADSP-21368/ADSP-21369, Rev. E, July 2009.

Texas Instruments: A New Audio File Format for Low-Cost, High Fidelity, Portable Digital Audio Amplifiers, White Paper, 2005.

Wellman, J., Kahmann, M., and Mathis, W.: TET-Watt – An AC Power Standard based on Class-D Topologies using ZePoC-Coding. Conference on Precision Electromagnetic Measurements, Korea, 2010.

Permissions

All chapters in this book were first published in ARS, by Copernicus Publications; hereby published with permission under the Creative Commons Attribution License or equivalent. Every chapter published in this book has been scrutinized by our experts. Their significance has been extensively debated. The topics covered herein carry significant findings which will fuel the growth of the discipline. They may even be implemented as practical applications or may be referred to as a beginning point for another development.

The contributors of this book come from diverse backgrounds, making this book a truly international effort. This book will bring forth new frontiers with its revolutionizing research information and detailed analysis of the nascent developments around the world.

We would like to thank all the contributing authors for lending their expertise to make the book truly unique. They have played a crucial role in the development of this book. Without their invaluable contributions this book wouldn't have been possible. They have made vital efforts to compile up to date information on the varied aspects of this subject to make this book a valuable addition to the collection of many professionals and students.

This book was conceptualized with the vision of imparting up-to-date information and advanced data in this field. To ensure the same, a matchless editorial board was set up. Every individual on the board went through rigorous rounds of assessment to prove their worth. After which they invested a large part of their time researching and compiling the most relevant data for our readers.

The editorial board has been involved in producing this book since its inception. They have spent rigorous hours researching and exploring the diverse topics which have resulted in the successful publishing of this book. They have passed on their knowledge of decades through this book. To expedite this challenging task, the publisher supported the team at every step. A small team of assistant editors was also appointed to further simplify the editing procedure and attain best results for the readers.

Apart from the editorial board, the designing team has also invested a significant amount of their time in understanding the subject and creating the most relevant covers. They scrutinized every image to scout for the most suitable representation of the subject and create an appropriate cover for the book.

The publishing team has been an ardent support to the editorial, designing and production team. Their endless efforts to recruit the best for this project, has resulted in the accomplishment of this book. They are a veteran in the field of academics and their pool of knowledge is as vast as their experience in printing. Their expertise and guidance has proved useful at every step. Their uncompromising quality standards have made this book an exceptional effort. Their encouragement from time to time has been an inspiration for everyone.

The publisher and the editorial board hope that this book will prove to be a valuable piece of knowledge for researchers, students, practitioners and scholars across the globe.

List of Contributors

F. Coccetti
Technische Universität München, Lehrstuhl für Hochfrequenztechnik, Munich, Germany

W. Dressel
Technische Universität München, Lehrstuhl für Hochfrequenztechnik, Munich, Germany

P. Russer
Technische Universität München, Lehrstuhl für Hochfrequenztechnik, Munich, Germany

L. Pierantoni
Dipartimento di Elettronica ed Automatica, University of Ancona, Ancona, Italy

M. Farina
Dipartimento di Elettronica ed Automatica, University of Ancona, Ancona, Italy

T. Rozzi
Dipartimento di Elettronica ed Automatica, University of Ancona, Ancona, Italy

H. T. Feldkaemper
Chair of Electrical Engineering and Computer Systems, RWTH Aachen University, Schinkelstr. 2, 52062 Aachen, Germany

H. Blume
Chair of Electrical Engineering and Computer Systems, RWTH Aachen University, Schinkelstr. 2, 52062 Aachen, Germany

T. G. Noll
Chair of Electrical Engineering and Computer Systems, RWTH Aachen University, Schinkelstr. 2, 52062 Aachen, Germany

C. Schlachta
Darmstadt University of Technology, Institute of Microelectronic Systems, Germany

M. Glesner
Darmstadt University of Technology, Institute of Microelectronic Systems, Germany

A. Glasmachers
Fachbereich Elektrotechnik und Informationstechnik, Universität Wuppertal, Fuhlrottstrasse 10, 42097 Wuppertal, Germany

J. Rauscher
Department of Microelectronics, University of Ulm, Germany

M. Tahedl
Department of Microelectronics, University of Ulm, Germany

H.-J. Pfleiderer
Department of Microelectronics, University of Ulm, Germany

C. Neeb
University of Kaiserslautern, Germany

M. J. Thul
University of Kaiserslautern, Germany

N. Wehn
University of Kaiserslautern, Germany

C. Saas
Munich University of Technology, Institute for Circuit Theory and Signal Processing, Arcisstr. 16, 80290 Munich, Germany

J. A. Nossek
Munich University of Technology, Institute for Circuit Theory and Signal Processing, Arcisstr. 16, 80290 Munich, Germany

B. Burdiek
Institut für Theoretische Elektrotechnik und Hochfrequenztechnik, Universität Hannover, Appelstr. 9A, 30167 Hannover, Germany

W. Mathis
Institut für Theoretische Elektrotechnik und Hochfrequenztechnik, Universität Hannover, Appelstr. 9A, 30167 Hannover, Germany

M. Schneider
Lehrstuhl für Allgemeine Elektrotechnik und Datenverarbeitungssysteme, RWTH Aachen, Schinkelstraße 2, 52062 Aachen, Germany

H. Blume
Lehrstuhl für Allgemeine Elektrotechnik und Datenverarbeitungssysteme, RWTH Aachen, Schinkelstraße 2, 52062 Aachen, Germany

T. G. Noll
Lehrstuhl für Allgemeine Elektrotechnik und Datenverarbeitungssysteme, RWTH Aachen, Schinkelstraße 2, 52062 Aachen, Germany

M. Ehlert
Micronas GmbH, Freiburg, Germany

M. Vollmer
University of Dortmund, Information Processing Lab, Germany

J. Götze
University of Dortmund, Information Processing Lab, Germany

J. Freitag
Institut für Theoretische Elektrotechnik, Appelstraße 9A, 30167 Hannover, Germany

W. Mathis
Institut für Theoretische Elektrotechnik, Appelstraße 9A, 30167 Hannover, Germany

N. Fichtner
Institute for High Frequency Engineering, Technische Universität München, Arcisstr. 21, 80333 München, Germany

U. Siart
Institute for High Frequency Engineering, Technische Universität München, Arcisstr. 21, 80333 München, Germany

Y. Kuznetsov
Moscow Aviation Institute (State Technical University) Volokolamskoe sh. 4, A-80, GSP-3, Moscow, 125993, Russia

A. Baev
Moscow Aviation Institute (State Technical University) Volokolamskoe sh. 4, A-80, GSP-3, Moscow, 125993, Russia

P. Russer
Institute for High Frequency Engineering, Technische Universität München, Arcisstr. 21, 80333 München, Germany

S. Schelkshorn
Technische Universität München Lehrstuhl für Hochfrequenztechnik
Fachgebiet Hochfrequente Felder und Schaltungen 80333 München, Germany

J. Detlefsen
Technische Universität München Lehrstuhl für Hochfrequenztechnik
Fachgebiet Hochfrequente Felder und Schaltungen 80333 München, Germany

S. Bertl
Technische Universität München, Fachgebiet Hochfrequente Felder und Schaltungen, Arcisstr. 21, 80333 München, Germany

A. Dallinger
Technische Universität München, Fachgebiet Hochfrequente Felder und Schaltungen, Arcisstr. 21, 80333 München, Germany

J. Detlefsen
Technische Universität München, Fachgebiet Hochfrequente Felder und Schaltungen, Arcisstr. 21, 80333 München, Germany

D. Lupea
Technische Universität Dresden, Institut für Verkehrsinformationssysteme, Mommsenstr. 13, D-01062 Dresden

U. Pursche
Technische Universität Dresden, Institut für Verkehrsinformationssysteme, Mommsenstr. 13, D-01062 Dresden

H.-J. Jentschel
Technische Universität Dresden, Institut für Verkehrsinformationssysteme, Mommsenstr. 13, D-01062 Dresden

T. von Sydow
Electrical Engineering and Computer Systems, RWTH Aachen University, Schinkelstr. 2, 52062 Aachen, Germany

H. Blume
Electrical Engineering and Computer Systems, RWTH Aachen University, Schinkelstr. 2, 52062 Aachen, Germany

T. G. Noll
Electrical Engineering and Computer Systems, RWTH Aachen University, Schinkelstr. 2, 52062 Aachen, Germany

L. Zhang
Otto-von-Guericke University of Magdeburg, IESK, PO Box 4120, D-39016 Magdeburg, Germany

U. Kleine
Otto-von-Guericke University of Magdeburg, IESK, PO Box 4120, D-39016 Magdeburg, Germany

J. Fischer
Institute for Technical Electronics, Technical University Munich, Theresienstrasse 90, D-80290 Munich, Germany

E. Amirante
Institute for Technical Electronics, Technical University Munich, Theresienstrasse 90, D-80290 Munich, Germany

A. Bargagli-Stoffi
Institute for Technical Electronics, Technical University Munich, Theresienstrasse 90, D-80290 Munich, Germany

D. Schmitt-Landsiedel
Institute for Technical Electronics, Technical University Munich, Theresienstrasse 90, D-80290 Munich, Germany

T. Mahnke
Institute for Integrated Circuits, Technical University of Munich, Germany

W. Stechele
Institute for Integrated Circuits, Technical University of Munich, Germany

M. Embacher
National Semiconductor GmbH, Fuerstenfeldbruck, Germany

W. Hoeld
National Semiconductor GmbH, Fuerstenfeldbruck, Germany

M. Vollmer
University of Dortmund, Information Processing Lab., Germany

J. Götze
University of Dortmund, Information Processing Lab., Germany

J. Fischer
Institute for Technical Electronics, Technical University Munich, Theresienstrasse 90, D-80290 Munich, Germany

E. Amirante
Institute for Technical Electronics, Technical University Munich, Theresienstrasse 90, D-80290 Munich, Germany

A. Bargagli-Stoffi
Institute for Technical Electronics, Technical University Munich, Theresienstrasse 90, D-80290 Munich, Germany

D. Schmitt-Landsiedel

Institute for Technical Electronics, Technical University Munich, Theresienstrasse 90, D-80290 Munich, Germany

B. Klehn
Infineon Technologies AG, Munich, Germany

M. Brox
Infineon Technologies AG, Munich, Germany

F. Kienle
University of Kaiserslautern, Germany

H. Michel
University of Kaiserslautern, Germany

F. Gilbert
University of Kaiserslautern, Germany

N. Wehn
University of Kaiserslautern, Germany

S. Drüen
TU München, Germany

K. Esmark
Infineon Technologies, München, Germany

W. Stadler
Infineon Technologies, München, Germany

H. Gossner
Infineon Technologies, München, Germany

D. Schmitt-Landsiedel
TU München, Germany

S. Henzler
Lehrstuhl für Technische Elektronik, Technische Universität München, Theresienstrasse 90, D-80290 Munich, Germany

J. Berthold
Corporate Logic, Infineon Technologies AG, Balanstrasse 73, D-81541 Munich, Germany

G. Georgakos
Corporate Logic, Infineon Technologies AG, Balanstrasse 73, D-81541 Munich, Germany

D. Schmitt-Landsiedel
Lehrstuhl für Technische Elektronik, Technische Universität München, Theresienstrasse 90, D-80290 Munich, Germany

A. Bargagli-Stoffi
Institute for Technical Electronics, Technical University Munich, Theresienstrasse 90, D-80290 Munich, Germany

E. Amirante
Institute for Technical Electronics, Technical University Munich, Theresienstrasse 90, D-80290 Munich, Germany

J. Fischer
Institute for Technical Electronics, Technical University Munich, Theresienstrasse 90, D-80290 Munich, Germany

G. Iannaccone
Dipartimento di Ingegneria dell'Informazione, Università degli Studi di Pisa, Via Diotisalvi 2, I-56122 Pisa, Italy

D. Schmitt-Landsiedel
Institute for Technical Electronics, Technical University Munich, Theresienstrasse 90, D-80290 Munich, Germany

A. Edelmann
FernUniversität in Hagen, Universitätsstr. 27/PRG, 58084 Hagen, Germany

S. Helfert
FernUniversität in Hagen, Universit¨atsstr. 27/PRG, 58084 Hagen, Germany

J. Jahns
FernUniversität in Hagen, Universitätsstr. 27/PRG, 58084 Hagen, Germany

F. Mukhtar
Institute for Nanoelectronics, Technische Universit¨at München, Germany

Y. Kuznetsov
Theoretical Radio Engineering Department, Moscow Aviation Institute, Russia

P. Russer
Institute for Nanoelectronics, Technische Universit¨at M¨unchen, Germany

C. Fischer
Daimler AG, Wilhelm-Runge-Straße 11, 89013 Ulm, Germany

M. Goppelt
University of Ulm, Institute of Microwave Techniques, Albert-Einstein-Allee 41, 89069 Ulm, Germany

H.-L. Blöcher
Daimler AG, Wilhelm-Runge-Straße 11, 89013 Ulm, Germany

J. Dickmann
Daimler AG, Wilhelm-Runge-Straße 11, 89013 Ulm, Germany

J. A. Russer
Lehrstuhl für Nanoelektronik, Technische Universität München, Arcisstrasse 21, Munich, Germany

F. Mukhtar
Lehrstuhl für Nanoelektronik, Technische Universität München, Arcisstrasse 21, Munich, Germany

A. Baev
Theoretical Radio Engineering Dept., Moscow Aviation Inst., Volokolamskoe shosse, 4, GSP-3, Moscow, 125993, Russia

Y. Kuznetsov
Theoretical Radio Engineering Dept., Moscow Aviation Inst., Volokolamskoe shosse, 4, GSP-3, Moscow, 125993, Russia

P. Russer
Lehrstuhl für Nanoelektronik, Technische Universität München, Arcisstrasse 21, Munich, Germany

O. Mitrea
Darmstadt University of Technology, Karlstr. 15, 64283 Darmstadt, Germany

C. Popa
University Politehnica of Bucharest, Bd. Iuliu Maniu 1–3, Bucharest, Romania

A. M. Manolescu
University Politehnica of Bucharest, Bd. Iuliu Maniu 1–3, Bucharest, Romania

M. Glesner
Darmstadt University of Technology, Karlstr. 15, 64283 Darmstadt, Germany

M. R. Kühn
Technische Universität M¨unchen, Fachgebiet Höchstfrequenztechnik, Arcisstr. 21, 80333 München, Germany

E. M. Biebl
Technische Universität München, Fachgebiet Höchstfrequenztechnik, Arcisstr. 21, 80333 M¨unchen, Germany

M. O. Olbrich
Technische Universität München, Fachgebiet Höchstfrequenztechnik, 80290 München, Germany

A. Gr¨ubl
now with Europäisches Patentamt

R. H. Raßhofer
now with TriQuint

E.M. Biebl
Technische Universität M¨unchen, Fachgebiet Höchstfrequenztechnik, 80290 M¨unchen, Germany

M. Schleyer
Fachgebiet Mikroelektronik, Technische Universität
Berlin, Germany

S. Leuschner
Intel Mobile Communications GmbH, Munich, Germany

P. Baumgartner
Intel Mobile Communications GmbH, Munich, Germany

J.-E. Mueller
Intel Mobile Communications GmbH, Munich, Germany

H. Klar
Fachgebiet Mikroelektronik, Technische Universität
Berlin, Germany

M. Streifinger
Technische Universität München, Fachgebiet
Höchstfrequenztechnik, Arcisstr. 21, 80333 München,
Germany

T. Müller
DaimlerChrysler Forschungszentrum, Wilhelm-Runge-
Str. 11, 89081 Ulm, Germany

J.-F. Luy
DaimlerChrysler Forschungszentrum, Wilhelm-Runge-
Str. 11, 89081 Ulm, Germany

E. M. Biebl
Technische Universität München, Fachgebiet
Höchstfrequenztechnik, Arcisstr. 21, 80333 München,
Germany

M. Camp
Leibniz University of Hannover, Institute of
Radiofrequency and Microwave Engineering, Appelstraße
9a, 30167 Hannover, Germany

Smart Devices GmbH & Co. KG, Schönebecker Allee 2,
30823 Garbsen, Germany

R. Herschmann
Leibniz University of Hannover, Institute of
Radiofrequency and Microwave Engineering, Appelstraße
9a, 30167 Hannover, Germany
Smart Devices GmbH & Co. KG, Schönebecker Allee 2,
30823 Garbsen, Germany

T. Zelder
Leibniz University of Hannover, Institute of
Radiofrequency and Microwave Engineering, Appelstraße
9a, 30167 Hannover, Germany

H. Eul
Leibniz University of Hannover, Institute of
Radiofrequency and Microwave Engineering, Appelstraße
9a, 30167 Hannover, Germany

M. Prochaska
University of Hannover, Institute of Electromagnetic
Theory and Microwave Technology, Appelstraße 9A,
30167 Hannover, Germany

A. Belski
University of Hannover, Institute of Electromagnetic
Theory and Microwave Technology, Appelstraße 9A,
30167 Hannover, Germany

W. Mathis
University of Hannover, Institute of Electromagnetic
Theory and Microwave Technology, Appelstraße 9A,
30167 Hannover, Germany

W. Kraus
Institute for Technical Electronics, TU-Munich, Germany

B. Stelzig
Institute for Technical Electronics, TU-Munich, Germany

T. Tille
Institute for Technical Electronics, TU-Munich, Germany

D. Schmitt-Landsiedel
Institute for Technical Electronics, TU-Munich, Germany

H. Li
SMS TI MT MS, Infineon AG, Postfach 80 09 49, D-81609
München, Germany

J. Eckmueller
SMS TI MT MS, Infineon AG, Postfach 80 09 49, D-81609
München, Germany

S. Sattler
CTT TS ADT, Infineon AG, Postfach 80 09 49, D-81609
München, Germany

H. Eichfeld
SMS TI MT MS, Infineon AG, Postfach 80 09 49, D-81609
München, Germany

R.Weigel
Lehrstuhl Technische Elektronik, Friedrich-Alexander-
Universität Erlangen-Nürnberg, Cauerstr. 9, 91058
Erlangen, Germany

B. Bechen
Fraunhofer Institute of Microelectronic Circuits and
Systems, Duisburg, Germany

A. Kemna
Fraunhofer Institute of Microelectronic Circuits and
Systems, Duisburg, Germany

M. Gnade
Fraunhofer Institute of Microelectronic Circuits and
Systems, Duisburg, Germany

T. v. d. Boom
Fraunhofer Institute of Microelectronic Circuits and Systems, Duisburg, Germany

B. Hosticka
Fraunhofer Institute of Microelectronic Circuits and Systems, Duisburg, Germany

M. Weber
Institute for Theoretical Electrical Engineering, Leibniz Universität Hannover, Hannover, Germany

T. Vennemann
Institute for Theoretical Electrical Engineering, Leibniz Universität Hannover, Hannover, Germany

W. Mathis
Institute for Theoretical Electrical Engineering, Leibniz Universität Hannover, Hannover, Germany